Multiwavelength Astrophysics

Multiwavelength Astrophysics

Edited by

France A. Córdova

Los Alamos National Laboratory

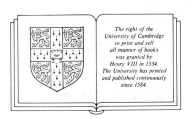

The right of the
University of Cambridge
to print and sell
all manner of books
was granted by
Henry VIII in 1534.
The University has printed
and published continuously
since 1584.

CAMBRIDGE UNIVERSITY PRESS

Cambridge

New York   New Rochelle   Melbourne   Sydney

Published by the Press Syndicate of the University of Cambridge
The Pitt Building, Trumpington Street, Cambridge CB2 1RP
32 East 57th Street, New York NY 10022, USA
10 Stamford Road, Oakleigh, Melbourne 3166, Australia

First published 1988

Printed in Great Britain at the University Press, Cambridge

British Library cataloguing in publication data

Multiwavelength astrophysics.
1. Astrophysics. Equipment & techniques
I. Córdova, France
523. 01'028

Library of Congress cataloguing in publication data available

ISBN 0 521 36197 4

Contents

Preface

In the past decade, a proliferation of guest observer facilities in almost every waveband, together with computerized astronomical data bases, have greatly facilitated correlative studies of most types of astrophysical source across the electromagnetic spectrum. So many resources go into multiwavelength observing, and into merging data from different facilities, that is productive to stand back and ask: What are we learning from multiwavelength studies? Do the results justify the large consumption of personal and technical resources? Is more multiwavelength activity, even to building space platforms with arrays of instruments covering the spectrum, warranted?

To answer these questions, we organized a workshop and planned a book on the subject. Both activities would emphasize the science learned from multiwavelength studies, and also include discussion of the use of large data bases, the methods of organizing campaigns, and future multiwavelength facilities. Implicitly, by dramatizing the results, and explicitly, by examining the problems with coordinating campaigns, the workshop and the book would point to ways to improve multiwavelength observations and data analysis.

The workshop, titled, "Multiwavelength Astrophysics," was held in Taos, New Mexico, August 9–14, 1987. It was hosted by Los Alamos National Laboratory and sponsored by the Laboratory's Institute of Geophysics and Planetary Physics (Director, Chick Keller) and its Earth and Space Science Division (Division Leader, W. Doyle Evans), and by the National Aeronautics and Space Administration. The meeting was the fourth Los Alamos workshop on Space Physics and Astrophysics. The scientific program was organized by myself (chairperson) and William Priedhorsky, with the help of the scientific organizing committee: Art Davidsen, Guenter Hasinger, John Hutchings, Timothy Kallman, Fumiyoshi Makino, Keith Mason, Tony Peacock, George Ricker, and Paula Szkody. Elaine Ruhe coordinated all the details of the workshop, including interacting on a personal level with each of the participants. She was assisted by Margaret Flaugh and Trisha Allen. The workshop's hosts at the Sagebrush Inn in Taos were Louise and Ken Blair.

The workshop covered many areas of astrophysics where multiwavelength studies have made a contribution: the Sun, late-type stars, O and B stars, cataclysmic variables, supernovae and their remnants, X-ray binaries, X-ray and γ-ray bursters, the interstellar medium, and active galactic nuclei. New designs for multiwavelength platforms were presented. Spontaneous groups of astronomers formed to examine ways of improving multi-facility observations and multiwavelength data analysis. So much enthusiasm was generated that many of the participants gathered again in a special session of the 1988 January meeting of the American Astronomical Society to further discuss ways to facilitate multiwavelength studies and present these ideas to a larger community. The result was a proposal to the *International Astronomical Union* to form a Working Group on the subject.

The goals of this book are similar to those of the Taos workshop: to comprehensively review a number of fields where multiwavelength observations have made a difference so that the benefit of such observations is clear; to show the range of data bases, analysis facilities, and observing instruments available for comparative waveband studies; to recommend ways to improve the multiwavelength approach; and, finally, to look in some detail at space instruments that are being proposed for multiwavelength studies. The book draws heavily on the inspiration and ideas of the Taos workshop. Indeed, many of the authors were participants. Yet the book goes beyond that meeting as it enlarges in a more comprehensive way upon the subjects discussed, and adds more examples of the application of multiwavelength tools.

This is not to say that the book is complete in its treatment of multiwavelength astrophysics; many phenomena where multiwavelength observations have had an impact are not covered at all; for example, solar flares. The reader will find, though, that the topics that are covered amply illustrate the effectiveness of multiwavelength observing efforts.

The results discussed in the following pages presage the importance of this type of research in the next decade, when the largest and most sensitive ground and space telescopes ever built will challenge astronomers with new discoveries. The eventual understanding of these phenomena may come from cooperative observing in every waveband, as it has for luminous X-ray sources. Today it is hard to envision *not* doing multiwavelength observations of almost any astronomical phenomenon. As D. Hartmann and S. Woosley write in their chapter, "Imagine the state in which the study of advanced stages of stellar evolution and supernovae would find itself were our entire data base limited to the neutrinos from the [SN 1987A] explosion."

We hope that this book will be a useful reference for astronomers and students interested in finding out what has been done, and what it is possible to do, in carrying out multiwavelength observations and doing correlative analyses. The book may also give some ideas to scientists proposing new instruments and analysis facilities, and those organizing new data bases.

I thank all of my co-authors for writing their comprehensive reviews in a timely way, and for making the editing of this book a very enjoyable enterprise. For uniformity of presentation, all of the chapters have been done using the same (TEX) format. To write their chapters, many of the authors learned this format for the first time. Jean Johnson at Los Alamos deserves praise for helping the rest of us non-TEXnicians.

There were many astronomers who carefully reviewed individual chapters; most are acknowledge in the chapters themselves, but I wish to thank them again for improving the quality of this book. I also wish to thank other reviewers who participated in a more anonymous way: Richard Epstein, Keith Mason, Chris Mauche, Gerrie Peters, and Barry Smith.

The editing of this book was supported, in part, with funding from the U.S. Department of Energy. It was supported in an altogether different way by my 'Foster family': Christian, Anne-Catherine, and Stephen.

France A. Córdova (Foster)   March 1988

Multiwavelength (λλλ) Astrophysics: An Overview

Martin C. Weisskopf

Space Science Laboratory, Marshall Space Flight Center

The realization of the capability, and the necessity, of performing λλλ astrophysics was the focus of the Taos Workshop on Multiwavelength Astrophysics and the results are well documented in the following chapters. The capability is a consequence of technological developments that continue to allow for new methods, techniques, and higher sensitivities over more and more of not only the electromagnetic spectrum, but also other channels of information as demonstrated by the recent detection of neutrinos from SN1987A. Of special significance for λλλ astrophysics is the space program which allows for observations beyond the obscuring effects of the earth's atmosphere and facilitates monitoring of time variable phenomena. The necessity for λλλ observing is underscored by the ever growing number of astrophysical problems that simply cannot be adequately addressed by means of detailed studies in a narrow spectral band. In the following chapters numerous examples of λλλ research are presented which span such diverse topics as non-degenerate stars including the Sun, the riddle of the origin of gamma ray bursts, and the study of interstellar matter within galaxies. These latter two topics typify the need for, and the diverse nature of, λλλ observations. In the study of the interstellar material within elliptical galaxies one can identify at least three characteristic temperature scales, $\sim 10^7$ K, $\sim 10^4$ K, and <100 K. To study material at these temperatures requires observations at X-ray, ultraviolet-visible, and infrared-radio wavelengths respectively. However, since the time scales associated with galaxy and galaxy-cluster phenomena are long, this research does not require absolutely simultaneous observations.

In contrast, the study of the emission processes and physical origins of gamma ray bursts, where the time scales are measured in seconds or less, demands truly simultaneous data. Observations must cover both the γ-ray/X-ray bands in order to completely characterize the burst spectra as well as a broad range of the longer wavelengths where the reprocessed burst radiation is emitted and which serves as a probe of both the geometry and the physical conditions of the plasma in the vicinity of the source of the burst phenomenon.

The increasing realization of λλλ astrophysics has been fostered by not only technology and scientific necessity but by other factors which are interesting to note. One of these factors is a gradual decline of the need to serve a lengthy apprenticeship within a given discipline in order to be an "observer." This phenomena is by and large a product of the space based programs, both in this country and abroad. For example, the International Ultraviolet Explorer and Einstein Observatories established significant Guest Investigator programs that formally solicited and sponsored proposals for their active use by the broader astronomical community. Data analysis software developed for the use of the Guest Observers removed the necessity for each scientist to understand all the technical details of these experiments. These phenomena will continue with the advent of the *Hubble Space Telescope (HST)*, the *RoentgenSatellit (ROSAT)* mission, the *Gamma-Ray Observatory (GRO)*, *Astro-D*, the *Advanced X-ray Astrophysics Facility (AXAF)*, the *X-ray Multiple-Mirror Telescope (XMM)*, the *Space Infrared Telescope Facility (SIRTF)*, and the *Infrared Space Observatory (ISO)*.

A second factor in the evolution of $\lambda\lambda\lambda$ astrophysics has been the absence of continuity in some of the space based disciplines, at least on a national basis. Thus, e.g., in the United States the existence of the enormous gap in time between the *Einstein Observatory* launched in 1978 and the *AXAF*, currently planned to be launched no earlier than 1995, has "forced" a significant fraction of American X-ray astronomers to turn to observations in other wavelength bands. A third factor, which has only just begun to contribute to the realization of $\lambda\lambda\lambda$ astrophysics, has been the rapid development of computer networks. These obviate the "need to be there" and encourage the development of standards and transportable software for the dissemination and analysis of astronomical data. The IRAF (Interactive Reduction and Analysis Facility) software being developed at Kitt Peak and *HST* Science Institute and MIDAS (Munich Image Data Analysis System) developed at the European Southern Observatory are prominent examples of emerging portable software standards.

On the other hand, and despite the advances that have taken place, it remains difficult to accomplish $\lambda\lambda\lambda$ astrophysics. The majority of difficulties and frustrations center about the mechanics of obtaining the relevant data. These are discussed in some detail in Part 2 of this work in the chapters, "Databases and Analysis Centers" and "Multiwavelength Observations Today." Among the most significant problems are: (1) The essential absence of standards and uniformity in the data one obtains from different observatories; (2) The essential absence of any central clearinghouse(s) for proposals that require data from different observatories (The latter is exacerbated, in part, in the United States by the existence of two funding agencies—one for ground-based and one for space-based research); (3) Frustrations resulting from the different observing constraints for different wavelength bands.

Despite the difficulties outlined above, progress has been made, as demonstrated in the following chapters. The future, too, looks bright. More powerful observational tools are coming on line, including ten meter telescopes, the Very Long Baseline Array, and future major space observatories. With the advent of these devices has come an awareness of the necessity for $\lambda\lambda\lambda$ astrophysics and the need to create the architecture to facilitate such research. This architecture is necessary as more and more emphasis will need to be placed on simultaneous observing campaigns because, as illustrated in the chapters on the Sun, active stars, cataclysmic variables, X-ray binaries, etc., understanding of the astrophysics requires this. In addition, one needs to consider not only the coordinated use of wavelength specific observatories, but also (space-based) missions designed specifically to address $\lambda\lambda\lambda$ problems, as discussed in the Part 2 chapter, "Future Space Instrumentation for Multiwavelength Astrophysics."

The direction for future advancement in $\lambda\lambda\lambda$ astrophysics is clear. With increased use of computer networks and the adoption of standard formats (e.g., FITS) and higher level portable software (e.g., IRAF) one can anticipate a time when data from the major observatories will be readily accessible to all astronomers at their home institutions. Data from these observatories needs to be prepared for broad dissemination with instrument signature removed. In addition, well documented and readily accessible analysis software and calibration data bases must also be made available. This will occur, if for no other reason than to justify the large investments made in the major programs.

Having the tools to perform $\lambda\lambda\lambda$ astrophysics is necessary but not sufficient for the reasons discussed above. Astronomers need to be able to use these tools easily and in a coordinated manner. The realization of this is also feasible and the groundwork is being laid. Examples are NASA's Astrophysics Data Program which allows astronomers to perform archival research across disciplines and under the umbrella of a single proposal process. An approach such as this will surely be utilized for the operation of future space observatories and will include provisions for coordinated observations. More joint observing programs with ground-based observatories are required. This, most likely, will require the most effort to organize. Conferences and workshops, such as the Taos Workshop, the Strasbourg Colloquium on the Coordination of Observing Projects, and the Astrophysics Data Systems workshops held in Maryland in the fall of 1987, which emphasize the policies and architecture required for the accomplishment of $\lambda\lambda\lambda$ astrophysics, increase confidence that great strides will be made over the ensuing years.

Part 1: The Scientific Impact of Multiwavelength Observations

The Hot Solar Envelope

R. Grant Athay

High Altitude Observatory

National Center for Atmospheric Research

1. Introduction

The Sun is a typical main sequence G star with a moderately thick convection zone. In stars of this spectral type, the convective instability ceases at depths below the visible photospheric layers. Overshooting convective eddies penetrate the photosphere, however, and give rise to the familiar granulation seen in the visual continuum. Comparisons of empirical models of the photosphere derived from changes in the spectral distribution of the photon energy flux across the solar disk to theoretical models reveal that transport of heat by the convective overshoot plays a significant role in the deep photosphere but diminishes to a negligible role in the upper photosphere. This gave rise to the conclusion, tacitly adopted for many years, that all vestiges of the convection disappear within the photosphere.

Meanwhile, the discovery that the Sun has a million-degree corona presented a difficult theoretical challenge. How could a star with an effective temperature near 5780 K give rise to a hot and very dynamic envelope at temperatures reaching in excess of 10^6K? The detailed answer to this question is still unclear. Nevertheless, in pursuing this answer we have come to recognize that the influences of the subphotospheric convection on the outer layers of the Sun extend far beyond the heat transport in convective overshoot itself. These extra influences are of subtle nature but their importance for the solar envelope is profound. The conceptual picture of how the subsurface convection drives the varied phenomena of the outer envelope now seems reasonably clear. Details of how the energy and momentum transfer occurs, however, are at best tentative, and more often totally lacking.

Stated in the broadest terms, convection couples with rotation to generate magnetic fields, which are subsequently concentrated by the convective action into flux tubes of kilogauss intensity. Continued agitation and bodily transport of these flux tubes by the convection further stress the magnetic field generating a flux of energy that is transported along the field and dissipated in the higher layers of the atmosphere. It is this dissipation of magnetic free energy that is believed to give rise to the hot solar corona with its associated mass loss in the solar wind, x-ray emission, and occasional violent bursts of activity in flares and flare-related events.

A second important result of the convection is the generation of pressure waves. The longer period waves are trapped and remain evanescent in the photosphere and below. Short period waves ($T \lesssim 150$ s) propagate, however, and may carry considerable energy flux above the photosphere. These waves are not believed to be of major significance in the corona but may be competitive with or even surpass the importance of magnetic free energy dissipation in the chromosphere, which is sandwiched between the photosphere and corona.

The magnetic flux tubes evidently are created in the convective region below the visible layers. However, in the photosphere, flux tubes of kilogauss strength still dominate the magnetic structure on all spatial scales extending from occasional large naked-eye sunspots to tiny elements only a fraction of an arcsec in diameter. The latter are sprinkled over the surface of the Sun in a mosaic pattern referred to as the network. The network border large supergranule cells of the order of $50''$ in diameter. These supergranules are not visible in brightness patterns in the lower photosphere but become so in the high photosphere and chromosphere. They are, however, clearly evident in velocity patterns even in the low photosphere.

The concentration of the field into kilogauss flux tubes bordering the supergranules is the result of the dominance of mechanical forces (gas pressure) over magnetic forces, i.e., in the large value of the plasma β. Since the scale height over which the gas pressure decreases is relatively small, the mechanical forces exerted on the field decrease rapidly with height allowing the field to expand to a more nearly potential configuration. As a result, in the corona, where the gas pressure is reduced by some six orders of magnitude below the photospheric value, the magnetic forces have become dominant (low β) and the field fills the entire volume.

Although we believe that we have identified some of the important links in the chain of events between convection and energy deposition in the chromosphere and corona, it is not clear that we have identified all of the essential ones. Also, those that we have identified are understood only in rather broad terms. The specifics of the dynamo process, the formation of kilogauss flux tubes, and the subsequent generation of magnetic free energy (or electric currents) and its ultimate transfer back to particle energy pose an array of very difficult problems, none of which is well understood. These challenging physical problems, together with our ability to observe the Sun in some detail, make the Sun of immense value as a laboratory for studying plasma physical processes. The problems encountered in the Sun have close parallels in magnetospheres of planets, other stars, pulsars, and galaxies. Just as the study of these objects contribute to our understanding of the Sun, so do solar studies contribute to the understanding of these and other astrophysical systems. The interaction of convecting plasma with magnetic fields to generate high-energy phenomena appears to be endemic to the majority of the known universe, and the Sun remains as one of the best means of studying this important chain of physical processes. Additionally, the variable radiation and particle outputs resulting from solar magnetic field-plasma interactions provide a variable driver for the Earth's magnetosphere, thermosphere, and lower atmosphere. Thus, the Sun serves as the laboratory through which we interpret much of what we observe both in the external universe and in our more immediate terrestrial environment.

As in other areas of astrophysics, the opening to the high-energy end of the electromagnetic spectrum provided by orbiting observatories has vastly increased our knowledge of the Sun. Beginning with the earliest x-ray and EUV images of the Sun obtained from rockets and satellites, our eyes were opened to the important and pervasive role of magnetic fields. Also, new understanding was developed for the nonthermal, high-energy events previously revealed by solar radio bursts, solar proton events, and geomagnetic and ionospheric storms. Similarly, visible light coronagraphs lifted out of the bright sky background of terrestrial observatories discovered an unsuspected complexity of activity in the corona. The spectacular and frequent coronal mass ejections discovered by *Skylab* are now recognized as being among the most energetic solar events.

Solar space research, however, must still be regarded as being in its infancy. We have explored some of the potential of space research but have not fully exploited any of it. In addition, one of the more important advantages of space research, viz., high spatial resolution, has hardly been explored at all. Solar granulation is structured on a scale comparable to the pressure scale height in the photosphere, H_p, which is of the order of 150 km. Magnetic flux tubes generated by the granulation and supergranulation are of similar scale or smaller in the low photosphere. A wealth of evidence suggests that as the magnetic fields of the photospheric flux tubes diverge to fill the bulk of the chromosphere and corona, the plasma is confined by the field and becomes highly structured on spatial scales of the same order as the flux tube diameter in the photosphere. Evidently, the buffeting of the flux tube by granulation ultimately impresses a fine structure on the plasma that persists even into the tenuous corona where the magnetic field energy is strongly dominant over the plasma energy. By implication, the energy processes propagating along the field to excite the plasma are similarly structured. The actual processes by which magnetic free energy is transmitted to the plasma, either as heat or as bulk kinetic energy, involve the much smaller size scales typical of magnetic reconnection and electric currents in plasmas. It is clear, therefore, that the physics of plasma excitation via the magnetic fields generated by convection inherently involves a variety of physical processes all of which occur on small spatial scales.

The proximity of the Sun readily allows the resolution of spatial scales down to about $\frac{1}{2}H_p$ (0.1″) through the regions of the spectrum where normal incidence optics can be used. This is sufficient to study the solar atmosphere on scales that are physically relevant and is the central goal of the High Resolution Solar Observatory currently under study as a future NASA mission. An international effort, known as the Large Earth-Based Solar Telescope, is being pushed in parallel with HRSO. Limitations of observing through the Earth's atmosphere, even with the use of adaptive optics, restrict the high resolution studies to small fields of view and short time periods. These restrictions severely limit the range of problems that can be addressed. The HRSO will be largely free of such limitations but encounters others that arise from the lack of variety and flexibility in the type of focal plane instruments that can be included. Thus, the two programs are more complimentary than competitive.

At radio wavelengths, the VLA already achieves a resolution of 0.1″ and is a powerful tool for studying the magnetic and plasma structure of the corona. High resolution imaging in the XUV is currently being tested using multilayer coatings to extend the use of normal incidence optics to shorter wavelengths. Also, the Pinhole Occulter Facility under study by NASA is designed for high resolution imaging (0.2″) in hard x-rays. These new techniques for imaging hold great promise for future studies with angular resolution improved over past and current observations by an order of magnitude in the corona and in the transition region between the chromosphere and corona.

It is not sufficient, of course, merely to image the small scale structure. Such images must be accompanied by spectroscopy that is capable of providing the requisite plasma diagnostics. This usually requires reasonably high spectral resolution. In addition, plasma and magnetic structures with small spatial scales often evolve rapidly in time requiring that observations be made at high time resolution. The combined requirements for high spatial, spectral, and temporal resolution place strong demands on the available photon flux even for the Sun. Fortunately, however, these demands can be satisfied to a reasonable degree in many applications, thanks to the proximity of the Sun.

Multiwavelength studies of solar phenomena already have a substantial history, but much is still required both in improving the quality of the data and in discovering effective new diagnostic methods. In a single chapter we cannot discuss all such needs. As a compromise, we have selected a class of problems related to the hot solar envelope. Even then, we must limit detailed discussion to selected areas and touch only briefly on the larger aspects of the problem.

The topic selected involves nonthermal, as well as thermal, phenomena, both of which involve physical processes whose study benefits enormously from observations over a broad wavelength range. The corona, for example, produces radiation extending from radio wavelengths to x-rays. The phenomena observed, or inferred from observations, range from radiation from low temperature (10^4K) to high temperature (10^6K) plasmas and from energetic electron and proton streams to coronal mass ejections and eruptive prominences. Some of these phenomena are best observed at XUV and x-ray wavelengths, whereas others are best observed at radio wavelengths and still others at visual wavelengths.

The study of solar flares represents another area of solar physics in which multiwavelength observations play an essential role. What one observes in a narrow or even moderate wavelength band reveals only a fraction of the phenomena occurring as an integral part of the flare event. Hard x-ray and γ-ray observations cast an entirely new light on the picture of flares developed from observations at visual wavelengths alone, while at the same time giving no hint of some of the important phenomena observed in the visual spectrum.

The *Solar Maximum Mission*, designed specifically for the study of flares, provides an excellent example of coordinated multiwavelength experiments focusing on a complex astrophysical problem. The battery of experiments extending from visual wavelengths through the XUV to x-rays and γ-rays are described in papers by Bohlin *et al.* (1980), Forrest *et al.* (1980), Orwig *et al.* (1980), Van Beek *et al.* (1980), Acton *et al.* (1980), Woodgate *et al.* (1980), and MacQueen *et al.* (1980). Results from these experiments have been published in numerous papers and are being summarized in a special issue of *Solar Physics* to appear in 1988.

2. The Solar Interior

The hot envelope surrounding the Sun has its origin in the solar interior. Within the central core of the Sun, helium nuclei are produced by fusion of protons accompanied by the release of energy. The layers immediately above the core are stable to convection because of the relatively small temperature gradient, and energy flows outwards by photon diffusion. Standard models place the extent of the radiative equilibrium core at about 76% of the solar radius. These models, however, ignore convective overshoot, and thus represent an upper limit on the extent of radiative equilibrium. Theoretical estimates of the extent of the convective overshoot regime, as well as estimates of the depth of the convection zone based on the depletion of light elements and solar p-mode frequencies suggest that the convection most likely reduces the radiative core to between 65% and 70% of the solar radius (cf. Press 1986).

Within the convectively unstable layers virtually all of the energy is carried by convection due to its efficiency in transporting heat coupled with the large opacity of the material to the photon flux. Convective stability returns in standard models near the base of the photosphere where the continuum optical depth is between 1 and 2. The granulation observed in the low photosphere lies in the overshoot region above the unstable layers. Average granules have a radius, r_g, of about 450 km. This is about three times the pressure scale height, H_p, at the top of the convectively unstable zone. However, H_p increases rapidly inward because of the reduced mean molecular weight associated with hydrogen ionization, and at a depth into the convection zone equal to r_g, H_p has increased to 350 km. Thus, r_g is roughly equal to H_p at a depth below the top of the convection zone equal to r_g.

The overshooting convection raises the temperature somewhat in the photosphere but does not carry sufficient energy flux to modify the temperature gradient by more than a small amount. At the top of the photosphere there is little or no remaining evidence of the granulation as such. On the other hand, the hot envelope overlying the photosphere vividly demonstrates that mechanical (non-radiative) energy flux in some form passes through the photosphere and dissipates in the tenuous outer layers. There is no doubt that the convective energy dissipates mainly as heat in and below the photosphere. However, some small fraction of the convective energy is dissipated in the generation of magnetic fields and a variety of wave modes. This remnant of the convective energy must also eventually escape the Sun. It does so by heating the atmosphere to produce enhanced radiation and the solar wind. In broad outline, therefore, the hot solar envelope is the result of the conversion of convective energy into magnetic and wave energy thence to heat in the outer layers of the atmosphere far above the top of the convection zone itself.

Magnetic field generation in the Sun is perhaps more properly described as a modulation of pre-existing fields. The dynamo action arises from motion of ionized plasma across existing magnetic field lines. As originally proposed by Parker (1955), the solar dynamo is believed to involve a cyclical buildup alternately of poloidal (N-S) field at the expense of torroidal (equatorial belt) field followed by regeneration of the torroidal field at the expense of poloidal field, etc. The conversion from torroidal to poloidal field is via helical motion or "helicity"–the so-called α-effect. Regeneration of the torroidal field results from differential rotation ($d\Omega/d\phi \neq 0$, where ϕ is latitude).

The characteristics of $\alpha - \Omega$ dynamos depend upon a number of parameters including, in addition to α and $d\Omega/d\phi$, their depth variation $d\alpha/dr)$ and $d\Omega/dr$ and the latitudinal variation $d\alpha/d\phi$. By suitable choices of parameters, one can mimic many aspects of the observed solar activity cycles (cf. Gilman 1986). Such an approach is inconclusive, however, since the convection itself will determine the values of each of the parameters, which means that they cannot be selected independently of a physical theory for the form of the convection. Unfortunately, convection in an ionized, rotating, compressible atmosphere is not well understood, and we are forced to rely on observations or theoretical estimates to provide information on the various parameters entering the model calculations.

Theoretical estimates of α (cf. Gilman review 1986) indicate much larger values than are used in the most successful dynamo models. On the brighter side, recent observations of solar oscillations are revealing the nature of $\Omega(r, \phi)$. Pressure waves trapped in the convection zone and photosphere are reflected by the chromospheric temperature rise. As they propagate downward the wave fronts refract along a curved trajectory that takes them back to the surface and a subsequent reflection. The index of refraction decreases with decreasing wave frequency, which allows the low frequency waves to penetrate deeper into the convection zone than waves of higher frequency.

Eigenvalues occur at frequencies ν_i in the wave spectrum whenever the ratio $2\pi R_\odot/L(\nu)$ is an integral number, where $R_\odot$ is the solar radius and $L(\nu)$ is the distance between successive surface reflections measured along a great circle. Thus, a power spectrum of the waves reveals discrete modes whose differential properties provide a means of probing different depths in the solar interior.

In particular, for an observer in a fixed reference frame, the rotation of the Sun splits the resonance frequencies ν_i into multiplets whose spacing depends upon $\Omega(r, \phi)$. Given a sufficiently precise definition of the wave spectrum, therefore, it is possible, in principle, to derive $\Omega(r, \phi)$ (cf. review by Brown, Mihalas, and Rhodes 1986). The results of such a derivation are more accurate in the outer layers occupied by the convection zone but still provide useful information well into the radiative core. Currently available results (Brown and Morrow 1987) are not sufficient to completely specify $\Omega(r, \phi)$ even within the convection zone. However, the data are consistent with $\Omega(r, \phi)$ being a function of ϕ only within the convection zone, i.e., with the same differential rotation at all depths. Within the radiative core the data are consistent with solid body rotation at the surface equatorial rate. It is clear, however, that better data are needed before we can be confident of the detailed form of $\Omega(r, \phi)$.

3. Overview of the Solar Atmosphere

Although it is convenient for purposes of discussion to divide the solar atmosphere into layers that are readily distinguished observationally, in this section we emphasize some of the physical processes that play major roles in determining the character of the atmosphere. The total energy flux from the Sun, $F_\odot$ ($= 6 \times 10^{10}$

ergs cm^{-2} s^{-1}), can be represented in the form

$$F_\odot = F_p + F_{ch} + F_{tr} + F_c \tag{1}$$

where the different terms represent the contributions, respectively, from the photosphere, chromosphere, transition region, and corona. The relative amplitudes (whose evaluations are discussed later) of the different terms, to order of magnitude, are

$$F_p = 10^4 \, F_{ch} = 10^5 \, F_{tr} = 10^6 \, F_c \, . \tag{2}$$

Thus, F_p accounts for some 99.99% of $F_\odot$ and only some 0.01% of the solar power is required to drive the hot envelope overlying the photosphere.

A second way of characterizing the distribution of the power output is in terms of temperature, T. The photosphere and chromosphere have $T < 10^4$K and the corona has $T > 10^6$K. The intermediate decades $10^4 - 10^5$K and $10^5 - 10^6$K represent the lower and upper transition region, respectively. We denote the energy fluxes from these regions by $F_{<4}$, F_{45}, F_{56} and $F_{>6}$ and express the relative contributions to order of magnitude, as

$$F_{<4} = 10^5 \, F_{45} = 10^6 \, F_{56} = 10^6 F_{>6} \tag{3}$$

In still a third categorization, we use F_n to denote the flux contributed by the layers in which neutral atoms outnumber free electrons and F_i to denote the flux from the ionized layers in which electrons outnumber neutrals. We then find $F_n = F_p + F_{ch}$ and $F_i = F_{tr} + F_c$ with

$$F_n = 10^5 \, F_i \, . \tag{4}$$

On the other hand, if we limit consideration to the chromosphere, transition region, and corona, and replace F_n by F_n^*, then $F_n^* = F_{ch}$ and

$$F_n^* = 10 \, F_i \tag{5}$$

The importance of these different ways of categorizing the rates of energy loss from the different atmospheric regimes will become apparent as we discuss the dominant mechanisms of energy loss.

Certain qualitative properties of the thermal structure of the solar atmosphere can be deduced without knowing the detailed nature of the energy input. The reason this can be done is that the energy output is primarily by radiation, and, as the temperature rises, the radiation loss passes through regimes of very different behavior. Throughout most of the photosphere radiation from free-free and free-bound transitions in the H$^-$ ion are dominant. The primary energy balance occurs via radiative cooling in the red and infrared portion of the spectrum and radiative heating in the violet.

In the upper photosphere and temperature minimum region, three important changes occur relative to the middle photosphere: (1) the heat input shifts from radiation to dissipation of mechanical energy flux; (2) the radiation output shifts from the H^- continuum to bound-bound transitions, primarily in singly ionized metals; and (3) the populations of the bound states begin to depart from thermodynamic equilibrium. The first two effects occur because the H^- opacity decreases in proportion to the square of the density, which acts both to rapidly decouple the ambient atmosphere from the intense radiation density and to increase the relative importance of the bound-bound transitions. The third occurs because the low continuum opacity allows even the line photons to escape from the atmosphere by diffusion in frequency into the line wings where the opacity is low.

The departure from thermodynamic equilibrium in the bound-bound transitions results in an inbalance between collisional excitations and de-excitations, and the primary cooling mechanism becomes excitation by electron collisions followed by radiative de-excitation and subsequent escape of the photon. For a given bound-bound transition, the rate of cooling, q_{ij} is given by

$$q_{ij} = c_{ij} N_e N_i e^{-h\nu_{ij}/kT}, \; ergs \; cm^{-3} \; s^{-1}, \tag{6}$$

where N_e is the electron density, N_i is the population density in the lower energy state of the transition, ν_{ij} is the photon frequency for the transition and c_{ij} depends on the collision cross-section. As an alternative form of equation (6), we replace N_i with $(N_i/N_{ai}) (N_{ai}/N_H) N_H$, where N_{ai} is the density of atoms of the specie containing level i and N_H is the hydrogen density, and we replace N_e with $(N_e/N_H) N_H$ to obtain

$$q_{ij} = c_{ij}A_{ai}(N_i/N_{ai})(N_e/N_H)N_H^2 e^{-h\nu_{ij}/kT} , \tag{6a}$$

where $A_{ai} = N_{ai}/N_H$ is the relative abundance of the element. Within the hot solar envelope both N_i/N_{ai} and N_e/N_H are functions of temperature only, which we designate by $f_i(T)$ and $f_e(T)$, respectively. It follows that when equation (6a) is summed over i and j, it has the resultant form

$$Q = Q_0 \, g(T)f_e(T)N_H^2 , \tag{6b}$$

where $g(T)$ combines the summed effects of $f_i(T)$ and $e^{-h\nu_{ij}/kT}$. For $T < 8000$ K, where hydrogen is ionizing rapidly, $f_e(T)$ is a much stronger function of temperature than is $g(T)$. Thus, for $T < 8000$ K the term that primarily controls the value of Q is $f_e(T)$. In other words, it is the electron density that controls the radiation loss rate.

The specific temperature structure that results from a balance between cooling by emission in spectral lines (equation 6b) and mechanical heating depends, of course, upon the nature of the heating as well as the cooling. However, some qualitative features of the temperature profile can be deduced simply from the manner in which N_e varies with temperature. The vast majority of electrons in the Sun either are bound to hydrogen nuclei or they are the product of hydrogen ionization. Furthermore, hydrogen ionization is a very strong function of temperature. To illustrate this, we consider the ratio $y = N_e/N_t$, where N_t is the sum total of all particles other than free electrons. At the solar temperature minimum (outside of sunspots), metals are mainly singly ionized and hydrogen is only negligibly ionized. This gives $y = 10^{-4}$. On the other hand, in the transition region and corona, hydrogen is essentially fully ionized which gives $y \approx 1$. Most of this change in y by a factor of 10^4 occurs in a relatively narrow temperature interval. For example, in the chromosphere, $y = 10^{-3}$ near 5400 K and $y = 0.5$ near 8000 K. The influence this exerts on the temperature profile is to create a plateau of relative flat temperature gradient in the region where most of the ionization occurs. Figure 1 illustrates this effect in an empirical reference model of the chromosphere (Vernazza, Avrett, and Loeser 1981), to which we have added a theoretical transition region model for a constant thermal conduction flux of 6×10^5 ergs cm^{-2} s^{-1} (see section E for discussion). In the range 6000-8000 K in the model illustrated, the hydrogen density decreases by a factor of 300 whereas N_e decreases by less than a factor of 3. A significantly steeper temperature rise would result in an increase in N_e with height. We emphasize that the temperature gradient within the plateau depends upon the form of the heat input. However, the extremely strong dependence of N_e on temperature ensures that a plateau will occur for a wide range of heat inputs (Athay 1981a).

As long as hydrogen is predominantly neutral, the radiation output can be changed continuously through a relatively wide range. It is for this reason that most of the heat energy in the hot envelope is lost by radiation while the gas is predominantly neutral. Once the hydrogen ionization exceeds about 50%, however, N_e is no longer explicitly dependent on temperature. Thus, the only means through which radiation loss can be adjusted to balance the heat input is through the dependence of collision rates on temperature, and, to a lesser extent, on changes in N_i that accompany temperature changes. Again, we see, in Figure 1, the reaction to this limited flexibility in the radiation loss, viz., the temperature rises steeply to a coronal value of over 10^6K. It rises so steeply, in fact, that thermal conduction through the transition region may provide the primary energy loss from the corona. As a consequence, the heat energy deposited in the corona by dissipation of mechanical energy may reappear mainly as radiation arising in the

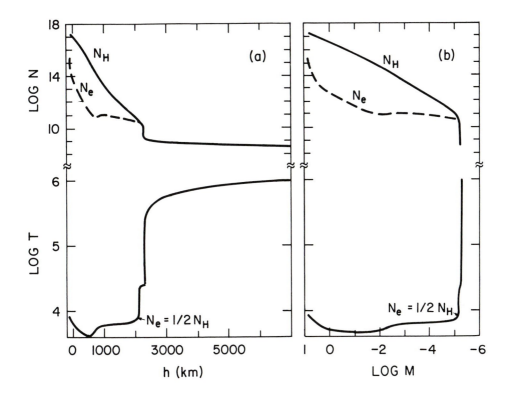

Figure 1 Plots of T, N_e and N_H versus geometrical height in km above the base of the photosphere (panel a) and the column mass, M, in gm cm^2 (panel b). The plots extend through the photosphere, chromosphere, and transition region. Below log T=4.3, the plots are taken from Vernazza, Avrett, and Loeser (1981), and above log T=4.3, they represent a constant thermal conduction flux of 6×10^5 ergs cm^{-2} s^{-1}.

10^4 to 10^5K temperature range. A lesser amount ($\approx 10\%$) flows outward by thermal conduction and the solar wind. The radiation flux from the corona itself is smaller still.

The steep temperature rise in the transition region and the resultant backflow of heat by conduction raises a number of challenging questions. These questions, together with new ones raised, are given even greater significance by observations of large velocity amplitudes and rapidly fluctuating radiation intensity from these same regions of the atmosphere. Whatever the answers to these questions may prove to be, it seems clear that the chromosphere and corona are inseparably joined both energetically and dynamically. The onset of the temperature rise to the corona appears to be strongly, if not totally, dependent on the form of the chromospheric heating. The transition takes place where hydrogen becomes mainly ionized, and this is directly coupled to chromospheric heating. Conduction from the corona may play a role, but surely the chromospheric heating plays a more important one.

In a later section, we will draw attention to dynamical phenomena indicating that chromospheric influences extend well into the transition region, illustrating that this region is not merely a thermal conduit for disposing of coronal energy.

The broad overview of the solar atmosphere presented in this section by-passes all of the essential physics of the heating as well as a number of unresolved conceptual difficulties. Similarly, it ignores the highly dynamic nature of the hot envelope as well as the magnetic structure and its influence. Thus, while we have attempted to explain the major features of the temperature profile, we do not mean to imply that we have solved any of the basic problems associated with either the heating or the dynamics of the atmosphere. In the following sections we will enumerate several of these problems and, where possible, indicate the observations that are needed to resolve them.

4. The Photosphere

The photosphere is observed in the continuum at wavelengths extending from 160 nm in the ultraviolet to about 30,000 nm at submillimeter wavelengths. Through the visual portion of the spectrum the strongest source of opacity is due to the H^- free-bound transition with a lesser but significant contribution from H^- free-free transitions. The free-bound opacity has a maximum near 800 nm, whereas the free-free opacity increases as λ^2. Beyond about 1350 nm the free-free opacity is dominant. Beyond the Balmer series limit near 360 nm, the free-bound continua of hydrogen and neutral and ionized metals become dominant over the H^- opacity.

For studies of solar convection, it is advantageous to look as deep into the Sun as possible. The minimum opacity, and hence where the escaping radiation has its deepest origin, occurs near 1600 nm. At this wavelength, the surface of optical depth unity occurs, in standard models, just below the top of the convectively unstable layers. It could be of some interest, therefore, to compare the granulation observed near 1600 nm to that observed somewhat higher up in the overshoot region. The differences are expected to be rather subtle since the best height separation one can achieve in the optical portions of the spectrum is about 40 km, which is an order of magnitude smaller than the mean granule size. Nevertheless, within that 40 km span the atmosphere changes from convective instability to stability and this should have observable consequences of interest. Such comparisons have not been made but are becoming feasible with current technology.

Since our interest in this chapter is in the hot envelope, we will mention only a few of the unresolved questions associated with the photosphere. The chief problems are those of MHD convection and energy transport in a partially ionized, compressible, radiating atmosphere. In relationship to the hot envelope, our interest focuses on the by-products of the convection in the form of the various wave modes that are generated by the turbulent motions and in the morphology and dynamics of the magnetic field.

The so-called 5-min oscillations on the Sun are actually composed of a band of trapped waves with periods extending down to about 3 min. The short period limit corresponds to the acoustic cut-off frequency near the beginning of the chromospheric temperature rise. Shorter period waves are undoubtedly present, but these propagate through the photosphere and dissipate in the overlying layers. Dissipation occurs because the conservation of $\rho\, V_a^2$, where ρ is density and V_a is the velocity amplitude, rather quickly increases V_a to shock amplitude as ρ decreases with height.

Short period waves are difficult to detect observationally. The photons escaping the atmosphere arise from a distribution of depths with a mean near $\tau = 1$ and an effective width, L, of about $2H_\tau$ or larger, where H_τ is the length over which τ changes by a factor of e. For short period waves, the wavelength, λ_s, decreases to the order of L or smaller and the wave signature is strongly reduced. This effect is pronounced for periods less than about 100 s.

Gravity waves are undoubtedly present in the photosphere, but these also are difficult to detect. Propagating gravity waves exist only at frequencies below the Brunt-Väisälä frequencies and in layers that are convectively stable. In the photosphere, this permits waves with periods in excess of about 550 s.

The Brunt-Väisälä frequency decreases sharply in the middle chromosphere due to hydrogen ionization. This leads to the possibility of a zone of trapped gravity waves extending from the photosphere into the middle chromosphere. Although several observers have reported evidence of gravity waves, no convincing demonstration yet exists. Also, it is not clear what to expect for such waves. The wave energy can be dissipated in the chromosphere by "breaking" followed by turbulent cascade similar to that observed for ocean surface waves, and it can be dissipated in the photosphere by radiation loss.

Radiative cooling times in the low photosphere are extremely short (< 10 s), and, as a result, radiative damping strongly influences both gravity waves and pressure (sound) waves. Nevertheless, the search for gravity waves and an evaluation of their role in energy transport to the hot envelope remains an important issue in solar physics. Similarly, the role of high frequency sound waves must be better understood. Wave energy still remains a prime candidate for chromospheric heating.

Magnetic field structure in the photosphere exhibits the spatial scales of both the granulation and the supergranules. Individual flux tubes are subarcsec (granule) in size, but they are distributed over the surface in a network pattern bordering the supergranule cells. The flux tubes are essentially vertical, most likely because of buoyancy effects. They are "rigid" in the sense that the field strength, B ≈ 1500 g, is sufficient to make $B^2/8\pi$ of the same order as the ambient gas pressure in the low photosphere. Thus, the convective action coupled with evacuation of the flux tubes has compressed the field elements to the point where the magnetic pressure resists any further compresssion.

By concentrating the field into small elements, the convection has increased the average of $B^2/8\pi$ over the surface of the Sun by a factor of about 10^3 compared to the value obtained for a uniform field of the same flux. The increased energy of the magnetic field is an important factor in the role the field plays in the outer atmosphere. For example, if the field is perturbed by a small amount $\frac{\Delta B}{B}$, the resultant perturbation in energy is $2\,B\Delta B/8\pi$, and if we take $\Delta B/B$ to be independent of B then $B\Delta B \propto B^2$. Thus, for a given magnetic flux, the total energy perturbation associated with a perturbation $\Delta B/B$ increases in proportion to B and is therefore larger when the field is concentrated in flux tubes. It follows that the role of the magnetic field as a means of transferring convective energy to heat in the hot envelope has been enhanced as a result of the concentration of the photospheric flux into elements of high field strength.

There is little doubt that the corona is heated by the magnetic field and that a sizable fraction of that heating is due to the field associated with the small flux tubes. In principle, therefore, the heating of the corona can be observed, indirectly, as fluctuations in the flux tubes excited by the convection. To estimate the magnitude of the fluctuations, we note that the average energy flux, F_c, required by the corona and transition region, is of the order of 10^6 ergs cm^{-2} s^{-1}. Since the flux tubes are known to cover only a fraction f of the solar surface, the required local energy flux at the site of the tubes is $F_c(f\epsilon)^{-1}$ ergs cm^{-2} s^{-1}, where ϵ is the efficiency factor. We equate this to the flux of magnetic energy to obtain

$$\frac{2B\,\Delta\,B}{8\pi}\,V_A \;=\; F_c(f\epsilon)^{-1}\;, \tag{7}$$

where the Alfvén velocity, V_A, is given by

$$V_A \;=\; \left(\frac{B^2}{4\,\pi\rho}\right)^{\frac{1}{2}} \tag{8}$$

and ρ is the ambient density. We then use equations (7) and (8) to evaluate $\Delta B/B$ for B = 1500 G, $f = 10^{-3}$ and $\rho = 10^{-7}$ gm cm^{-3}, which is characteristic of the low photosphere. The result gives $\Delta B/B \approx 10^{-2}\,\epsilon^{-1}$.

If ϵ is of order unity, the fluctuations in B responsible for the coronal heating are so small that they would be extremely difficult to measure. However, this constraint might be too severe and a value of $\epsilon \approx 10^{-1}$ may be more realistic. This would increase $\Delta B/B$ to the 10% level. In addition, if the magnetic perturbations heat the chromosphere as well as the corona, the energy flux increases by a factor of ten and we expect $\Delta B/B \approx 10^{-1} \epsilon^{-1}$. Thus, the range of expected values for $\Delta B/B$ range from 10^{-2} to of order unity and a reliable determination of the magnitude of $\Delta B/B$ would provide an extremely useful guide to the nature and extent of the magnetic heating.

One of the curiosities of solar magnetic field structure is the frequent prominence of the lines dividing positive and negative polarities. In normal potential field configurations, one might expect that these "neutral lines" would be located at right angles to the field lines in regions of gradual change in the field structure, and that their primary claim to uniqueness would be in the vanishing, or small, slope of the field lines relative to the vertical direction. This is far from what is sometimes observed, however. The neutral lines often lie in regions of steep gradients in field strength. Furthermore, the field direction often has a strong component paralleling the neutral line indicating that the field is strongly sheared. The shear is evident in a variety of ways. It is directly observed at photospheric levels with vector magnetographs (Hagyard *et al.* 1984). In the chromosphere, it is apparent in the plasma structure through the patterns of elongated fibrils that tend to point towards the neutral line when they are some distance away but often turn to parallel the neutral line as they approach it. In such cases, long, dark Hα filaments frequently lie along and above the photospheric neutral ine. The shear persists into the transition region where it is observed in the plasma flow pattern as a strong velocity shear line overlying the photospheric magnetic neutral line.

The magnetic shear evident at the neutral line is part of an evolutionary pattern. In a somewhat typical case, the neutral line in a young region of newly emerging flux is unsheared. Both the field lines and the Hα fibrils and filaments cross the neutral line more or less at right angles. Such a pattern may persist for a few days before the shear develops. The transition from the unsheared to the sheared state (if it occurs) appears, at times at least, to be relatively abrupt, possibly occurring in a matter of a few hours (Foukal 1971). Detailed observations of such changes are lacking, however.

It seems likely that the shear in the magnetic field concentrates near the neutral line because this is where the field has the least rigidity, i.e., the lowest field strength. The cause of the shear is not at all clear, however. One possibility is the relative displacements (motions) of the poles (footpoints) of opposite polarities. No one has demonstrated, however, that such displacements are sufficient to produce shear of the magnitude that is observed. In particular, it seems very difficult to induce a rapid onset of shear by this mechanism. A second possibility often suggested for producing shear is the intrusion of new magnetic flux to create a highly non-potential field pattern that subsequently relaxes to a lower energy non-potential state that includes shear.

Still a third possibility to consider is that the shear arises from fluid flows in the photosphere, or below, that differentially transport the field lines, but not necessarily their footpoints. If, for example, the horizontal components of the magnetic field initially are $B_x \neq 0$ and $B_y = 0$, then an induced velocity field in which $V_y \neq 0$ and $\partial V_y / \partial X \neq 0$ (vertical vorticity) will distort the magnetic field such that B_y is no longer zero. Motions of this type ought to be very common in the solar atmosphere, as in any turbulent atmosphere, and provided they exist on the scale of supergranules or larger, they should be regarded as possible sources of magnetic shear.

The creation of magnetic shear along a neutral line appears to be a key ingredient to the occurrence of two-ribbon flares. Here we note only that this provides an excellent example of how phenomena at the photospheric level or below couple rather directly with high energy events in the corona. The proper study of such cause-and-event relationships demands observational capabilities extending from the visual spectrum to x- and γ-rays.

5. The Chromosphere

The chromosphere was originally defined by the nature of the spectrum seen at the extreme solar limb at the time of total eclipse. Specifically, the chromosphere denoted the region in which emission lines of neutral and singly ionized elements dominated the optical spectrum. It now is more customary to define the chromosphere in terms of temperature structure. Thus, for our purposes, we define the base of the chromosphere as the beginning of the temperature rise, i.e., the temperature minimum. We define the top of the chromosphere as the top of the temperature plateau formed by hydrogen ionization. Above this level, the temperature gradient increases markedly, and the temperature rise occurs essentially at constant gas pressure. As noted earlier, this transition in temperature gradient occurs when hydrogen is approximately half ionized. Thus, the top of the chromosphere marks the transition from mainly neutral gas to mainly plasma. As thus defined, the chromosphere extends from roughly 500 to 2000 km above the base of the photosphere, the gas pressure decreases from approximately 10^3 to 10^{-1} dynes cm^{-2} and the temperature increases from approximately 4300 to 8000 K. Notably, however, the electron density decreases by only about one order of magnitude, and the largest fraction of this occurs in the first 150 km. Figure 1 illustrates the mean model as given by Vernazza, Avrett, and Loeser (1981). This is a purely empirical model obtained by synthesizing a wide variety of data. The level at which hydrogen is half ionized is marked in the figure.

At the temperature minimum, the dominant radiation loss has already shifted from the continuum to the lines. Through the initial temperature rise, the radiation loss, and hence the heating rate, both in ergs gm^{-1} s^{-1} and in ergs cm^{-3} s^{-1} increases outwards. Across the temperature plateau, the radiation loss (heating) in ergs gm^{-1} s^{-1} also reaches a plateau and remains approximately constant until the top of the chromosphere is reached. The corresponding loss rate in ergs cm^{-3} s^{-1} decreases approximately exponentially with a scale height equal to the density scale height.

The most complete computations of the energy loss rate consistent with the empirical model (Anderson and Athay 1988) give a total net radiation loss rate for the chromosphere and temperature minimum region of 1.1×10^7 ergs cm^{-2} s^{-1}. The largest contribution to the total comes from Fe group elements ($\approx 45\%$). Other strong contributors are Ca ($\approx 31\%$), and Mg ($\approx 16\%$). In each case, the primary loss comes from the singly ionized ion. Because of the large proportion of neutral hydrogen, the early Lyman lines are still sufficiently opaque to suppress significant radiation losses in these lines.

One of the main sources of uncertainty in the above quotation of radiation losses is in the temperature minimum region. The loss rate in this region is particularly sensitive to relatively small changes in temperature, and, unfortunately, the empirical model is not defined with sufficient accuracy to allow much confidence in the computed rate of radiation loss. On the other hand, the fact that the energy loss rate rises rather sharply through the low chromospheric layers suggests that the contribution from the temperature minimum region is most likely of secondary importance. We cannot be certain of this, however, until a better empirical model of this region is available.

An interesting problem arises at the top of the chromosphere. Across the temperature plateau energy balance is achieved by ionizing hydrogen to maintain an approximately constant energy loss at a rate of 4.5×10^9 ergs gm^{-1} s^{-1} (Anderson and Athay 1988). Once hydrogen is mainly ionized, however, this mechanism of energy balance is no longer possible, and energy balance, if indeed a balance exists, must change rather markedly. As an average, of course, the energy input rate must change along with the change in radiation output.

The total energy radiated from the portion of the hot solar envelope at temperatures above 10^4K is approximately 6×10^5 ergs cm^{-2} s^{-1}, about half of which is in the Lyman-α line. The mass of the atmosphere above the chromosphere is 6×10^{-6} gm cm^{-2}. Thus, on average, the radiation loss rate above 10^4K is 10^{11}

ergs gm^{-1} s^{-1}. This is over an order of magnitude higher than the loss rate across the chromospheric plateau. It follows that the average heating rate per gm in the corona and transition region is over an order of magnitude higher than in the chromosphere. This represents a real change in the divergence of the energy flux since the column mass at the base of the corona is nearly the same as the column mass at the top of the chromosphere. Evidently, the higher temperature and/or increased ionization either enhances the dissipation of the energy flux heating the chromosphere or taps an entirely different energy flux that passes freely through the chromosphere. The latter alternative is a distinct possibility, as is clear from the following argument recently advanced by Anderson and Athay (1988).

The energy flux carried by a sound wave is given in ergs cm^{-2} s^{-1} by

$$F_w = \frac{1}{2} \rho V_a^2 V_s ,$$

(9)

where ρ is the density, V_a is the velocity amplitude, and V_s is the sound speed. Since we expect the wave energy to dissipate when V_a approaches V_s, we may approximate F_w by replacing V_a with V_s. We then note that V_s is approximately constant within the temperature plateau, and hence, that the divergence of F_w is controlled by the divergence of ρ. Thus, we may write

$$\nabla \cdot F_w = \frac{1}{2} V_s^3 \, \nabla \cdot \rho , ergs \ cm^{-3} \ s^{-1} ,$$

(10)

which we may rewrite as

$$\nabla \cdot F_w = \frac{1}{2} \frac{V_s^3 \rho}{H_\rho} , ergs \ cm^{-3} s^{-1} ,$$

(11)

where H_ρ is the density scale height. To convert to ergs gm^{-1} s^{-1}, we divide equation (11) by ρ to obtain

$$\frac{\nabla \cdot F_w}{\rho} = \frac{1}{2} \frac{V_s^3}{H_\rho} .$$

(12)

We next replace V_s with

$$V_s = \left(\frac{\gamma P}{\rho} \right)^{\frac{1}{2}}$$

(13)

and H_ρ with

$$H_\rho = \frac{kT}{\mu m_g}$$

(14)

to obtain

$$\frac{\nabla \cdot F_w}{\rho} = \frac{1}{2} \gamma g \left(\frac{\gamma kT}{\mu m} \right)^{\frac{1}{2}} , ergs \ gm^{-1} \ s^{-1} ,$$

(15)

where m is the hydrogen mass. As a mean for the chromospheric temperature plateau, we adopt $\gamma = 1.1$ (due to hydrogen ionization), $T = 7000$ K and $\mu = 1$. The resulting maximum energy flux is 1.2×10^{10} ergs gm^{-1} s^{-1}, which is remarkably close to the energy flux required to match the empirical model, but is a factor of eight below that required for the corona. This suggests two conclusions: (1) sound waves are a good, if not likely, candidate for heating the chromosphere; and (2) they are not a candidate for heating the corona. The latter conclusion was reached by Athay and White (1978) from a similar consideration.

If the chromosphere is heated by short-period sound waves, we should expect to observe velocity amplitudes near the sound speed, which is given to good approximation by $V_s = .091(\gamma T)^{\frac{1}{2}}$. Although there is no real consensus on observed velocity amplitudes in the chromosphere, there is agreement that the non-thermal component of line broadening gives so-called "microturbulence" velocities, ξ, that approach the sound speed in the temperature plateau. Table 1 summarizes values of ξ derived by different authors. Hirayama and Irie's (1984) results are based on widths of metal lines observed at eclipse and those of Tripp *et al.* (1978) are based on a synthesis of Si II emission line profiles observed in the far ultraviolet. The results given by Beckers and Canfield (1975) are distilled from a variety of earlier publications containing rather disparate results. None of the results shown in Table 1 is determined to an accuracy better than about 10%, and in most cases, the uncertainty is even larger.

Table 1
Non-Thermal Line Broadening Velocities,
ξ, in km s^{-1}

Height	Hirayama and Irie (1984)	Beckers and Canfield (1975)	Tripp et al. (1978)	$V_s/\gamma^{\frac{1}{2}}$
800	4	1.7	2	6.9
1000	6	2.7	6	7.0
1500	9	5.3	7.2	7.2
2000	12	8.3	7.8	7.8

The plateau in the VAL model extends from approximately 1000 to 2000 km. Tripp *et al.* examined a range of parameters and selected a model with $\xi = V_s(\gamma = 1)$ within the temperature plateau, which accounts for the exact agreement shown in the table. It is notable that throughout the plateau, the average of the three sets of observational results give ξ within 30% of $V_s/\gamma^{\frac{1}{2}}$. Clearly, we need to determine ξ with more precision, but it already seems clear that measured values of ξ are broadly consistent with heating by short-period sound waves. This does not, of course, rule out other heating mechanisms. On the other hand, the approximate equality between the radiation loss rate and the wave energy flux obtained by setting $V_a = V_s$ together with the approximate equality of V_s and ξ supports the assumption that heating by sound waves is of major importance.

Chromospheric oscillations observed in UV emission lines of Si II by OSO-8 (Athay and White 1979a) show regular oscillations in the five-minute band. The velocity amplitudes, however, are much below the sound speed. This eliminates the five-minute oscillations as a major source of chromospheric heating but does not exclude the higher frequency waves.

Chromospheric structural detail looks quite different in different spectral lines. Part of this is due to the increased temperature sensitivity of the collision rates at shorter wavelengths, but a larger part is due to differences in the gradients in the atomic level populations from which the lines arise. In strong lines of abundant metals, the excitation potentials are low and level populations decrease exponentially with a scale height about equal to the density scale height H_ρ. In such lines, one observes a relatively narrow range of height roughly equivalent to $2H_\rho$.

A much wider range of heights can be observed in a few spectral lines that arise from atomic energy levels lying several electron volts above the ground state. The rising temperature through the chromospheric layers causes the populations of these excited levels to decrease much more slowly with height than the ground state populations, thereby broadening the layers that are observed. The Hα line of hydrogen provides the best example of such behavior. In the VAL average Sun model, for example, the entire layer extending from the top of the chromosphere (as defined here) to within 150 km of the temperature minimum show a variation in the population density of the n = 2 level in hydrogen by only a factor of 8. By comparison the total hydrogen density varies through a range of 10^4. Fortunately, the absolute values of the n = 2 population give sufficient Hα opacity to make the chromosphere readily observable but not so large as to obscure the underlying layers. It is for these reasons that Hα images of the chromosphere reveal a wealth of structural detail not seen in any other line. An example is shown in Figure 2.

Chromospheric structure largely reflects the magnetic field geometry. In strong field regions (sunspots and plages), the field is nearly vertical and the Hα intensity shows a pattern of finely divided bright dots. In weaker field regions, the field develops a horizontal component and the Hα intensity reveals patterns of elongated fibrils. Away from active regions, the fibrils are organized on the scale of the supergranules with clusters of fibrils converging at the network bordering the supergranule cells. Within active regions, the fibril patterns reveal a larger scale organization consistent with the overall pattern of the magnetic field. A frequently occurring and striking feature of the active region fibril pattern is the curving of the fibrils as they turn to parallel the long dark filaments. The dark filaments, in turn, overlie the magnetic neutral lines in the photosphere. The fibril pattern paralleling the dark filaments is consistent with other evidence indicating that the field is strongly sheared near the neutral line.

Magnetic shear is not present at all neutral lines, or at all times on a given neutral line. Similarly, not all neutral lines have overlying dark filaments. The dark filaments seen on the disk come in three classes, all associated with neutral lines: (1) active region filaments that are low-lying and thus inconspicuoius at the limb; (2) arch filament systems associated with newly emerging flux; and (3) quiescent filaments bordering or well away from active regions. The latter are typically elevated above the chromosphere and appear as prominences at the limb. They do not usually show evidence of strong shear. Similarly, the arch filaments appear as short dark threads crossing the neutral line more-or-less at right angles again suggesting that the field is not strongly sheared. The active region filaments, however, are characteristically associated with strong shear (Foukal 1971). (For a recent review of filaments see Schmieder 1988).

Chromospheric spicules seen at the solar limb as jet-like eruptions rising out of the upper chromosphere with speeds of about 20 km s^{-1} appear to lie over the network. The spicules are oriented close to the vertical and are believed to follow the more nearly vertical field lines emerging from the central regions of the photospheric flux tubes. Otherwise, the spicules have similar dimensions (approximately 1$''$ diameter) and lifetimes (10 min.) to the fibrils. It may be, in fact, that the principal distinction between spicules and fibrils is their orientation. This, however, is conjecture and may prove to be incorrect.

Spicules are of major interest in solar physics for a variety of reasons. Estimates of spicule densities and temperatures (Beckers 1972) give densities near those in the upper chromosphere and temperatures somewhat above 10^{4}K. Moreover, the density decreases as the spicule rises with a scale height that is much larger than the scale height in the upper chromosphere. This suggests that the spicule likely has its origin in the upper chromosphere.

The upwards mass flux in spicules is large when averaged over the Sun, the mass flux in spicules exceeds the solar wind flux by some two orders of magnitude and is sufficient to replace the coronal mass at the rate of about 50 coronal masses per day. More importantly, the rate at which work is expended against gravity in lifting the spicule mass, when averaged over the surface, is estimated at 4×10^5 ergs cm^{-2} s^{-1}, which can also be expressed as approximately 10^{11} ergs gm^{-1} s^{-1}. We recall that this is the same as the average energy flux that heats the corona.

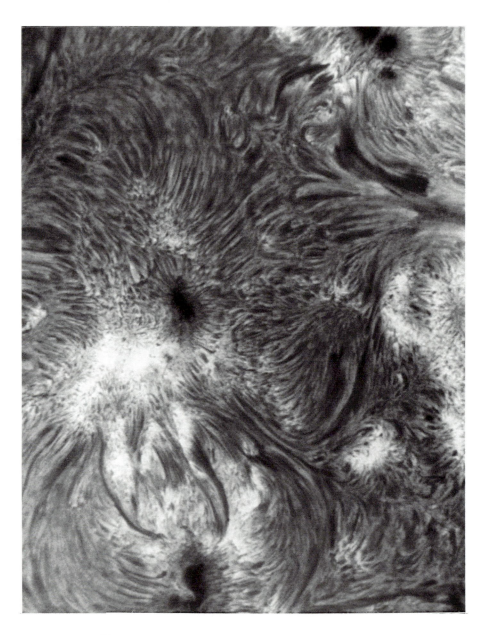

Figure 2   An $H\alpha$ image of a solar active region. The bright plage and sunspot umbrae (darkest compact features) have strong magnetic flux with mainly vertical field lines. Elsewhere, the elongated fibrils and larger dark filaments reveal more nearly horizontal field patterns. (Photo courtesy of Big Bear Solar Observatory, California Institute of Technology).

The material ejected upwards in spicules must eventually return to its level of origin and the work expended in lifting the material must eventually reappear as heat energy and subsequently as radiation flux from the chromosphere and/or transition region and corona. In this regard, we note that the spicule ejections are observed at temperatures near 10^5K (Dere *et al.* 1987) along with a more general downflow (Doschek, Feldman, and Bohlin 1976) of material with approximately the same mass flux as estimated from the Hα ejections. It appears, therefore, that the spicule material is heated to at least 10^5K in the process of ascending. However, no downflow is observed near 10^6K (Doschek, Feldman, and Bohlin 1976) so there is no reason to believe that the spicule matter reaches coronal temperature.

The near equality of the energy flux in spicules and the radiation flux from the transition region raises questions as to how the two are related. Is there a single primary energy flux at the top of the chromosphere that is expended equally between lifting the spicules and heating the corona and transition region? Does the major heating itself involve the lifting of spicule matter as a precursor? Or, are there two separate mechanisms of heating – one that includes the lifting of spicules and one that goes more directly into heat energy? We cannot yet answer these questions.

Among the other issues yet to be resolved are the following: the mechanism by which spicules are accelerated, the nature and origin of the chromospheric motions associated with line broadening, the wave modes present in the chromosphere, the origin of magnetic shear, the release of shear energy to create flares, magnetic field topology in and near dark filaments, the amplitude of plasma fluctuations associated with fibrils, and the origin of helium line emission at chromospheric levels. Each of these is a topic of long-standing interest and must be resolved before we can claim to understand the observed phenomena and physics of the chromosphere.

In order to resolve many of the above issues, we need detailed observations in a variety of lines formed at different chromospheric levels. Most of the more useful lines for this lie in the uv and EUV. In particular, the emission lines of H, He I, He II, C I, C II, O II, Si II, and Fe II deserve careful attention.

6. The Transition Region

The choice of the term "transition region" to denote the layers sandwiched between the chromosphere and corona is perhaps unfortunate in that it does not give sufficient recognition to the importance and variety of the phenomena occurring there. The transition region is unique in a variety of ways. It is not only the region of steepest temperature gradients, but also exhibits the most variability and the largest non-thermal velocities. The temperature rise from 8000 K to 10^6K occurs in less than one pressure scale height.

The transition region is effectively thin in all spectral lines, which means that all photons resulting from collisional excitations of atoms and ions escape the transition region and represent a real energy loss. One need only sum up the energy radiated in the different lines, therefore, in order to determine the rate of energy loss, hence rate of heating. The rate of energy loss, q_{ij}, in a given line is expressed by equation (6).

To obtain the radiation flux emerging from the atmosphere it is necessary to integrate equation (6) over depth. In practice, it is customary to take an average of $N_e N_i e^{h\nu_{ij}/kT}$ over height and replace the integrated form of equation (6) with

$$\frac{q_{ij} \ e^{h\nu_{ij}/kT}}{c_{ij} \ A_i} \ = \ N_e^2 \ \Delta h/\Delta lnT \tag{16}$$

where A_i is the relative abundance of the ion and $\Delta h/\Delta lnT$ is the inverse of the logarithmic temperature gradient near the temperature T_{ij} for which N_i is a maximum. The quantity q_{ij} is obtained by observation and the remaining terms on the left-hand side of the equation are either known constants or can be evaluated from T_{ij}. To determine T_{ij}, one assumes a balance between collisional ionization and radiative recombination, which uniquely defines T_{ij} as well as ΔlnT_{ij}.

In deriving q_{ij} from observations, the best we can achieve is the average value of the true q_{ij} over the area corresponding to the observed line intensity. If the emission is spatially discontinuious, which some evidence suggests, we should write

$$q_{ij} \ (obs) \ = \ f \ q_{ij} \ , \tag{17}$$

where f is a fill factor representing the fraction of the surface within the resolution element that is emitting.

The left-hand side of equation (16) with q_{ij} replaced by $q_{ij} \ (obs)$ is known as the differential emission measure, which we denote by $E(T)$. Since $E(T)$ depends only on atmospheric parameters, a curve defining $E(T)$ can be derived from observations combining a number of spectral lines formed over a range of values of T_{ij}. Similarly, summing equation (6) over all transitions that emit within a given temperature range yields a curve

$$Q(N_e, T) \ = \ Q_0 \ q(T) \ N_e^2, ergs \ cm^{-3} \ s^{-1} \tag{18}$$

giving the total radiation loss rate in ergs cm^{-3} s^{-1} as a function of temperature. Plots of $E(T)$ and $q(T)T^{-2}$ are shown in Figure 3. The latter represents the radiation loss for constant gas pressure.

Since we have replaced q_{ij} by $q_{ij}(obs)$ in defining $E(T)$, we need to multiply equation (16) by f before equating $E(T)$ to the right-hand side. Thus, we obtain

$$E(T) \ = \ N_e^2 \ f \ \Delta h/\Delta lnT \ . \tag{19}$$

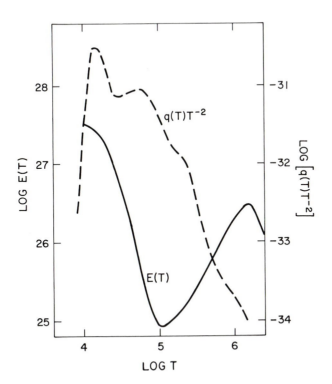

Figure 3 Plots of the emission measure $E(T)$ and radiation loss function $q(T)T^{-2}$ for constant pressure. The peak in $q(T)T^{-2}$ near log T=4.2 is due to the Lyman-α line.

If we interpret $\Delta lnT/\Delta h$ as a logarithmic temperature gradient and note that the classical thermal conduction flux is given by

$$F_T = K_0 \, T^{5/2} \, \frac{dT}{dh} \,, \tag{20}$$

we can then solve equations (17) and (18) in terms of the ratio of the two unknowns F_T and f to obtain

$$\frac{F_T}{f} = \frac{K_0 \, T^{7/2} \, N_e^2}{E(T)} \,. \tag{21}$$

It is known from empirical chromospheric and coronal models that the mean electron pressures at the top of the chromosphere and at the base of corona at 10^6K are practically the same. This is entirely consistent with empirical models of the tansition region, which indicate that $P_0 = N_e T$ is essentially constant. Thus, equation (21) can be rewritten as

$$\frac{F_T}{f} = \frac{K_0 \, P_0^2 \, T^{3/2}}{E(T)} \qquad (22)$$

The plot of $E(T)$ in Figure 3 is closely approximated between 10^5 and 10^6K by

$$E(T) = E_0 \, T^{3/2} \, . \qquad (23)$$

This combines with equation (22) to give

$$\frac{F_T}{f} = \frac{K_0 \, P_0^2}{E_0} \, , \qquad (24)$$

and we have the interesting result that in the upper half of the transition region F_T/f is approximately constant. Using $K_0 = 1.1 \times 10^{-6}$, $P_0 = 4 \times 10^{14}$ from standard models and $E_0 = 1.6 \times 10^{17}$ from the plot in Figure 3, we find $F_T/f = 10^6$ ergs cm^{-2} s^{-1}. The apparent constancy and magnitude of F_T/f have dominated our concepts of the transition region ever since the first realization over three decades ago that the conductive flux could be directly related to observations. It is still not clear what the results mean in terms of the structure and physics of the transition region. In the following, we consider two alterntive inter pretations: one in which $f = 1$ and a second in which $f << 1$.

By setting $f = 1$, we conclude that the conduction flux is roughly constant at 10^6 ergs cm^{-2} s^{-1} through the temperature range 10^6 to 10^5K. This is a highly appealing result since the conduction flux is sufficient to supply all of the energy needed to balance the radiation loss from the transition region. Furthermore, the energy radiated between 10^5 and 10^6K is only 1×10^5 ergs cm^{-2} s^{-1}, so F_T should indeed be nearly constant. Because of the natural appeal of such a model, most of the extant models of the transition region adopt this approach. An important conclusion from models of this type is that essentially all of the unexpended mechanical energy flux at the top of the chromosphere passes into the corona and is dissipated at temperatures in excess of 10^6K. Furthermore, the vast majority of this energy is subsequently conducted back through the transition region where it is lost by radiation.

Models of the above class, as well as static coronal loop models that are closely related, have the appeal of being relatively simple in concept and intuitively attractive. On the other hand, they face two strong objections. All models of this type (cf. Athay 1981b) lead to predicted emission measures, $E(T)$, that decrease monotonically with decreasing temperature. This is a direct and unavoidable consequence of the strong temperature dependence of the conductive flux (equation

20). Additionally, there is strong observational evidence that $f \ll 1$, which violates the basic assumption on which the models are based. The latter conclusion is reached (Feldman, Doschek, and Mariska 1979; Feldman 1983, 1987) from a variety of arguments based on observational data. The most direct illustration that $f \ll 1$ is obtained by comparing the mean values of Δh inferred from equation (14) to the observed width, Δh^*, of the emitting region at the limb. Consistent with the definition of f used here

$$f = \frac{\Delta h}{\Delta h^*} \tag{25}$$

and is found to be less than 10^{-2} at temperatures near mathattwelve10^5K.

Inhomogeneous structure in the corona has been studied by a variety of methods. The most recent (Orrall and Rottman 1986) has been a comparison of the average electron density, $< N_e >$, inferred from electron scattering intensity to the quantity $< N_e^2 >$ inferred from emission line intensities. The quantity X defined by

$$X = < N_e^2 > / < N >^2 \tag{26}$$

is found to be large compared to unity ($\simeq 10$) both in coronal holes and active regions. This is compatible with a number of earlier studies (cf. Feldman, Doschek, and Mariska 1979). It is not clear just how to relate X to the fill-factor f. However, inferred values of Δh based on emission measures near 10^6K are of the order of 3000 km. By comparison, the effective thickness of the emission line corona (intensity scale height) observed at the limb is bout 3×10^4 km. Thus, from equation (25) we would obtain $f = 10^{-1}$. This estimate of f is rather crude, but taken together with the estimates of X, it is sufficient to demonstrate that f is much larger near 10^6K than at 10^5K.

If we accept that $f \ll 1$, then we face just the opposite conclusions from the models with $f = 1$, viz., the mean conductive flux is too small to supply the heat input necessary to balance the radiation loss from the transition region. Thus, most of the unexpended mechanical energy at the top of the chromosphere must be dissipated directly in the transition region. One should then interpret the $E(T)$ curve as being fixed by the energy dissipation as a function of temperature.

Two important factors in transition region models not included in the preceding discussion are the heating of spicule material and its subsequent cooling. Similar and well correlated spicule structure is observed at temperatures from near 10^4 to 10^5K (Dere *et al.* 1987) and, in addition, most of the plasma at temperatures near or below 10^5K shows widespread descending motion. It is clear, therefore, that much of the rising spicule material reaches temperatures in excess of 10^5K.

The hot descending spicule material carries an enthalpy flux given by

$$F_{en} = \frac{5}{2}nkTVf(T) .$$ (27)

In the transition region $nkT \approx 0.2$ and at 10^5K the observed velocity is near 10 km s^{-1}. Assuming that the observed velocity actually represents $Vf(T)$, we find $F_{en} \approx 5 \times 10^5$ ergs cm^{-2} s^{-1}, which, again, is remarkably close to the observed radiation flux originating in the transition region below 10^5K. It is also of interest that the enthalpy flux carried downward by the hot plasma near 10^5K is closely equal to the rate at which energy is expended in lifting the spicules against gravity (see the preceding section on the chromosphere). A possible scenario suggested by Pneuman and Kopp (1978) is that the mechanical energy flux coming through the chromosphere first accelerates and heats the spicule matter, and, somewhat later, the falling, heating matter cools by radiation loss. The trouble with this model is that it does not explain the $E(T)$ curve below 10^5K any better than does the conduction model. Similarly, a combination of conduction and downflow does not improve the situation (Athay 1981b).

What we are left with, therefore, are two identified sources of energy that are ample to match the total radiation output in the transition region but neither of which seems capable of satisfying the observed $E(T)$ curve. Although several attempts have been made to explain the shape of $E(T)$ for $T < 10^5$K including heating of small magnetic loops by finely divided electric currents (Rabin and Moore 1984), fluctuating heating and cooling cycles (Athay 1984a) and systems of cool loops (Antiochos and Noci 1986), none has gained wide acceptance. Thus, we are forced to admit that we really know very little about the nature of the transition region, and more especially so in the temperature range below 10^5K.

The eruptions of spicules from the top of the chromosphere and the return downflow of 10^5K plasma is ample evidence of the dynamic nature of the transition region. This is not the only evidence, however. The small scale motions reflected in the non-thermal component of line width continue to increase into the transition region. Results from the *Skylab* mission (Mariska, Feldman, and Doschek 1978) for T between 1.6×10^4 and 2×10^5K can be represented reasonably well by the relation

$$\xi = 0.4 \, V_s$$ (28)

In addition, the NRL HRTS data for lines formed near 10^5K show a frequent occurrence of high speed ejecta (Brueckner and Bartoe 1983) accelerating upwards to velocities as high as 400 km s^{-1}. This is approximately twice the sound speed in the corona but approximately equals the estimated Alfvén speed near 10^5K.

Even at relatively low spatial resolution, the transition region near 10^5K shows wide ranges in brightness from point to point on the disk. Bright regions frequently exceed the average by a factor of 10^2 and the fainter region within the supergranule cells are often below average by a factor of 10^{-2}.

In some active regions, there is almost a continuous display of "bursts" with the characteristics of small flares (Emslie and Noyes 1978; Athay 1984b; Porter *et al.* 1987). Although the brightness patterns in the transition region show persistent enhanced emission in the network and in active regions consistent with the patterns observed in the chromosphere, the actual brightness at a given location fluctuates on time scales of a few minutes with amplitudes often exceeding 100%. The network structure is very pronounced in the 10^4 to 10^5K regime, but fades at higher temperatures and is no longer recognizable at 10^6K. This is usually interpreted as a spreading of the magnetic field to completely fill the corona.

Extensive observations of Doppler shifts in CIV lines formed near 10^5K using the OSO-8 satellite exhibit oscillations in the 5-minute band a major fraction of the time. However, the oscillations are of low amplitude and indicate an energy flux (assuming they are due to propagating pressure waves of only 1×10^4 ergs cm^{-2} s^{-1} (Athay and White 1979b). This is consistent with our earlier arguments that heating by sound waves is unimportant in the transition region and corona.

The very steep temperature gradients in the transition region, if indeed they exist, provide a number of challenging theoretical and interpretational problems. If we set the fill factor to unity near 10^5K, the temperature scale height, Δh, falls below 10 km, which is less than one mean-free-path for the more energetic particles. This leads to a situation in which electrons in the high energy tail of the thermal distribution function can penetrate downward into regions of markedly lower temperature. Thus, the local electron distribution function is non-maxmillian. Studies of the resulting distribution function together with the influence of the enhanced high energy tail on ionization equilibrium and thermal conduction by Shoub (1983) and others show that major effects are present at the temperature gradients and densities indicated for the transition region.

Further work is needed in these important problems. However, we must also establish the true temperature gradients by pursuing observations in the transition region at high spatial resolution.

The issue of transition region structure is a critical factor in resolving a number of outstanding problems including the magnitude of the thermal conduction, local energy balance, and velocity amplitudes. Without resolving the fine structure, we can only speculate on its nature and influences. Other issues that are critical to an understanding of the transition region include those associated with spicules. How is the spicule material heated, and how high does the temperature rise? How far does the spicule material penetrate into the corona? These questions are important to understanding the role of the spicules as a heat source (Athay and Holzer 1982), the observed form of $E(T)$ and the true nature of the velocity field.

One of the more intriguing questions concerning the transition region is whether there is an equilibrium state. The marked variability and dynamic nature of the plasma suggests that it may be continually out of equilibrium. The very limited ability of the plasma to adjust its radiation loss to a specific energy input suggests a similar conclusion. If such is indeed the case, we need to adopt new approaches to the type of data collected and its interpretation (Athay 1984a).

Studies of the transition region necessarily rely heavily on uv and EUV observations. The wide temperature span requires that many ions be studied ranging from neutral hydrogen and the singly ionized abundant metals through multiply ionized states such as Mg X. Since approximately half of the radiation from the transition region is in the Lyman-α line, particular attention should be devoted to studying the region producing Lyman-α.

7. The Corona

It has been clear since the *Skylab* era that the corona is largely controlled by the magnetic field. On the largest scale, the brighter regions of the corona correspond to areas in which the magnetic field within about $1.5R_\odot$ of the photosphere is closed on the sun. Darker regions, known as coronal holes, correspond to areas where the field opens to interplanetary space.

Coronal holes, as shown in Figure 4, are observed as areas of sharply depleted emission due to reduced density and lower coronal temperature. They were first discovered at the limb in observations of forbidden emission lines. It was not until they were observed on the solar disk in soft x-ray emission that their uniqueness was realized, however. Subsequent study has revealed some of the evolutionary aspects of coronal holes related to the sunspot cycle. More importantly, the coronal holes have been identified as the source of high speed solar wind streams and as the mysterious solar M-regions responsible for recurrent geomagnetic storms (cf. review by Hundhausen 1977).

Although coronal holes are most readily observed on the disk in soft x-ray and EUV emission lines, such observations have been relatively limited in duration. Much was learned from *Skylab* data, but the observing period of *Skylab* lasted for only a few months, which is about the lifetime of a large coronal hole at low solar latitude. Since the larger coronal holes have a long lifetime relative to the solar rotation period, their evolution can be observed in limb data with extended temporal coverage. As a result, most of what is known concerning the temporal evolution of coronal holes has been derived from synoptic limb observations

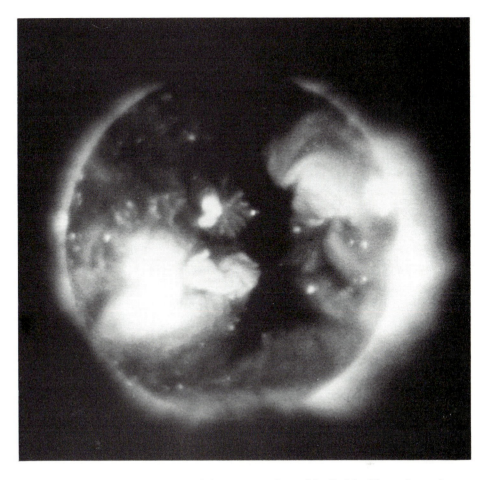

Figure 4　A soft x-ray picture of the corona taken with *Skylab*. The polar regions are to the top and bottom. A large dark coronal hole extends from one pole across the equatorial belt. (Photo courtesy of American Science and Engineering.)

of the K-corona. The K-corona is produced by electron (Thomson) scattering of photospheric radiation and can be detected in polarized light at the ground from high altitude sites with clear air. From studies of 12 years of data, Hundhausen, Hansen, and Hansen (1981) demonstrated that polar coronal holes are largest near sunspot minimum and decrease in area as activity increases becoming absent or nearly so at sunspot maximum and the year or two immediately following. Low latitude holes arise either as separate entities or as extensions of polar holes. They are observed at all phases of the sunspot cycle. The total area covered by coronal holes appears to maximize near sunspot minimum and decreases by a large factor at maximum activity (Fisher and Sime 1984). Most, if not all, of the modulation is accounted for by changes in the polar holes and their low latitude extensions, and it is unclear how the low latitude holes that are not extension of the polar holes vary with the activity cycle. It is the low latitude holes that give rise to the high speed wind streams observed in the ecliptic plane at 1 A.U. and that are identified with the M-region (Hundhausen, Hansen, and Hansen 1981). The tendency of the strong recurrent M-regions to be most evident in the declining phase of the sunspot cycle apparently results from increased lifetimes of the low latitude holes during this epoch.

In a study of one coronal hole located over a solar pole, Munro and Jackson (1977) found that the boundaries of the coronal hole diverged rapidly with distance above the solar surface.

At the base of the corona, the hole as seen in soft x-rays extended some 20 ° in latitude in either direction from the solar north pole. As seen at $2.0R_\odot$ in the K-corona, however, the hole extended approximately 60 ° in latitude in either direction from the pole. This rapid fluting of the polar holes very likely results from the strong predominance of closed field regions in low solar latitudes so that far from the sun the field, even at low latitudes, is dominated by the open fields in the polar regions. It may be, therefore, that low latitude coronal holes diverge much more slowly with distance from the surface. The rate of which the field flares plays an important role in determining the asymptotic speed of the solar wind and may be a factor in producing the high speed streams (Kopp and Holzer 1976).

Somewhat curiously, the coronal holes seem to have little influence on the underlying transition region and corona. The only apparent direct indication of the holes in the transition region and chromosphere is in the lines of He I and He II. The EUV He II emission lines exhibit the coronal holes in rather moderate contrast. In addition, there is indirect evidence of the holes in that the spicules are somewhat taller and larger and the transition region lines decrease less rapidly with height above the limb.

The helium lines are distinct from other transition region lines in the way they are excited. A majority of the excitation is by electron collisions (Seeley and Feldman 1985, Athay 1988), but there is an added component excited by coronal soft x-rays. The soft x-rays create alpha particles that recombine with free electrons to produce enhanced emission in the He II lines, especially in the Lyman-α line at 30.4 nm. The radiation in this line, in turn, photoionizes He I and electron recombinations to He II enhance the He I emission. At the location of coronal holes where the coronal x-ray flux is strongly reduced, this process of helium excitation is reduced. The result is observed in the 30.4 nm He II line as reduced brightness and in the 1083 nm line of He I as reduced absorption. The effect in the 1083 nm line is often rather subtle, however, and the coronal hole imprint is not readily identified. Also, the imprint in the transition region and chromosphere is over an area comparable in extent to the area of reduced soft x-ray emission in the corona, and is fixed therefore by the size of the coronal hole in the low corona.

A number of studies of coronal holes have been carried out using the 1083 nm line of He I as a proxy measure of the holes (cf. Sheeley and Harvey 1978). It is not clear what confidence can be placed in such studies. The holes are identified as regions of reduced absorption in the He I lines. The absorption itself is patchy in character over the entire solar surface and can vary from place to place for reasons that have nothing to do with coronal holes. Moreover, the soft x-ray flux commonly has a highly irregular distribution along the borders of coronal holes and the radiation falling on the chromosphere will necessarily be irregular also. There seems little doubt that the presence of major coronal holes can be detected using the He I absorption. Quantitative measures of the associated coronal hole area are suspect, however.

In closed field regions, the classical large scale feature is the coronal streamer, an example of which is illustrated in Figure 5. The lower, bulbous shaped portion of the streamer does not usually extend above about 2.5 $R_\odot$. This portion straddles the photospheric neutral line and frequently surrounds a quiescent prominence. The structural patterns observed within and surrounding the bulbous portion of the streamer are consistent with those of a long magnetic arcade that follows the meanderings of the neutral line separating large regions of opposite polarity outside of active regions. The long thin rays extending above the lower bulbous portion of the streamer are believed to follow lines of force that open to outer space. A current sheet separates the oppositely directed lines of force on either side of the central ray. This suggested magnetic field picture has been pieced together over many years of study from computed coronal magnetic field patterns based on photospheric measurements and from observed magnetic sector structure in the interplanetary solar wind.

In the active regions, field lines are mostly closed either between sunspots of opposite polarity or between a sunspot and a nearby plage. Patterns vary from one region to another and with the evolution of individual regions. Thus, no specific pattern for the coronal structure over active regions has been recognized. However, the corona over active regions is usually hotter and denser than in the streamers and, as a result, the active regions are characterized by unusually strong emission.

Figure 5 An eclipse photograph of the corona through a neutral density filter with a radial transmission gradient design to suppress the bright inner corona relative to the outer corona. A large coronal hole in the south polar region is bordered on either side by coronal streams. The small bright features near the lunar limb are prominences. (Photo courtesy of High Altitude Observatory.)

The large scale structure of the corona follows the pattern set by the large scale magnetic field. Within this pattern, the local coronal plasma is highly structure on small spatial scale. Evidently coronal heating is distributed in a highly non-uniform and finely divided manner. This gives rise to individual brighter loops and rays tracing out the particular small flux elements experiencing enhanced heating. Such structures are especially evident in soft x-rays (Figure 4) and in EUV emission lines (Figure 6).

Much effort has gone into modeling both static loops and loops with plasma flow along the loop axis. The literature is too extensive to discuss here, and the reader is referred to a recent review by Zirker (1986). In essence, the problem of modeling static loops is not basically different from modeling static plane parallel transition regions. The heat input is subject to redistribution by thermal conduction along the loop and is lost either by radiation or conduction through the ends of the loop. An assumed hydrostatic pressure balance imposes a relationship between the maximum temperature and height of the loop, and the total radiation loss is related to the maximum temperature and plasma density. Clearly, one must be careful not to over specify the conditions for the loop otherwise a stable equilibrium may not exist. In a plane parallel atmosphere, the effective height of the atmosphere is set by the temperature, which, in turn, is fixed by the heating function, and a stable equilibrium may require proper freedom at the boundaries. The same is true of loops, and it is not surprising, therefore, that instability results when the loop parameters and boundary conditions are over specified.

On a still smaller scale than the loops and rays are the coronal bright points seen in soft x-ray and EUV images. These are evident in Figures 4 and 6 and have been identified with small magnetic bipoles in the photosphere. The bright points occur at all solar latitudes but most often in active latitudes. Similarly, they occur throughout the sunspot cycle. Their rate of birth varies only slowly throughout the sunspot cycle but out of phase with the birth of large active regions (Golub, Harvey, and Webb 1986).

The small bipoles with which the bright points are associated may be newly emerging ephemeral regions, relatively long-lived dipoles or chance encounters of two poles of opposite polarity not previously connected. About 50% are associated with the dipoles, 30% with ephemeral regions and 20% with chance encounters (Golub, Harvey, and Webb 1986).

In view of the evidence for pronounced coronal structure on both large and small spatial scales, it seems probable that the majority of the coronal plasma is contained in small relatively isolated features and that the diffuse ambient background corona is of secondary importance. A number of observers (cf. Orrall and Rottman 1986) have concluded from comparisons of observed intensities in the election scattering continuum and in various emission lines that the "irregularity factor" $X = < Ne^2 > / < Ne >^2$ is much greater than unity in the inner corona and increases outwards. (See preceding section on the transition region.)

A new era in coronal physics began with the discovery of coronal mass ejections, CMEs. The CMEs were first observed with *Skylab* and later by the Air Force P-78 satellite and *Solar Maximum Mission*. The larger and brighter ones are also observed with the ground-based K-corona meter in Hawaii. It was soon recognized that CMEs come in a variety of sizes, forms and speeds, ranging from large, high-speed ejections to the small, indistinct and sometimes slow ejections. In the less distinct cases, it is difficult to establish a clear set of criteria that define a CME,

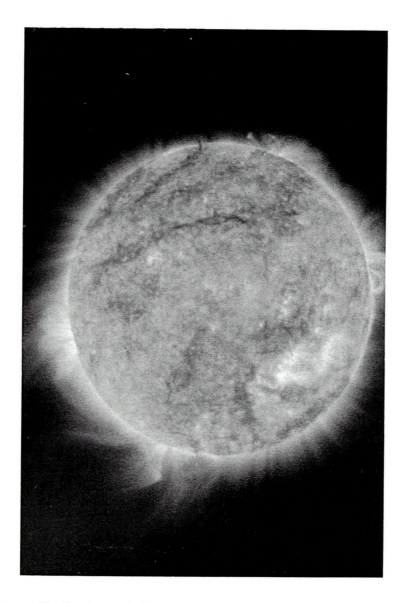

Figure 6   The first image of the corona in soft x-rays obtained October 23, 1987, with a normal-incidence Cassegrain telescope with multilayer optics. The bandpass is $17.1 - 17.5$ nm, and the photons are mainly in lines of Fe IX and Fe X formed at temperatures of $0.95-1.3\times10^6$K. Angular resolution is $1 - 1.5$ arcsec.

Coronal holes can be seen near the two poles in the upper left and lower right quadrants. A smaller coronal hole is visible near the bright active regions at high southern latitudes. The elongated dark features correspond to dark filaments observed in Hα. Closed loops and open rays are seen beyond the limb and a number of bright points are seen on the disk. (Photo courtesy of Arthur B. C. Walker, Jr., Joakim Lindblom, and Troy W. Barbee, Jr., Center for Space Science and Astrophysics, Stanford University; Richard B. Hoover, NASA-George C. Marshall Space Flight Center.)

and different observers use somewhat different criteria. The larger events, however, are readily and easily recognized. An example is shown in Figure 7.

Large CMEs consist of three rather distinctive components (Illing and Hundhausen 1985). The innermost of these consists of dense knots of prominence material that are the remnants of an erupting quiescent prominence (dark filament) and may remain visible in Hα out to several solar radii. The outer most of the principle CME features is a shell of denser than average coronal material visible in white light as a bright arch surrounding the prominence material at some distance away. The third component is sometimes less distinct and may not always be present. It lies inside the bright coronal arch but precedes and surrounds the prominence material. This pattern is closely similar to the familiar structure observed in the lower portions of coronal streamers, viz., a prominence surrounded by a dark cavity, which, in turn, is encased by brighter coronal material.

After studying a large number of events, Hundhausen (1988) and his co-worker describe a large CME as the eruption of a large-scale magnetic system carrying with it the imbedded plasma and sweeping up the overlying coronal material in its outward expansion. For example, a coronal magnetic arcade at the base of a streamer may become unstable and expand outwards carrying the encased prominence and its surrounding cavity as it expands. Such a picture appears to be consistent with a major fraction of the well-observed CMEs.

The leading edge of the CMEs represented by the top of the coronal arch moves with speeds varying from highly subsonic (< 50 km s^{-1}) to super-alfvenic (> 1000 km s^{-1}). The fastest undoubtedly produce strong shocks and the intermediate speeds undoubtedly produce slow shocks. It dos not seem that such shocks commonly appear as distinctive features in the K-corona observations, but they may be associated with type II radio bursts (Gary *et al.* 1984).

From a tabulation of solar events possibly related to 78 CMEs observed between March and September 1980 (near sunspot maximum) by Webb (1987), we find the following associations: 56 with prominences, 50 with radio bursts, 50 with x-ray enhancements and 36 with flares. A subset of this group of CMEs with velocities in excess of 250 km s^{-1}, which has 37 events, shows as possible associations 26 with prominences, 27 with radio bursts, 27 with x-rays and 23 with flares. The ratios are not significantly different between the subset and the full set. *Skylab* data for 1973-74 (about 1.5 years prior to sunspot minimum) showed only 42% associated with prominences and 10% associated with flares (Wagner 1984). The difference between these results and those at sunspot maximum could result from the reduced frequency of flares and prominences near sunspot minimum. Part of the difference, however, may be due to the use of different criteria for the

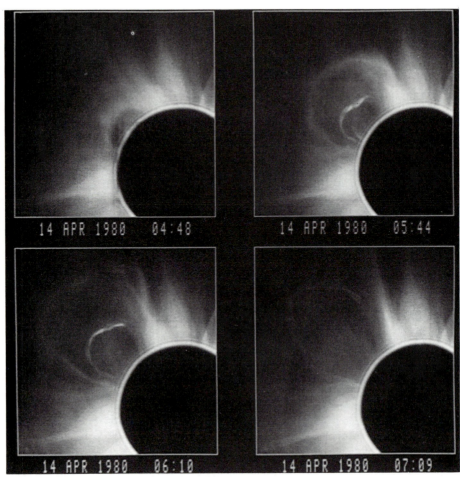

Figure 7 A sequence of images of a coronal mass ejection observed in the visual continuum with the *SMM* satellite. The second and third images of 05:44 and 06:10 clearly exhibit the prominence trailing behind the surrounding coronal arch moving outwards at a speed of approximately 270 km s^{-1}.

two studies. This possibility is reinforced by the findings that 48% of the CMEs observed by *Skylab* were not associated with any other known solar event, whereas possible associations were identified for all 78 of the CMEs observed in 1980.

R. Harrison (1986) finds from a study of three CMEs with closely related flare and x-ray associations, that the beginning of the CME is accompanied by soft x-ray emission but that the flares occur several tens of minutes later, and, further, that the flares occur at one foot point of the associated coronal loop. He also notes that previously reported flare CME associations follow the same pattern. This is consistent with the previously established associations between flares and erupting dark filaments, which show that the dark filament eruption precedes the associated flare ribbons, which, in turn, lie some distance either side of the dark filament. It thus seems clear that flares are the result of the event leading to the CME rather than the origin of it.

Hundhausen (1988) finds from a uniform study over the period 1980 to 1987 (with significant gaps) that the frequency of CMEs decreased from 0.9 per day at sunspot maximum to about 0.1 per day at sunspot minimum in 1986 then increased again to about 0.5 per day in 1987 with the beginning of the new cycle. These data, together with the *Skylab* data show, however, that the correlation of CME frequency with the sunspot cycle is more complicated than simply a direct correlation with sunspot number. Both this conclusion and the associations noted in the preceding paragraph are consistent with the description of a typical large CME as an eruption of a coronal magnetic arcade.

Heating in the solar corona is still poorly understood. The prevalence of coronal loops together with the strong correlation of the brighter regions of the corona with the regions of the corona with closed field regions is a clear indication that the heating is due to the magnetic field. It is evident, also, from the small diameters of coronal loops that the heating fluctuates on small spatial scales.

Parker (1983a,b) has suggested that the general heating in closed magnetic regions is due to the release of energy associated with small-scale tangential discontinuities in the field that are necessarily present in a dipole field that is continually agitated by small-scale convection. In his proposed model, there is no equilibrium state and dissipation is more or less continuous. Others (Hollweg 1984; Davila 1986) have suggested resonance absorption of Alfvén waves.

One of the more urgent needs in coronal physics is to determine how the energy input is distributed, both radially and spatially over the surface of the sun. Unfortunately, we do know even the order of magnitude of the average input. If most of the transition region energy is supplied by conduction from the corona, then the average energy required in the corona is approximately 10^6 ergs cm^{-2} s^{-1}. If, on the other hand, thermal conduction through the transition region is not the dominant energy loss from the corona, the average heat flux required for the corona may be as low as 10^5 ergs cm^{-2} s^{-1}.

The nature of the solar wind is sensitive to how the heat energy is distributed radially (Holzer 1982). Energy supplied in the subsonic regime inside of the critical point tends to determine the mass flux, whereas that supplied beyond the critical point tends to determine the asymptotic wind speed. On this basis, it appears that heating beyond the critical point is needed in order to explain the observed wind speeds.

To the best of our current knowledge, we suggest that coronal heating is finely structured in directions perpendicular to field lines but varies only slowly in directions parallel to field lines. In bright coronal loops, the heat flux may be locally as high as 10^7 ergs cm^{-2} s^{-1} (Withbroe and Noyes 1977), but, on average, is one or two orders of magnitude lower. Unfortunately, these rather broad generalizations are of little help in identifying the specific mechanism by which the energy is transported and dissipated.

Other outstanding issues in the corona include plasma flows, stability (thermal and dynamic), mass ejections, particle acceleration, relative abundances of elements, temperature structure, etc.

The principal coronal emission lines are in the EUV and soft x-ray wavelengths. On the other hand, the white light K-corona observed at the limb has provided a wealth of important data on coronal structure and dynamics, and the primary monitoring of coronal activity has been done with radio telescopes.

References

Acton, L. W., Culhane, J. L., Gabriel, A. H., and 21 co-authors 1980, *Solar Phys.*, **65**, 53.

Anderson, L. A., and Athay, R. G. 1988, submitted for publication.

Antiochos, S. K., and Noci, G. 1986, *Ap. J.*, **301**, 440.

Athay, R. G. 1981a, *Ap. J.*, **250**, 709.

Athay, R. G. 1981b, *Ap. J.*, **249**, 340.

Athay, R. G. 1984a, *Ap. J.*, **287**, 412.

Athay, R. G. 1984b, *Solar Phys.*, **93**, 123.

Athay, R. G. 1988, *Ap. J.*, June 1, in press.

Athay, R. G., and Holzer, T. E. 1982, *Ap. J.*, **255**, 743.

Athay, R. G., and White, O. R. 1978, *Ap. J.*, **226**, 1135.

Athay, R. G., and White, O. R. 1979a, *Ap. J. Suppl.*, **39**, 333.

Athay, R. G., and White, O. R. 1979b, *Ap. J.*, **229**, 1147.

Beckers, J. M. 1972, *Ann. Rev. Astron. and Ap.*, **10**, 73.

Beckers, J. M., and Canfield, R. C. 1975, *Motions in the Solar Atmosphere*, AFCRL-TR-75-0592 (Hanscomb, AFB, MA: Air Force Cambridge Research Laboratories).

Bohlin, J. D., Frost, K. J., Burr, P. T., Guha, A. K., and Withbroe, G. L. 1980, *Solar Phys.*, **65**, 5.

Brown, T. M., Mihalas, B. W., and Rhodes, E. J. 1986, in *Physics of the Sun,* ed. P. A. Sturock, T. E. Holzer, D. M. Mihalas, and R. K. Ulrich, (Dordrecht: Reidel), p. 177.

Brown, T. M., and Morrow, C. A. 1987, *Ap. J. Lett.* **314**, L21.

Brueckner, G. E., and Bartoe, J.-D. F. 1983, *Ap. J.* **272** 329.

Cook, J. W., and Socker, D. G. 1987, *Science,* **232**, 1267.

Davila, J. M. 1986, in *Coronal and Prominence Plasmas*, NASA Conf. Pub. 2442, ed. A. I. Poland, (Washington, D.C.: Government Printing Office), p. 445.

Dere, K. P., Bartoe, J.-D. F., Brueckner, G. E, Cook, J. W. and Socker, D. G. 1987, *Science,* **232**, 1267.

Doschek, G. A., Feldman, U., and Bohlin, D. 1976, *Ap. J. Lett.*, **205**, L17.

Emslie, A. G., and Noyes, R. W. 1978, *Solar Phys.*, **57**, 373.

Feldman, U. 1983, *Ap. J.*, **275**, 367.

Feldman, U. 1987, *Ap. J.*, **320**, 426.

Feldman, U., Doschek, G. A., and Mariska, J. T. 1979, *Ap. J.*, **229**, 369.

Fisher, R., and Sime, D. G. 1984, *Ap. J.*, **285**, 354.

Forrest, D. J., Chupp, E. L., and 11 co-authors 1980, *Solar Phys.*, **65**, 15.

Foukal, P. 1971, *Solar Phys.*, **19**, 59.

Gary, D. E., Dulk, G. A., House, L., Illing, R., Sawyer, C., Wagner, W. J., McLean, D. J., and Hildner, E. 1984, *Astron. and Ap.*, **134**, 222.

Gilman, P. A. 1986, in *Physics of the Sun*, ed. P. A. Sturrock, T. E. Holzer, D. M. Mihalas, and

R. K. Ulrich, (Dordrecht: Reidel), p. 95.

Golub, L., Harvey, K., and Webb, D. 1986, in *Coronal and Prominence Plasma*, NASA Conference Publ. 2442, ed. A. I. Poland, (Washington, D.C.: Government Printing Office), p. 365.

Hagyard, M. J., Smith, J. B., Jr., Teuber, O., and West, E. A. 1984, *Solar Phys.*, **91**, 115.

Harrison, R. A. 1986, *Astron. and Ap.*, **162**, 283.

Harvey, K. L. 1987, *Ap. J.*, **323**, 380.

Hirayama, T., and Irie, M. 1984, *Solar Phys.*, **90**, 291.

Hollweg, J. V. 1984, *Ap. J.*, **277**, 392.

Holzer, T. E. 1977, *Journ. Geophys. Res.*, **82**, 23.

Hundhausen, A. J. 1977, in *Coronal Holes and High Speed Wind Streams*, ed. J. B. Zirker (Boulder, Colo.: University of Colorado Press), p. 225.

Hundhausen, A. J. 1988, in preparation.

Hundhausen, A. J., Hansen, R. T., and Hansen, S. F. 1981, *Journ. Geophys. Res.*, **86**, 2079.

Illing, R. M. E., and Hundhausen, A. J. 1985, *Journ. Geophys. Res.*, **90**, 275.

Kopp, R. A., and Holzer, T. E. 1976, *Solar Phys.*, **49**, 43.

MacQueen, R. M., Csoeke-Poeckh, A., Hildner, E., House, L., Reynolds, R., Stanger, A., TePoel, H., and Wagner, W. 1980, *Solar Phys.*, **65**, 91.

Mariska, J. T., Feldman, U., and Doschek, G. A. 1978, *Ap. J.*, **226**, 698.

Munro, R. H., and Jackson, B. V. 1977, *Ap. J.*, **213**, 874.

Orrall, F. Q., and Rottman, G. 1986, in *Coronal and Prominence Plasma*, NASA Conference Pub. 2442, ed. A. I. Poland, p. 395.

Orwig, L. E., Frost, K. J., and Dennis, B. R. 1980, *Solar Phys.*, **65**, 25.

Parker, E. N. 1955, *Ap. J.*, **122**, 293.

Parker, E. N. 1983a, *Ap. J.*, **264**, 635.

Parker, E. N. 1983b, *Ap. J.*, **264**, 642.

Pneuman, G. W., and Kopp, R. A. 1978, *Solar Phys.*, **57**, 49.

Porter, J. G., Moore, R. L., Reichmann, E. J., Engvold, O., and Harvey, K. L. 1987, *Ap. J.*, **323**, 380.

Press, W. H. 1986, in *Physics of the Sun*, ed. P. A. Sturrock, T. E. Holzer, D. M. Mihalas, and R. K. Ulrich (Dordrecht: Reicel), p. 77.

Rabin, D., and Moore, R. 1984, *Ap. J.* **285**, 359.

Schmieder, B. 1988, in *Dynamics and Structure of Solar Prominences*, ed. E. Priest and J. L. Ballester (Dordrecht: Reidel), in press.

Seeley, J. F., and Feldman, U. 1985, *Mon. Not. Roy. Astron. Soc.*, **213**, 417.

Sheeley, N. R., and Harvey, J. W. 1978, *Solar Phys.*, **59**, 159.

Shoub, E. C. 1983, *Ap. J.* **266**, 339.

Tripp, D. A., Athay, R. G., and Peterson, V. L. 1978, *Ap. J.* **220**, 314.

Van Beek, H. F., Hoyng, P., Lafleur, B., and Simnett, G. M. 1980, *Solar Phys.*, **65**, 39.

Vernazza, J. E., Avrett, E. H., and Loeser, R. 1981 *Ap. J. Suppl.*, **45**, 635.

Wagner, W. J. 1984, *Ann. Rev. Astron. and Ap.*, **22**, 267.

Woodgate, B. E., Tandberg-Hanssen, E. A., and 12 co-authors 1980, *Solar Phys.*, **65**, 73.

Webb, D. F. 1987, *Table of Solar Activity Associated with Coronal Mass Ejections Observed by the SMM Coronagraph/Polarimeter in 1980*. NCAR Technical Note 297 (Boulder, Colo.: High Altitude Observatory).

Withbroe, G. L., and Noyes, R. W. 1977, *Ann. Rev. Astron. and Ap.*, **15**, 363.

Zirker, J. B. 1986, *Solar Phys.*, **100**, 281.

Active Late-Type Stars

Jeffrey L. Linsky

Joint Institute for Laboratory Astrophysics
National Bureau of Standards and
University of Colorado
Boulder, Colorado

1. What Is Stellar Activity?

Stellar activity has become a generic term referring to a set of phenomena that occur when the basic assumptions of a classical stellar atmosphere are not valid. These assumptions are:

1. Significant energy is transported from the stellar core to space only by radiation and convection. In a classical atmosphere, mechanical energy can be produced by convective energy transport and oscillations (radial or nonradial), but the inevitable conversion of mechanical energy to heat is presumed not to alter significantly the thermal structure of the stellar atmosphere.

2. To a good approximation the atmosphere is in hydrostatic equilibrium; that is, there are no significant systematic flows, such as winds, although low-speed circulation patterns induced by rotation can occur.

3. Magnetic fields, if present, are sufficiently weak that they do not control the flow of matter or alter the energy balance in the stellar atmosphere.

These classical assumptions lead to the very useful Vogt–Russell theorem (see Lang 1980), according to which the mass, age, and initial chemical composition of a star determine its effective temperature, radius, and gravity, and hence its location in the H-R diagram. This theorem is very powerful in predicting the evolution of most stars, in particular stars without large mass loss rates or complex internal structures, but it often leads to highly inaccurate predictions concerning atmospheric structure. The classical assumptions of no mechanical heating, hydrostatic equilibrium, and no significant magnetic fields predict a single distribution of temperature with height across the surface of a star and thus a relatively uniform emitted spectrum across the stellar surface that is independent of time, except for small variations due to gravity darkening at the poles, meridional circulation patterns, and the cellular nature of convection. When the assumption of no mechanical heating is strictly obeyed, the atmosphere is a photosphere in which radiative–convective equilibrium imposes a monotonic decrease in temperature with height, except perhaps for a small inversion due to nonLTE ionization of H^- (the Cayrel effect). If we include mechanical heating due to the damping of purely acoustic (i.e. nonmagnetic) compressional waves, then the outer layers of a stellar atmosphere can be hotter than the photosphere. Such layers are called chromospheres, transition regions (TR), and coronae depending upon their temperature distribution and local energy balance (see Linsky 1980). Atmospheres heated by acoustic waves have relatively small emission measures (and thus relatively faint ultraviolet and X-ray emission) and little horizontal and temporal variation. Thus the classical assumptions lead to a rather uninteresting picture of an inactive (or quiescent) star with little in the way of horizontal or temporal variability, weak UV and X-ray emission, and no significant differences in the atmospheric properties of stars with the same mass, age, and chemical composition.

By contrast, we know that regions of strong, closed magnetic fields (i.e. magnetic flux tubes) are the basic structural elements of solar activity, because they are the locations of enhanced nonradiative heating, variable phenomena on time scales of seconds to months, and systematic flows. Very different energy balance and atmospheric properties characterize the flux tubes compared with regions of weak fields. While the Sun is a very useful prototype of an active star, the degree of its activity is puny by comparison with other stars of late spectral type. For example, large concentrations of magnetic flux can modify the photospheric energy balance to produce relatively dark structures (called sunspots) that cover up to 0.2% of the solar surface, whereas the analogous starspots in RS CVn-type systems and dMe stars can cover 10–50% of the stellar surface. On such stars, rapid increases (flaring) in the X-ray, ultraviolet, and microwave portions of the spectrum can be 10^3–10^5 times more energetic than solar flares.

Stellar activity encompasses those phenomena that occur when locally strong magnetic fields modify the local momentum and energy balance in a stellar atmosphere to produce observable effects. Some aspects of stellar activity are variable on time scales of milliseconds to days due either to intrinsic variations in the plasma (i.e., flaring or variable heating rates) or to rotation into and out of view of relatively long-lived features with large brightness contrast compared to the mean atmosphere (i.e. dark spots in the photosphere or bright active regions in the chromosphere and corona). Other aspects of activity are spatial structures like polar spots and analogs of the solar chromospheric network (a cellular pattern observable as enhanced emission in chromospheric emission lines), which manifest themselves as a long-term enhancement or decrease in emission. A third aspect of activity is a rather steady enhancement in emission from the chromosphere and corona due to the large heating rates associated with the conversion of magnetic fields or wave modes into heat by the dissipation of shocks or acceleration of electrons that become thermalized. All of these aspects of stellar activity likely vary during the stars' magnetic cycle (e.g. Wilson 1978, Baliunas and Vaughan 1985).

2. Some Basic Questions Of Stellar Activity Research

Various aspects of stellar activity have been studied for many years. Some useful milestones are the excellent photographic spectroheliograms in the Ca II H and K lines obtained by Hale and Ellerman (1903, 1904) showing the solar chromospheric network and plages, the discovery of bright cores in the Ca II H and K lines formed in stellar chromospheres (Schwarzschild and Eberhard 1913; see also reviews by Bidelman 1954, Linsky and Avrett 1970, Linsky 1977, 1980), the first observation of a white light stellar flare on the M-dwarf star UV Cet (Joy 1949), the first evidence for starspots on the RS CVn-type binary system AR Lac (Kron 1947), and the first conclusive evidence for stellar activity cycles (Wilson 1978). The pace of this research has accelerated rapidly with the availability of sensitive telescopes operating at radio (i.e. the VLA, Arecibo, and the VLBI networks), ultraviolet (*IUE*), X-ray (*HEAO 1*, *HEAO 2*, and *EXOSAT*), and optical wavelengths, which permit us to define more clearly the goals of stellar activity research.

What Types Of Stars Are Active?

Specific diagnostics of different activity phenomena are generally useful for only a limited range of stars. Thus the complete range of stars that are active in one way or another must be pieced together from results obtained using all of the available diagnostics. As more fully described in Section 3, the domain of activity includes pre-main sequence stars, main sequence stars from spectral type A9 to late M, single subgiant and giant stars of spectral types early F to K1, and stars in a variety of close-binary systems including the RS CVn, W UMa, Algol, and FK Com types. Table 1 summarizes the activity diagnostics so far detected for each class. It is important to keep in mind that the absence of detection (designated by the symbol N in Table 1) can mean either (1) the diagnostic is likely present but not detectable by current techniques, (2) the diagnostic has never been seen in any member of the class and is probably not present, or (3) there is no published systematic search for activity using the diagnostic. An example of the first category is the rapidly rotating stars for which broad line profiles preclude measurements of the magnetic Zeeman splitting, even though photometric evidence for large dark spots indicates that strong fields should be present.

What Stellar Parameters Control Activity?

Rutten (1984) and Schrijver (1987a) have shown that there is a minimum (or basal) value for the surface fluxes of chromospheric emission lines (e.g. Ca II H+K, Mg II h+k, and the Si II 1812 Å multiplet) as a function of B–V color and luminosity class, and that the surface fluxes of active stars can be as much as 200 times larger than the basal flux. The range in surface flux of higher temperature

Table 1 Evidence For Stellar Activity

Star type	Mag. Spots	Fields	Cycles[a]	Plages	Enhanced UV[b]	X-ray	Radio[c]	Flares Detected Optical	UV	X-ray	Radio[c]
Sun	Y	Y	Y	Y	Y	Y	Y	Y	Y	Y	Y
A-type dwarfs	N	N	N	N[d]	N[d]	N	N	N	N	N	N
F-type dwarfs	N	N	N	N	Y	Y	N	N	N	N	N
G-type dwarfs	R	Y	Y	Y	Y	Y	N	N	N	N	N
K-type dwarfs	R	Y	Y	Y	Y	Y	N	N	N	R	N
M-type dwarfs	Y	Y	Y	Y	Y	Y	Y	Y	Y	Y	Y
F-early K III–IV	N	N	N	N	Y	Y	N	N	N	N	N
T Tauri	Y	N	N	Y	Y	Y	Y	Y	Y	Y	Y
Naked T Tauri	?	N	N	N	Y	Y	N	N	N	N	N
RS CVn	Y	?	Y	Y	Y	Y	Y	?	Y	Y	Y
W UMa	Y	N	N	Y	Y	Y	Y	N	N	Y	N
Algol	N	N	N	Y	Y	Y	Y	N	N	Y	Y
FK Com	N	N	N	Y	Y	Y	N	N	N	N	N
peculiar AB stars	N	Y	N	N	N	Y	Y	N	N	N	N

Y = Diagnostic commonly detected in stars of this class.
N = Diagnostic not yet detected in any member of this class.
　　For some classes there have been no systematic searches for this diagnostic.
R = Diagnostic rarely seen.
[a] Long-term cycles detected in chromospheric emission or starspot coverage.
[b] Ultraviolet emission line flux observed to be larger than basal flux values.
[c] Nonthermal radio emission.
[d] Lα emission and X-rays detected from Altair (A7 V).

features like C IV and soft X-rays for stars of a given B–V color is even larger, although the existence of basal flux levels is not yet established. These results confirm the inadequacy of the classic stellar atmosphere assumptions and point to the importance of parameters other than spectral type (or color) that control stellar activity. Two potentially important parameters are stellar age and rotation rate (or orbital period).

Many studies have demonstrated the monotonic decrease in flux of chromospheric and coronal indicators with stellar age as indicated by cluster membership or by lithium abundance. Although Skumanich (1972) originally proposed that chromospheric Ca II emission decays as the power law $t^{-\frac{1}{2}}$, Simon, Herbig, and Boesgaard (1985) showed that the decline with age is better fit by an exponential law with e-folding times shorter for the high-excitation transition region (TR) lines than for the low-excitation chromospheric lines.

Emission from the corona, TR, and chromosphere increases rapidly with rotation, at least for stars redder than (B–V)=0.45 (spectral type F5 V). Various functional dependencies have been proposed including a law of the form $L_x \sim (v \sin i)^2$ suggested by Pallavicini *et al.* (1981) and a more complex relation in which L_x/L_{bol} depends linearly on equatorial rotational velocity for periods less than 12 days and then decays exponentially for longer periods (Walter 1981, 1982; Caillault and Helfand 1985). There is also evidence for saturation at high rotational velocities and young ages (Vilhu and Rucinski 1983; Caillault and Helfand 1985). Basri, Laurent, and Walter (1985) and Drake, Simon, and Linsky (1988) have discussed different rotation–age relations for ultraviolet emission lines and for nonthermal microwave emission, respectively.

Since stellar rotation decreases with age on the main sequence, age and rotation rate (or rotational period) are statistically interchangeable parameters, and without additional information one cannot determine which parameter controls stellar activity. The required additional information is provided by the RS CVn systems with $P_{orb} \lesssim 20$ days. For these systems tidal forces produce synchronous rotation (i.e. $P_{rot} = P_{orb}$), so that the rotation rate (either P_{rot} or v_{rot}) does not depend on stellar age. For these systems and the shorter-period W UMa systems, the functional dependences of coronal and chromospheric surface fluxes on rotation rate are the same as for the main sequence single stars (e.g. Basri 1987). Rotation, rather than age, therefore must be the controlling parameter.

The next step in identifying the controlling activity parameter was made by Mangeney and Praderie (1984), who showed that for main sequence stars of all spectral types $L_x \sim (R_o^*)^{-1.2}$, where R_o^* is the effective Rossby number, the ratio of the rotational period to the convective turnover time deep in the convective zone where the $\alpha\omega$-dynamo is presumed to operate. Subsequently, Noyes *et al.* (1984) showed that the normalized Ca II H+K line flux ($R'_{HK} = F'_{HK}/\sigma T^4_{eff}$, where F'_{HK} is the surface flux in the H and K lines corrected for the photospheric contribution) depends smoothly with small scatter on the Rossby number in the stellar convective zone. Hartmann *et al.* (1984) showed that the corresponding normalized Mg II h+k line flux for late-type dwarf stars is also a smooth function of the Rossby number, and Simon *et al.* (1985) showed that a similar relation holds for the normalized C IV flux for young late-type dwarf stars. Basri (1987) also showed that the Mg II and C II surface fluxes each vary with Rossby number functionally in the same way for both cool dwarfs and RS CVn systems. Thus stellar age is only an auxiliary parameter, whereas the Rossby number, which characterizes the strength of the dynamo that amplifies magnetic fields (Durney and Latour 1978), appears to be the important activity parameter.

Until recently, reliable photospheric magnetic field parameters could be measured only for the chemically peculiar and slowly rotating B and A-type stars, because Zeeman analysis of absorption line profiles in polarized light only measures fields with simple geometries. The complex magnetic geometries of stars similar to the Sun frustrate such methods and require a different approach. Robinson, Worden, and Harvey (1980) demonstrated that the Zeeman broadening of absorption lines in unpolarized light can be used to measure typical field strengths and fractional filling factors even for complex field geometries. This technique has been refined and applied by Saar *et al.* (1986) and Saar and Linsky (1985) to optical and infrared spectra of dwarf stars of spectral type G0–M5. An important result is that non-spot photospheric field strengths scale as $P_{gas}^{1/2}$, where P_{gas} is the photospheric gas pressure, independent of rotation rate (Saar 1987). On the other hand, the photospheric magnetic flux increases with both rotation rate and Rossby number and saturates at large Rossby numbers (Linsky and Saar 1987), consistent with the Skumanich and MacGregor (1986) dynamo model. This result completes the argument that stellar activity is magnetic in character, and the fundamental parameters are those that control the dynamo generation of magnetic flux.

What Are The Physical Mechanisms Responsible For Activity?

Many studies relate the observables of stellar activity to such stellar parameters as the Rossby number, rotation rate (and period), age, spectral type, and gravity. The fundamental physics of stellar activity, however, occurs not on a global scale but rather on very small scales. This physics describes mechanisms of energy transfer from mechanical and magnetic modes into heat on rapid (i.e. flare) or slow time scales. These heating processes include the dissipation of acoustic energy in shocks and electrodynamic processes in magnetic regions, such as the shock dissipation of magnetohydrodynamic wave energy, anomalous joule dissipation, and heating by magnetic reconnection. Theories of these heating processes have been reviewed by Kuperus, Ionson, and Spicer (1981) and by Spicer, Mariska, and Boris (1986), among others.

While an essential goal of stellar activity research is to identify the important energy transfer (or heating) mechanisms, one observes only the end product of the energy transfer as enhanced radiative output and the time scale for heating or cooling, whichever is longer. Understanding the energy budget of the outer atmosphere of a star involves understanding the energy budget of its constituent components, such as closed magnetic field structures (i.e. loops), open field structures, nonfield regions, and transient phenomena such as flares and emerging flux regions. In each of these structures heating occurs in a plasma with a wide range of temperatures, thus the radiative output occurs in X-rays, ultraviolet, optical, infrared, and radio regions of the spectrum. Energy can also be transferred by non-radiative modes such as waves, expansion, and turbulence that can be observed as line broadening or asymmetries. Energy transferred by thermal conduction or by charged particle beams is difficult to observe directly but is eventually converted to heat and thus observable radiation.

Measuring the energy budget of active phenomena and inferring the associated physical processes thus requires coordinated observations at many wavelengths, often including broad-band photometry and high-resolution spectroscopy. Highly transient phenomena like flares require simultaneous observations with high time resolution, whereas the study of relatively long-lived phenomena such as starspots, magnetic fields, and chromospheric active regions (and their interrelations) require contemporaneous but not strictly simultaneous observations. Each type of observation brings unique information on complex phenomena occurring at many levels and thus at many temperatures in an atmosphere.

In subsequent sections I summarize the diagnostic capabilities of different types of observations and then summarize what multiwavelength observations are telling us about the physics of stellar activity for two classes of active stars.

3. Multiwavelength Probes Of Multilayer Activity

Optical And Infrared Photometry

I begin by describing the use of broad-band photometry in the optical and infrared, because these data provide information on the location and size of starspots and accurate timing for flares with respect to which other aspects of stellar activity may be compared spatially and temporarily. Since the bulk of radiation from late-type stars emerges from the photosphere and appears in the optical and infrared, the ability to detect starspots and flares against this bright background requires that the brightness contrast relative to the photosphere be large. Two virtues of photometry are that only modest instruments on small telescopes are required and that monitoring observations to study long-term trends as well as flaring during multiwavelength observing campaigns can be automated.

Periodic variability in the optical light curve (see Figure 1), often called the photometric wave or distortion wave, has been cited for the last 30 years as evidence for an asymmetrical distribution of dark starspots on active stars. Recent reviews of this work include those of Rodonò (1983, 1986) and Vogt (1983). Such work has led to the following conclusions:

1. Stellar spots, like solar spots, are dark and cool relative to the stellar photosphere. This conclusion cannot be deduced from a light curve alone, because a distribution of bright features on the stellar disk could produce the same light curve. Conclusive evidence that the spots are dark and cool comes from the reddening of color indices and strengthening of spectral indicators of cool

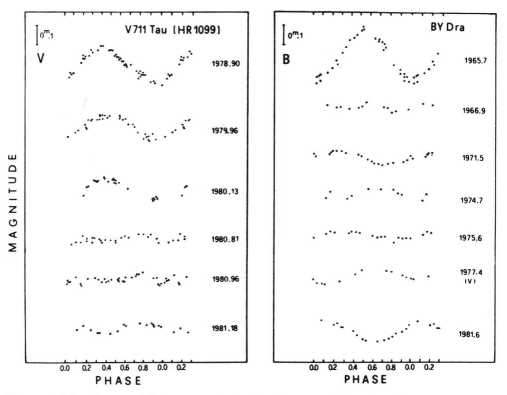

Figure 1 Selected seasonal light curves of the RS CVn binary V711 Tau (V light) and of the flare star BY Dra (B light) showing the characteristic photometric waves and their remarkable variations. The epoch of observation (year) is indicated for each light curve. (From Rodonò 1983).

matter such as TiO (e.g. Ramsey and Nations 1980) coincident with minimum visual brightness. In addition, Doppler imaging techniques (e.g. Vogt and Penrod 1983), to be described later, confirm that the spots are dark. For stars cooler than the Sun, typical spot temperatures are 3600 ± 200 K, and the ratio of spot to photosphere temperature appears to be characteristic more of sunspot umbrae than penumbrae (Rodonò 1983), although the absence of spatial resolution makes this determination difficult.

2. Spots exist on essentially all classes of active stars. An extensive literature exists concerning spots on active M dwarfs (e.g. BY Dra and UV Cet-type stars) and RS CVn binaries. In addition, sinusoidal variations in the optical light curves, indicative of an asymmetrical distribution of spots, have been observed in dwarf stars in the Hyades (Radick *et al.* 1987) and Pleiades (van Leeuwen and Alphenaar 1982) clusters and in T Tauri stars (Bouvier, Bertout, and Bouchet 1986; Herbst *et al.* 1986). It appears that a sufficient condition for appreciable spottedness is that a star have a deep convective

zone and an equatorial rotation velocity exceeding 5 km s^{-1} (Bopp 1981); binarity is significant only to the extent that tidal forces impose rapid rotation. V471 Tau–a synchronous binary (dK + WD)–is a good example of a rapidly rotating spotted K dwarf (Young *et al.* 1983).

3. Spot fractional coverage of the stellar surface extends from the detection limit (several tenths of 1%) often to values as large as 10–30%. Compared to the Sun, where the spot coverage is generally less than 0.2%, the spot coverage of active stars can be two or more orders of magnitude larger. The maximum degree of spottedness yet proposed for any star is 51% of one hemisphere of II Peg in September–November 1986 (Byrne 1987). In fact, the true spot coverage could be even greater, because the amplitude of the optical light curve reflects the *asymmetry* in hemispheric spot distribution, not the absolute value.

4. The distribution of spots (or spot groups) across the surface of an active star can be modelled from the optical light curve. Models for M dwarfs and RS CVn systems assuming circular spots typically show one large spot at high latitude or at the pole (e.g. Rodonò *et al.* 1986), contrary to solar spots, which are concentrated in the middle latitudes. These models are not unique, however, as the inclination of the rotational axis to the line of sight is generally not known (except for double-line spectroscopic binaries), the visual magnitude of the unspotted (i.e. immaculate) star is not known but often presumed to be the maximum visual brightness ever observed, the spots need not be circular, and there is a tradeoff between spot size and assumed brightness. Nevertheless, spot models should be approximately valid, and they are useful for comparison with the location of other activity phenomena.

5. The size and location of spots (or spot groups) change on time scales of months to many years (see Figure 1). Several different types of spot cycles have been identified (e.g. Catalano 1983; Rodonò 1983), including the typically 5–10 year period for spots to migrate once around an RS CVn star, the period for the wave amplitude to decrease and reappear (a measure of spot asymmetry), the time for individual spot groups to disappear, and the time for the degree of spottedness to undergo one cycle, roughly 60 years for BY Dra stars (Hartmann *et al.* 1981). Thus several different types of spot cycles appear to operate simultaneously, but Vogt's (1983) compilation of data indicates that for stars with rotational periods greater than about 7 days the cycles have solar-like periods (about 10 years), whereas stars with shorter rotational periods have systematically longer spot cycles.

6. These results provide circumstantial evidence that spots on active stars are regions of strong magnetic fields in analogy with sunspots. There are as yet

no direct measurements of magnetic fields in these spots, because the spots
are dark and thus contribute very weakly to the composite stellar light.
Nevertheless, the circumstantial case is very strong, and I will presume that
the spots are indeed regions of very strong fields.

Transient enhancement of the optical continuum is an important aspect of
the very complex phenomenon called a flare. Broad-band optical enhancements
were the first aspects of flares observed in UV Cet and other M dwarf flare stars
and are observed with great regularity, whereas optical flares on the Sun (called
white light flares) are rare events with only some dozen events well studied (see
review by Rust 1986). The difference is almost certainly a contrast effect, since
the solar photosphere ($T_{eff} = 5770$ K) is very bright in the visible, whereas the
cooler photospheres of M dwarfs ($T_{eff} = 3000$–4000 K) are much fainter. Thus M
dwarf flares are easily detected in the Johnson U band but difficult to observe in
the infrared. For example, U band enhancements as large as 4–5 magnitudes have
been detected, whereas enhancements of the I band flux are only a few hundredths
of a magnitude.

Optical light curves for M dwarf flares typically have fast rises to maximum
light (1–1000 s), slow decays (e-folding times of 1–100 min), and the light curves
often are complex, with several peaks commonly seen (see review by Byrne 1983).
Although the chromospheric emission lines (Balmer lines and Ca II) brighten simultaneously with the optical continuum, they generally decay with a much longer
time scale (Bopp and Moffett 1973). The number of flares as a function of U-band
peak flux is difficult to determine, because the many faint flares are hard to detect
and the very few giant flares require long observing runs to collect a meaningful
sample. Nevertheless, Kunkel (1973) estimated that the time-averaged optical flare
luminosity is about 0.004 L_{bol} for the prototype star UV Cet and never exceeds
0.01 L_{bol} for any member of the class. Large solar flares are commonly observed
over or near sunspots, but there is only a small tendency for M dwarf flares to
occur preferentially at times of maximum spottedness.

M dwarf flares generally have very blue broad-band colors that are independent of flare luminosity and time until late in the decay phase. Since there is no
consensus concerning the mechanism of the optical flares on M dwarfs, some useful insight can be acquired from detailed studies of solar white light flares, which
probably are analogous phenomena. During the initial impulsive phase of solar
white-light flares, the optical continuum commonly shows a strong Balmer discontinuity, whereas during the decay phase the continuum is generally flat with little
or no Balmer discontinuity. Rust (1986) argues that the first type of spectrum
is due to heating in the lower chromosphere producing hydrogen free-bound and
free-free emission, whereas the second type of spectrum is probably H$^-$ continuum
emission produced by heating at the top of the photosphere. In his analysis of the
well-observed July 1, 1980 white light flare, Rust argues that proton beams cannot

be responsible for the white light emission, because they are not well correlated temporally with the white light emission, and that electron beams are not sufficiently energetic, except perhaps during the impulsive stage. Instead, the strong soft X-ray flux (peaking at 1.8 keV) was sufficiently luminous, had good temporal correlation with the white light emission, and could penetrate to the upper photosphere where it would be absorbed, ionize hydrogen, and produce sufficient H^- emission to explain the white light luminosity. If this explanation is also valid for M dwarf flares, it could be tested by simultaneous optical and X-ray observations. Also, preflare dips and rises commonly seen in the U band and infrared could be due to enhanced absorption in the H^- continuum, which would be produced by X-ray heating, or by the impact of infalling matter, as originally proposed by Grinin (1976).

Optical enhancements during flares on RS CVn and related spectroscopic binary systems are difficult to observe, because at least one member of such systems has a solar-like spectrum and thus a bright optical continuum. Thus, Baliunas *et al.* (1984) detected no change to ±0.02 mag in the optical flux of λ And during the 5–6 November 1982 flare, and Chambliss *et al.* (1978) detected no changes in V magnitude as large as ±0.005 mag during the giant February 1978 flare on V711 Tau during which the microwave flux rose to 1 Jy. On the other hand, increases of 0.1 mag might have been detected by the FES (approximately the V band) on *IUE* coincident with enhanced ultraviolet and radio emission during the October 3, 1981 flare on V711 Tau (Linsky *et al.* 1988). If this result is valid, then the flux in the V band alone during this flare exceeded that of all the ultraviolet emission lines by an order of magnitude. Thus accurate measurements of the optical emission coincident with that in other bands are needed to assess the energy budgets of flares on RS CVn and other active stars.

Optical Spectroscopy

Both low and high-resolution spectroscopy can provide important information on active phenomena in the photosphere and lower chromosphere.

Vogt and Penrod (1983) developed a technique called Doppler imaging that exploits the correspondence of wavelength position in a rotationally-broadened line profile with spatial position of a relatively dark or bright feature on the stellar disk. Thus intensity distortions in a line profile are one-dimensional maps of the surface brightness distribution on a star, and a sequence of high quality line profiles obtained at many phases can be inverted to produce a two-dimensional image of the star. In a recent refinement of this technique, Vogt, Penrod, and Hatzes (1987) incorporated the principle of maximum entropy image reconstruction into their profile inversion procedure and demonstrated that reasonably accurate images can be obtained even when the S/N of the data is only several hundred and the assumed inclination of the rotation axis is in error by ±20°.

A virtue of the Doppler imaging technique is that it is an independent method for obtaining the distribution of starspots across a stellar surface that can be used to test the accuracy and uniqueness of spot modelling based only upon visual photometry. A good test case is the RS CVn-type system V711 Tau = HR 1099 for which $v \sin i = 38$ km s^{-1} and the inclination angle $i = 33°$ both are well known. This system was observed in September–October 1981 by Rodonò *et al.* (1986), who obtained an image of the K1 IV star by modelling the optical light curve, and by Vogt and Penrod (1983), who made a Doppler image of the same star. The results of these two independent approaches are summarized in Figure 2. Both methods conclude that there is a high latitude spot that crosses the central meridian near phase 0.20 and a low latitude spot (centered near $+10°$) that crosses the central meridian at phase 0.54. They also concluded that each spot covers roughly 4% of the surface of the K star, but the Doppler imaging technique determined that the spots are not circular. The striking agreement between two independent techniques shows that each can determine the location and size of large starspots on an active star provided that high quality data are available and that the inclination angle is known roughly.

As previously mentioned, the complex magnetic geometries of late-type stars have frustrated attempts to infer properties of photospheric magnetic fields by analysis of absorption lines observed in opposite senses of circular polarization. However, new procedures can infer the field strength and filling factor from the enhanced Zeeman broadening of high Landè g absorption lines compared to low Landè g lines in unpolarized light. As described by Saar (1988), the small Zeeman broadening of the line wings observed close to the continuum can provide reliable estimates of typical field strengths provided that the S/N$\geq$100, the rotational velocity is less than about 10 km s^{-1}, and weak line blends can be cancelled to high precision by comparison with profiles of the same line in inactive stars of the same spectral type. This procedure provides no information on field strengths in the dark starspots, as they contribute negligibly to the integrated starlight even when the spots are large, but it does measure the parameters of the network and plage fields (if the Sun is a good prototype). These fields are generally 1200–1500 Gauss on the Sun and are presumed to have a small dispersion about a mean value on convective stars cooler than the Sun.

On the Sun, the nonspot fields tend to be clumped in active regions that are typically bright when observed in the cores of chromospheric emission lines. These active regions (often called plages) commonly surround large sunspots. Thus multiwavelength observing campaigns to study stellar activity should include Zeeman broadening measurements, chromospheric emission line fluxes, and Doppler imaging (and/or optical photometry) data to study the geometrical interrelations of spots, magnetic fields, and active regions.

Wilson (1978) pioneered the use of accurate spectrophotometry with a 1 Å wide passband centered on the emission cores of the Ca II H and K lines to moni-

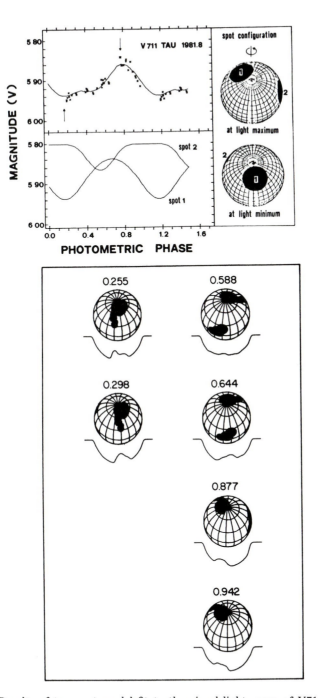

Figure 2 (Top) Results of two-spot model fit to the visual light curve of V711 Tau in 1981.8. Arrows indicate phases of minimum and maximum light. (From Rodonò *et al.* 1986). (Bottom) Results of Doppler imaging analysis of V711 Tau also in 1981.8. Phases are given above stellar image and line profiles below. (From Vogt and Penrod 1983).

tor the of chromospheric activity of late-type stars. His goal was to search for cyclic behavior analogous to the familiar solar 11 year sunspot cycle (22 year magnetic cycle). He noted that the Ca II flux data for many stars were "noisy", because the day-to-day fluctuations were far larger than the expected instrumental noise. Subsequently Vaughan *et al.* (1981), Baliunas *et al.* (1983), and others have found that these fluctuations often are periodic, presumably caused by an asymmetric distribution of chromospheric plages. By this technique, one can accurately measure the rotational periods of late-type dwarfs and giants, which often cannot be determined by any other method. Thus Ca II flux data in multiwavelength campaigns provide a rough measure of the fractional area of the stellar surface covered by bright chromospheric plages and the longitudes where the major plages are centered.

The fluxes and profiles of chromospheric emission lines, in particular the hydrogen Balmer lines and the Ca II H and K lines, have been measured for many active stars during and outside of flares. Linsky (1980, 1983) reviewed the data for late-type stars outside of flares, and Worden (1983) and Giampapa (1983) reviewed the literature for M dwarf flares. Linsky (1984) and Bopp (1983) summarized the chromospheric emission line data for RS CVn systems.

I would like to call attention to several important aspects of the data. Schrijver (1987a) showed that there is a minimum emission line surface flux, which he calls the basal flux, that characterizes very inactive dwarf and giant stars as a function of B–V color. The Ca II H and K excess flux (the amount above the basal flux for the stellar B–V) correlates well with fluxes of lines formed in different layers of a stellar atmosphere. The solar Ca II flux is just above the basal flux. For active stars while not flaring, the surface fluxes of the Ca II H and K lines can be up to 10 times larger than the basal flux. In moderate resolution spectra of the dMe and pre-main sequence (PMS) stars, the Ca II line cores can be much brighter than the surrounding (faint) continuum, whereas for the hotter RS CVn systems the cores of these lines are rarely as bright as the continuum. The Balmer lines are in emission by definition in the dMe stars but are in absorption in the less active dM stars, which, nevertheless, must have a chromosphere for these lines to be present (Cram and Mullan 1979). In the RS CVn systems, the Hα profiles usually are in absorption but "filled-in" with chromospheric emission. For dMe stars there is considerable evidence from monitoring campaigns for an anticorrelation of chromospheric emission line fluxes with optical brightness, as one would expect if chromospheric active regions overlie starspot complexes. For RS CVn systems, however, the evidence is ambiguous. Bopp (1983) notes, for example, that the anticorrelation is commonly seen with the Ca II line flux, but Dorren *et al.* (1981) in their extensive Hα study of V711 Tau found no evidence for an anticorrelation during the 1977–1978 observing season, some evidence for an anticorrelation in their 1979 data, and more scatter at all times in the Hα flux data than photon statistics would indicate.

Since chromospheric line fluxes strengthen greatly during flares, they indicate when flares are in progress, although they decay more slowly than the optical continuum. The Ca II lines and Hα lines of dMe stars become stronger in emission, while in RS CVn systems the Hα line profiles generally change from filled-in absorption lines to broad and often complex emission lines. There is an excellent correlation between the variations of Hα emission and radio flux during flares (Fraquelli 1982, 1984). Thus the variable Hα flux in RS CVn systems noted by Dorren *et al.* (1981) might indicate continuous low-level flaring. This high degree of short-term variability might also explain why the Hα line flux generally does not show a clear anticorrelation with the degree of spottedness. Simultaneous monitoring of the Hα flux (and profiles), radio emission, X-ray emission, and optical photometry are needed to sort out this problem. Evidence supporting the hypothesis that Balmer line flux variability is caused by flaring includes the simultaneous *EXOSAT* and Balmer line measurements of UV Cet (Butler *et al.* 1986), which show good time correlations between rapid enhancements in the X-ray and Balmer line flux. Rapid variability of the Ca II H and K line flux observed by Baliunas *et al.* (1981) in λ And and ε Eri also might be due to low-level flaring. This hypothesis could be tested by multiwavelength observing.

During flares on dMe stars, the Balmer lines often become broad (observable out to ±10 Å) like during solar flares, but there is little data yet that the Hα line develops enhanced blue wing emission early in flares and enhanced red wing emission late in flares, as is commonly seen in solar flares. The broadening of the Balmer lines and their merging together high in the Rydberg series is due to the Stark affect and is commonly used to estimate electron densities.

Ultraviolet Spectroscopy

Although observations with the *Copernicus* satellite and with balloon-borne and rocket-borne spectrographs provided a few glimpses of the most intense ultraviolet emission lines emitted by a few bright stars, almost all that has been learned about the ultraviolet spectra of late-type stars, in particular active stars, has been obtained with the *IUE* satellite, which was launched in 1978. The *IUE* has become a powerful tool for studying stellar activity, because it has sufficient sensitivity to study weak emission lines in a great many active stars at low dispersion, and it can observe the brighter emission lines at a resolution of $\lambda/\Delta\lambda = 10,000$ in many active stars. *IUE* also is able to record the whole 1200–3200 Å spectral region with only two observations, and scheduling is sufficiently flexible to permit long-term monitoring and simultaneous observations with other observatories. The *IUE* time allocation committees deserve commendation for their cooperation in supporting coordinated observations, but only a few such observing campaigns have been implemented successfully.

Table 2 Comparison of Ultraviolet Spectroscopic Capabilities

Spacecraft	Dates Operational	Spectral Range (Å)	Spectral Resolution ($\lambda/\Delta\lambda$)	Faintest Hot Star Observable (m_v)
Copernicus (OAO-3)	1972–1982	912–3000	20,000[a] 5,000[a]	6 6
IUE	1978–present	1175–3200	10,000 200-500	11 16
HST/HRS	1989?–	1100–3200 1100–3200 1100–1700	100,000[a] 20,000 2,000	13 15 18
HST/FOS	1989?–	1150–3200 1850–3200 1150–2500	1,300 400 250	21 22 22
HST/FOC	1989?–	1175–3000	2,400	17–21
HUT	1989?–	425–1850	300–500	17
EUVE	1992?–	80–700	100	12–17.5
LYMAN	1994?–	912–1250 1200–2000 100–2000	30,000 10,000 1,000	17 18 21

[a]Very limited spectral coverage at one time.

Because of it limited sensitivity, *IUE* cannot observe fainter active stars, such as PMS stars, at high dispersion or stellar flares with high time resolution. The Goddard High Resolution Spectrograph and the Faint Object Spectrograph on HST will provide greatly improved sensitivity, time resolution, and spectral resolution, but at the price of decreased simultaneous spectral range and scheduling flexibility. I am concerned that the anticipated large oversubscription rate and the scheduling complexity of this low Earth orbit spacecraft will frustrate most attempts to monitor active stars for long periods or to set up multiwavelength observing campaigns. A proposed second generation instrument, the Space Telescope Imaging Spectrograph will greatly extend the spectral range that can be observed at one time, and the LYMAN Far Ultraviolet Spectroscopic Explorer will extend the useful spectral range from 1200 Å down to at least the Lyman continuum edge (912 Å). The capabilities of these instruments are summarized in Table 2.

IUE data have provided important insights concerning the physics of stellar activity for late-type main sequence and evolved stars and binary systems, as well as PMS sequence stars. Three important recent reviews are those of Jordan and Linsky (1987), Imhoff and Appenzeller (1987), and Dupree and Reimers (1987).

The 1200–3200 Å region covered by the *IUE* spectrographs is rich in spectral features (see Jordon and Linsky 1987). These include numerous emission lines of C I, O I, Si II, and Fe II formed at temperatures of 4000–6000 K at the base of the chromosphere; the hydrogen Lα line and C II 1335 Å resonance lines formed at the top of the chromosphere; and lines of Si III, C III, Si IV, O IV, C IV, and N V formed at 30–150,000 K in the TR. Figure 3 illustrates the low and high-resolution 1200–2000 Å spectra obtained by *IUE* for the star β Dra (Brown *et al.* 1984). Many of these lines are collisionally excited resonance lines for which the surface flux (flux per unit area on the star) can be related simply to the emission measure $Em = \int_{\Delta T} N_e^2 dh$ for the small range of temperature ΔT where most of the line flux is emitted. Some spectral features are intersystem lines that are collisionally excited but depopulated by both line radiation and collisions (proportional to the electron density). Examples of such intersystem lines are Si III 1892 Å, C III 1909 Å, and O III 1666 Å (see Figure 3), all formed near 60,000 K, and lines in the C II 2325 Å and Si II 2335 Å multiplets formed at 7–15,000 K. Ratios of these intersystems lines to each other and to resonance lines formed at about the same temperature provide valuable measures of the electron density, pressure, and thus the emitting volume at the temperatures where the lines are formed. Other lines are not collisionally excited but rather are produced by fluorescent processes in which photons from a strong emission line (typically Lα , Lβ, Mg II, or O I) create a high population of the excited state of the fluorescent line. Examples include Lβ pumping of the O I 1304 Å multiplet, the O I 1641 Å, and the C I 1994 Å lines (all shown in Figure 3), Lα pumping Fe II lines, and O I pumping of lines of CO and S I. Fluorescent lines tend to dominate the spectra of low-gravity K and M stars for which the chromospheric pumping lines are very optically thick, the electron densities (and thus collisional excitation rates) are low, and there is little hot plasma. The He II 1640 Å line is commonly classified as a recombination line resulting from X-ray photoionization of He$^+$.

The presence in the ultraviolet spectrum of emission lines formed over a wide range of temperatures (4,000–150,000 K) and excited by different physical processes permits one to infer the distribution of emission measure with temperature and, with some assumptions concerning the geometry, the distribution of temperature and density with height. One technique involves plotting for many stars the surface fluxes (or the ratio of surface flux to bolometric flux) in one line against the corresponding quantity in another line formed at a different temperature in the same star. The first such flux-flux diagrams by Ayres *et al.* (1981) showed power-law distributions with slopes of unity when TR line fluxes are plotted against TR line fluxes, slopes of 1.5 when TR line fluxes are plotted against chromospheric line fluxes (e.g. Mg II or Ca II), and slopes of 3 when X-ray fluxes are plotted

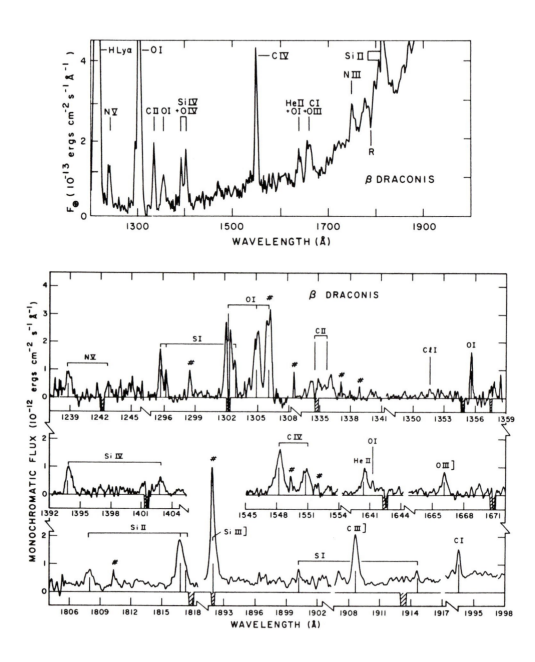

Figure 3 (Top) Low-resolution, short-wavelength spectrum of β Dra (G2 Ib–II) with prominent emission lines indicated. (Bottom) Selected high-resolution emission lines of β Dra. (From Brown *et al.* 1984).

against chromospheric line fluxes. Oranje *et al.* (1982) found similar results for a larger sample of stars observed by IUE, and Bennett *et al.* (1984) found a similar set of power-law slopes in flux-flux diagrams for active and quiet regions on the Sun observed by the Solar Mesospheric Explorer satellite. These results point to a general property of stellar activity: the relative distribution of nonradiative heating with plasma temperature depends on a single "active parameter", apparently independent of stellar effective temperature, gravity, and binarity.

The nature of this activity parameter was clarified by Schrijver's (1987a) discovery that the power-law fits in the flux-flux diagrams are significantly improved by first subtracting basal fluxes from the surface fluxes in the chromospheric emission lines (e.g. Ca II, Mg II, and Si II) and then plotting surface fluxes in the TR lines and X-rays against only the excess surface fluxes in the chromospheric lines. Schrijver (1987b) argued that the basal fluxes are likely due to the dissipation of acoustic waves and are thus not magnetic in character on the basis that (1) in the Sun basal flux levels are observed in the centers of supergranule cells where the magnetic fields are very weak, and (2) the dependence of the observed Mg II basal fluxes on stellar effective temperature and gravity is similar to values computed by Bohn (1984), who included only acoustic wave heating. On the other hand, the chromospheric excess fluxes and the total TR and X-ray fluxes appear to depend only upon the stellar rotation rate and the Rossby number (see Section 2) and thus are very likely magnetic in character.

A second technique, commonly called emission measure analysis, was developed by Pottasch (1964) and Jordan and Wilson (1971) and is described in detail by Jordan and Brown (1981). This technique can be used to derive the spatially-averaged distribution of emission measure with temperature from optically-thin, collisionally-excited emission lines formed in the TR and corona (see Figure 4). With the additional assumptions of hydrostatic equilibrium and plane-parallel geometry (for dwarf stars) or spherical geometry (for giants), the distribution of pressure and temperature with height can be derived from the emission measures, provided that the density at one temperature is determined from a density-sensitive line ratio (see Figure 4). Jordan and Linsky (1987) summarized the atmospheric models computed for active and quiet dwarfs, giants, PMS stars, and hybrid chromosphere stars using this technique. The TR pressures determined for dwarf stars lie in the range 0.03-2.0 dyne cm^{-2}, which is the same range as solar atmosphere models of the centers of supergranule cells, where magnetic fluxes are weak, to bright plages, where magnetic fluxes are large. Also the derived TR temperature gradients (dT/dh) as a function of temperature appear to have the same shape in all active stars, suggesting that the same physical processes control the thermal structure of active stars. Jordan *et al.* (1987) found that for active dwarf stars the derived TR heating rates are inconsistent with acoustic heating but could be explained as heating by MHD waves with modest magnetic fields.

Optically thick chromospheric resonance lines including Mg II h and k, Ca II

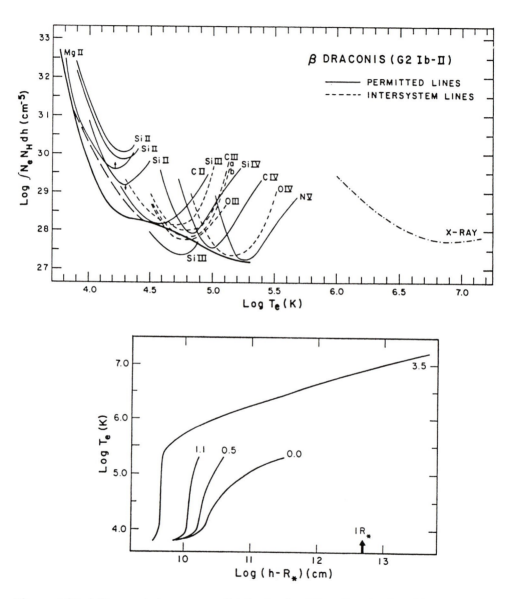

Figure 4 (Top) Mean emission measure distribution for β Dra. Permitted and intersystem lines are shown as full and dashed lines respectively. (Bottom) Variation of tempearture with height for the computed models of the outer atmosphere of β Dra. The different models are identified by the value of the electron pressure at 2×10^5 K, P_o, which has units of 10^{14} cm^{-3} K. The model with $P_o = 3.5 \times 10^{14}$ cm^{-3} K is extended to coronal temperatures. (From Brown *et al.* 1984).

H and K, the O I 1304 Å multiplet, and hydrogen Lα have been computed using various nonLTE codes in order to infer the chromospheric temperature and density structures in late-type dwarf and giant stars. Prior to 1972 these line profiles were computed using the complete redistribution assumption, but subsequent models have generally used the more realistic partial redistribution technique developed by Mihalas and collaborators. Linsky (1985) reviewed the development of the partial redistribution technique and summarized the stellar models computed using it. For a clear statement of the physical principles underlying this technique see Hubeny (1985). In the last few years codes have been developed to compute line profiles in expanding spherically symmetric chromospheres by solving the partial redistribution radiative transfer equations in the comoving frame (Drake and Linsky 1983). Both the earlier complete redistribution and later partial redistribution modelling techniques lead to the general result that the chromospheric emission lines brighten with increasing chromospheric temperature gradient (dT/d[log(m)]), where m is the mass column density, which in turn leads to higher densities (and thus collisional excitation rates) as a function of line optical depth. To a lesser extent these lines also brighten as the temperature minimum at the top of the photosphere moves inward. Both changes in the atmospheric temperature distribution are consequences of enhanced nonradiative heating, which is an essential component of stellar activity. Solar models (e.g. Shine and Linsky 1974; Shine *et al.* 1975) and stellar models (e.g. Kelch *et al.* 1979) are examples of this general result. Also, Cram and Mullan (1979) found that the Hα line and higher members of the Balmer series first become deep absorption lines and then become bright emission lines in dMe stars as the chromospheric temperature gradient is steepened and as the amount of material in the lower TR is increased in their models.

High-resolution *IUE* spectra have provided important but unexpected information on the dynamics of stellar activity. Stencel *et al.* (1982) first called attention to a systematic redshift of TR lines compared with low excitation lines formed at the base of the chromosphere. This suggestion was subsequently confirmed by Ayres *et al.* (1983, 1988) for a number of active stars. In particular, they showed that redshifts of the optically thin intersystem lines of Si III and C III relative to low-excitation chromospheric lines indicates that the higher temperature material is flowing downward relative to the cooler material. This result is interesting because a similar phenomenon is observed on the Sun, where high spatial resolution allows us to locate the downflowing material in regions of high magnetic flux and thus presumably in magnetic flux tubes (Dere 1982). Again we find evidence for the bright emission lines (indicators of activity) being associated spatially with regions of high magnetic flux. The cause of the downflow is not yet clear, but it could be explained by a combination of mass flux conserving flows in closed flux tubes if the downflowing plasma is denser than the upflows.

I previously called attention to the use of photometric light curves and Doppler imaging techniques for inferring the location of dark spots on the photospheres of

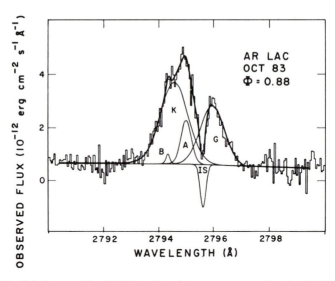

Figure 5 The Mg II k line profile of AR Lac fitted by components for the K0 IV star (component K), the G2 IV–V star (component G), two plages on the K0 IV star (components A and B), and the interstellar absorption feature (component IS). Walter *et al.* (1987) have obtained a Doppler image of AR Lac by modelling a set of Mg II line profiles obtained at many orbital phases.

active stars. In a similar way, monitoring the integrated fluxes of ultraviolet emission lines as a function of rotational phase and Doppler imaging techniques using the profiles of ultraviolet emission lines can locate individual active regions on the surfaces of active stars. The first technique, often called <u>rotational modulation</u>, requires only low-resolution spectra and is often a component of multiwavelength observing campaigns. Examples of this technique will be given in Section 4, but it is important to monitor active stars for at least two rotational periods to distinguish between transient events (flares) and long-lived events (active regions).

The first systematic application of the <u>Doppler imaging</u> technique to a chromospheric emission line was by Walter *et al.* (1987). They analyzed *IUE* high resolution spectra in the Mg II k line for the RS CVn-type system AR Lac (G2 IV–V + K0 IV) obtained at eight rotational phases in October 1983. This technique requires a sequence of high-resolution spectra with high signal-to-noise, well distributed in rotational phase, to provide accurate measurements of the longitude of individual active regions and cruder information on their latitudes and areas. An example of the Mg II line profiles that Walter *et al.* (1987) analyzed is shown in Figure 5. This observed profile was modelled as five gaussians representing global emission from the G2 IV–V star, the K0 IV star, interstellar Mg II absorption, and two active regions (plages) that lie on the equator of the K0 IV star. Subsequently Neff and Neff (1987) analyzed 18 Mg II spectra also of AR Lac obtained in September 1985. They determined the locations and sizes of two plages near the equator and one at high latitude on the K0 IV star, and a very inactive region and a flare on the G2 IV–V star. Inclusion of Doppler imaging analysis of a chromospheric emission line in future coordinated multiwavelength observing campaigns will provide plausible estimates for the location and size of emitting regions at levels of an active stellar atmosphere only observable in X-rays and radio radiation, from which such spatial information cannot be inferred.

Rapid increases in the fluxes of ultraviolet emission lines and continua have been detected as flares from dMe, RS CVn, and other active stars. These data are generally obtained as part of multiwavelength observing campaigns and will be discussed in Section 4.

X-ray Photometry And Spectroscopy

The most useful spectral region for studying the 10^6–10^8 K thermal plasmas in stellar coronae is the X-ray region extending from 0.1 keV (124 Å) to 10 keV (1.24 Å). Calculations of the emergent spectrum for solar abundances (e.g. Mewe and Gronenschild 1981) show that emission lines dominate the spectrum for T$< 2 \times 10^6$ K and that the bremsstrahlung continuum dominates at higher temperatures. As the plasma temperature increases, the bulk of the X-ray emission is shifted to higher energy so that even broadband detectors can infer crude temperatures.

Solar X-ray observations provide important clues about the fundamental structures and processes likely to occur in the coronae of active stars. In their comprehensive review Vaiana and Rosner (1978) emphasized that high-resolution broadband X-ray images from Skylab and from rocket payloads reveal a highly structured corona consisting of relatively dark coronal holes and bright loop-like structures clumped into active regions. The holes are open magnetic field regions that are dark due to low electron densities and temperatures, whereas the closed field loops are bright due to high densities and temperatures. The field geometry cannot be measured directly but rather is inferred from extrapolation of measured photospheric fields. Thus the solar corona is in reality an ensemble of mini-coronae each with its own physical properties (temperatures, densities, heating rates) and isolated from adjacent regions by strong magnetic fields that inhibit thermal conduction of energy. In addition, the solar corona is dynamic, with flows and brightness changes on a wide range of time scales.

X-ray observations of active stars have been reviewed by Linsky (1981), Golub (1983), and Rosner, Golub, and Vaiana (1985). Most of what we have learned about stellar coronae has been gleaned from the *Einstein (HEAO 2)* satellite, with additional data from *EXOSAT* and *HEAO 1*. A useful way of sorting out which parameters of an active stellar corona can be measured is summarized in Table 3. In general, one should use the lowest spectral resolution necessary for a given measurement in order to take advantage of higher sensitivity and imaging capabilities.

Table 3 Generic Types of Measurement and Required Spectral Resolution

Measurement	Spectral Resolution (E/ΔE)			
	1–3	10–30	100–300	$\geq 10^3$
Target identification	X			
Timing and monitoring	X			
Temperature (T)		X		
Emission measure (EM)		X		
Range of T			X	
EM(T)			X	
Electron density (n_e)			X	
Volume			X	
Flow velocity				X
Doppler imaging				X
Velocity separation of binaries				X

Broad-band imaging (energy resolution E/ΔE <3) is ideal for target identification, timing measurements, and monitoring for X-ray variability. *Einstein* High Resolution Imager (HRI) and Imaging Proportional Counter (IPC) were powerful instruments because their high angular resolution provided low background and minimal source confusion. The High Resolution Camera (HRC) and Advanced CCD Imaging Spectrometer (ACIS) proposed for *AXAF* will be far more powerful instruments than the HRI and IPC because of their enhanced angular resolution and sensitivity. The IPC and HRI detected coronal emission from dwarf stars of spectral types F, G, K, and M. Aside from the hot OB stars for which the X-ray emission is likely from shocks in their winds, the hottest known X-ray emitting star is Altair (A7V) (Schmitt *et al.* 1985) and the coolest are the late M dwarfs (Johnson 1983). For a given spectral type the range in L_x values is a factor of 1,000–10,000 with the youngest and most rapidly rotating generally being the more active stars. Among the F dwarfs log(L_x) can be as large as 30.0, and for the young dMe stars log(L_x) $\leq$ 29.6. For comparison, a typical value for the Sun is log(L_x) = 27.7. T Tau stars can be as luminous as log(L_x) = 31.7, and RS CVn systems can be as luminous as log(L_x) = 31.8 (Walter and Bowyer 1981). The distances out to which these active stars should be observable by *AXAF* are listed by Linsky (1987) and can be as large as 40 kpc. Broad-band imaging has been used to monitor active stars for variability and flaring, often as a component of multiwavelength observing campaigns (see Section 4). Ambruster *et al.* (1987) found for late-M dwarfs that most of the X-ray flux outside of large flares is variable on time scales of minutes to an hour, suggesting that the coronae of active stars are heated by transient events called microflares. Although the IPC lacked sufficient spectral resolution to measure temperatures accurately, Golub (1983) was able to deduce that coronal temperatures rise from log(T_{cor}) = 6.3 for the older M dwarfs with low L_x to roughly $\log(T_{cor}) = 7.3$ for the young M dwarfs with high L_x. During flares, T_{cor} typically increases substantially. The poor fit of a single temperature plasma to the IPC pulse height spectra for active M dwarfs suggests the presence of at least two thermal components.

Low-resolution Spectroscopy ($E/\triangle E = 10–30$), as was achieved with the *Einstein* Solid State Spectrometer (SSS) and will be obtained with the *AXAF* ACIS, permits more precise measurements of coronal temperatures and the corresponding emission measures. The practical analysis requires matching of the observed shape of the flux distribution and unresolved emission blends with theoretical spectra folded through the instrumental response. One important result obtained with the SSS by Swank *et al.* (1981) is that coronae of RS CVn systems and Algol appear to have plasma predominately at two temperatures, with a cooler component at $\log(T_{cor}) = 6.6–7.0$ and a hotter component at $\log(T_{cor}) = 7.4–8.0$. A similar result was obtained by the SSS for the dMe star AD Leo (Swank and Johnson 1982).

Moderate-resolution spectroscopy ($E/\triangle E = 100–300$) generally requires either transmission or reflection gratings such as the Objective Grating Spectrometer (OGS) on *Einstein*, the Transmission Grating Spectrometer (TGS) on *EXOSAT*, and several transmission grating options for *AXAF*. The *AXAF* Quantum Colorimeter (QC) also will obtain such spectra. Instruments with this resolution are capable of resolving spectral lines formed at different temperatures and thus can be used to infer the EM(T) distribution for the corona, just as *IUE* ultraviolet spectra have been used to infer EM(T) for the TR and chromosphere. Also, density-sensitive line ratios in the helium, boron, and carbon isoelectronic sequences are available to infer densities and thus emitting volumes of the hot or flaring plasma. Unfortunately few observations are available from these instruments due to their low throughput, although Schrijver and Mewe (1986) used the *EXOSAT* TGS to obtain 10–200 Å spectra of Capella and σ^2 CrB, and Mewe *et al.* (1982) obtained spectra of Capella with the *Einstein* OGS. The enormously higher throughput of the grating spectrometers on *AXAF* will enable moderate-resolution spectra with 100 second time resolution of bright flares on dMe and RS CVn systems (Linsky 1987).

High-resolution spectroscopy ($E/\triangle E \geq 1000$) is needed to measure the wind velocities of stellar coronae and to Doppler image coronal active regions. No such data for active stars presently exist, but high-resolution spectrometers on *AXAF* will open up this new frontier of X-ray spectroscopy.

Radio Observations

The radio spectral region is extremely useful for studying both the hot thermal plasma and the nonthermal electrons in the coronae of active stars. Both incoherent and coherent emission processes are used to explain the observed radio

Table 4 Characteristics of Radio Emission Mechanisms

Mechanisms	Source Size	T_B or T_{eff} (K)	Circular Polarization	Time Variability	Stars where observed
Thermal bremsstrahlung	large ($R \gg R_\star$)	$\sim 10^4$	low (near zero)	low ($\sim$years)	Sun (non-magnetic regions), OB stars, K and M giants
Gyroresonance emission	large ($R \gtrsim R_\star$)	$\approx 10^7$	low	low?	Sun (magnetic regions), dMe nonflare?
Gyrosynchrotron or synchrotron	moderate ($R \lesssim R_\star$)	10^8 to a few 10^{10}	moderate ($< 30\%$)	moderate (min to hr)	Sun (microwave bursts), dMe, RS CVn (core-halo), OB, Bp
Cyclotron maser	small ($R \ll R_\star$)	high (to 10^{20})	often high (to 100%)	high (ms–s spikes) (s–hr outbursts)	Sun (microwave spike bursts), dMe flares
Plasma radiation	small ($R \ll R_\star$)	high (to 10^{17})	moderate to high (10–90%)	high (s–min bursts) (min–day outbursts)	Sun (decimeter radiation), dMe flares

luminosities, polarization properties, and spectral and temporal properties of the observed emission. Excellent recent reviews of the different emission mechanisms include those of Kuipers (1985), Dulk (1985, 1987), and Melrose (1987). In this section I briefly describe each of these mechanisms and indicate on which active stars they likely occur, but I refer the reader to the cited reviews for more detailed descriptions. The distinguishing characteristics of each mechanism are summarized in Table 4, which is based on a similar table in Dulk (1987).

Thermal bremsstrahlung is the free-free radiation emitted when thermal electrons are accelerated by the electric fields of ions. This emission from chromospheres and coronae must always be present, although often at a level far below detection threshholds. Since the free-free opacity is proportional to the square of the wavelength, the atmospheric height corresponding to optical depth unity ($\tau_\lambda = 1$) increases with wavelength. For the Sun, $\tau_\lambda = 1$ occurs in the chromosphere at millimeter wavelengths, so that the brightness temperature $T_B \leq 10^4$ K in this spectral region. At centimeter wavelengths outside of active regions, T_B increases with increasing wavelength, and the TR is optically thin so that $T_B = T_{cor}\tau_\lambda$ is much less than the coronal temperature. In the presence of a magnetic field, optically-thin thermal bremsstrahlung emission can have a small degree of circular polarization. Bremsstrahlung is the generally accepted emission mechanism for the extended winds of many (but not all) O stars and cool giants and supergiants, but this mechanism cannot explain the centimeter-wavelength emission from active dwarfs or close binaries. The reason for this is that the bremsstrahlung radio flux predicted on the basis of the observed thermal free-free X-ray emission is at least an order of magnitude smaller than observed (Gary and Linsky 1981; Gary 1985), except for the long-period RS CVn system Capella (Drake and Linsky

1986).

When nonrelativistic electrons are accelerated by magnetic fields, they radiate at the electron gyrofrequency $\Omega_e = eB/m_e c = 1.8 \times 10^7 B$ Hz and at its harmonics ($\Omega = s\Omega_e$), where s is an integer. The term gyroresonance emission describes emission at or very near the resonance frequency when the condition $s^2(2kT/m_e c^2) \ll 1$ or $s^2 T \ll 6 \times 10^9$ K is valid. This condition is generally satisfied in stellar coronae when $T < 10^8$ K and $s < 10$. If the coronal magnetic fields were constant, then gyroresonance radiation would occur only at or near the resonance frequency, it would be slightly anisotropic, and it would have a low degree of circular polarization (a consequence of high opacity at the resonant frequencies). However, the magnetic field must diverge above the photosphere, so that along a line of sight into the corona there will generally be some location at which the resonance condition is satisfied and the narrow emitting region near this level will be optically thick. Thus gyroresonance emission is usually optically thick continuum emission with a brightness temperature $T_B = T_{cor}$. Radio emission at 1–10 cm wavelengths from the solar corona above spots is usually gyroresonance emission at $s = 3$ (e.g. Felli, Lang, and Willson 1981). Gary and Linsky (1981) computed gyroresonance emission fluxes from dMe stars. They concluded that 6 cm emission from UV Cet (dM5.5e) could be explained by this mechanism, but the emitting region must be at least as large as the whole star, and either the coronal temperature exceeds 2.5×10^7 K or the photospheric fields (covering the whole star) are roughly 25,000 Gauss. Thus the mechanism could contribute to the observed emission, but it probably is not the dominant mechanism.

With increasing electron energy, the microwave emission per electron increases rapidly and the number of emitting electrons required to explain an observed flux level decreases rapidly. Gyrosynchrotron emission is radiation from either thermal or nonthermal electrons when $s = 10$–100. This typically occurs for mildly relativistic electrons (Lorentz factor $\gamma = (1 - (\frac{v}{c})^2)^{-1/2} \leq 3$), which could be thermal ($T = 10^8$–10^{10} K) or a power law distribution of electron energies. Emission by this process is incoherent with low circular polarization ($<10\%$) and $T_B < 10^{10}$ K. Higher energy nonthermal electrons ($\gamma \gg 1$) emit synchrotron radiation for which the circular polarization can be as large as 30%. The effective temperature of the electrons $T_{eff} = E_o/k$, where E_o is the energy of the emitting electron. $T_B = T_{eff}$ for optically thick emission, and $T_B = T_{eff}\tau_\lambda$ for optically-thin emission. For the optically-thick case, $T_B = 10^{10}$ K corresponds to 1 MeV electrons.

To determine which emission process, one must know the electron energy distribution, source size, or spectrum shape. For example, Linsky and Gary (1983) showed that the 6 cm flux levels of a few mJy, typically observed from such dMe stars as UV Cet and YY Gem, can be explained by gyrosynchrotron emission from a relatively small number of electrons (only 10^{-3} the number of ambient electrons) for which $T_{eff} > 10^8$ K. These electrons could be confined to a few magnetic loops in a flaring (nonthermal) or a postflaring (thermal) state. Mutel *et al.* (1985)

and others used VLBI techniques to resolve the emitting source in RS CVn and Algol systems. They find that during the initial stage of flares the radio emission is confined to a compact core with dimensions less than a few 10^{11} cm. The source is optically thick at centimeter wavelengths, with $T_B = T_{eff} = (1\text{–}6)\ 10^{10}$ K. They deduced that this is gyrosynchrotron emission from mildly relativistic electrons ($\gamma = 2\text{–}10$) located in compact loops (B $\approx$ 100 Gauss) where particles are being accelerated. At the later stages of a flare the volume of relativistic electrons expands to the dimensions of the binary system and becomes optically thin at centimeter wavelengths, with $T_B \approx 10^9$ K. Mullan (1985) proposed that these nonthermal electrons are accelerated by turbulence-induced electric fields that are sufficiently strong to cause nearly all the electrons in a loop to runaway to high energies, where radiation rather than collisions eventually thermalizes their energies. Gyrosynchrotron emission might also explain nonthermal radiation from O and B stars, Bp and Ap stars, and microwave bursts on the Sun and dMe stars.

The previously described processes cannot explain radio emission during flares when $T_B > 10^{12}$ K and the emission is nearly 100% circularly polarized, narrow band, and/or highly variable. Under these conditions a coherent process such as an electron cyclotron maser or plasma radiation likely occurs. These processes are described by Dulk (1985), Melrose (1987), and others. Here I will merely say why radio flares on dMe stars must be coherent emission processes. If one takes the emitting area to be the whole stellar surface, then $T_B = 10^{13} - 10^{16}$ K for meter-wave emission from many dMe stars during flares (e.g. Mullan 1985), and these brightness temperatures are lower limits if the flares are smaller in size. These high brightness temperatures are confirmed by evidence for spike rise times faster than 0.2 sec observed during a flare on AD Leo (Lang *et al.* 1983), indicating an area of less than 0.03 of the stellar surface and thus $T_B \geq 10^{13}$ K. Many flares, such as the AD Leo event just mentioned, are observed to be essentially 100% circular polarized. In addition, Bastian and Bookbinder (1987) obtained dynamic spectra of flares on UV Cet that indicate narrowband emission. It is interesting that while flares on dMe stars and the Sun commonly show the symptoms of coherent emission, there is no evidence for similar phenomena during flares on close-binary systems such as the RS CVn systems. Mullan (1985) presented theoretical arguments that could explain why dwarf stars and subgiants (such as RS CVn systems) exhibit such different flare phenomena.

4. Examples Of Multiwavelength Studies Of Active Stars

I summarized in Section 3 the spectroscopic and photometric tools available to the astronomer for understanding the phenomena of stellar activity. These tools can be used to determine which phenomena are present and to estimate the physical properties of the plasmas taking part. The spectroscopic tools are essential in modelling the plasmas to infer the heating processes. In this section I give examples of how several of these tools together have been used to obtain a broad perspective on active phenomena at all levels in a stellar atmosphere.

RS Canum Venaticorum Systems

As originally defined by Hall (1976), RS Canum Venaticorum (RS CVn) systems consist of detached binaries with an F or G-type hotter component of luminosity class V or IV, and typically a cooler early K-type secondary of luminosity class IV. The orbital periods are 1–14 days and strong Ca II H and K emission is observed from one or both stars. Systems with longer (or shorter) period often are called long (or short) period RS CVn systems, while the short-period contact systems are named W UMa systems after their prototype. Systems with orbital periods less than 20 days usually are synchronous or very nearly so due to tidal forces, whereas longer period systems are generally asynchronous, as tidal forces have not had sufficient time to enforce synchronism. Popper and Ulrich (1977) proposed a plausible evolutionary scenario for these systems.

Many observers have studied the rich phenomenology of these systems at wavelengths from X-rays to the radio. These phenomena are reminiscent of what is observed on the Sun, but they are vastly more energetic and often cover a much larger area of the stellar surface. A likely cause of this high degree of activity is the tidally-induced rapid rotation, which can be as large as 70 km s^{-1} at the equator compared to typical values of less than 5 km s^{-1} for single early-K subgiants. Recent reviews include those of Hall (1981), Rodonò (1983), Catalano (1983), and Linsky (1984). Multiwavelength observing campaigns are particularly needed to address two major questions concerning activity in these systems.

(1) What is the geometrical relationship of spots (or spot groups) and active regions in the chromosphere and corona? In particular, do active regions occur only above large spots where the magnetic fields are large, or are they uncorrelated spatially with large spots in the photosphere. Spots are identified either by photometry or by Doppler imaging using absorption line profiles, while active regions are identified either by rotational modulation of ultraviolet or X-ray fluxes or by emission-line Doppler imaging techniques. Many coordinated observations, well distributed in rotational phase over at least one and preferably several orbits, are required for each technique in order to determine whether spots and active regions are spatially correlated. Spurious correlations can appear for observing campaigns lasting less than one rotational period or with poor phase sampling as flares can persist for hours to days and it is difficult to distinguish between flaring and the emergence of a long-lived active region during a short observing campaign. Needless to say, it is difficult to acquire large blocks of satellite observing time and around the world ground-based telescope time to obtain the required observations, so only a few observing campaigns have been attempted with mixed success.

The most conclusive example of a correlation between spots and plages was obtained by Rodonò *et al.* (1987) during a coordinated *IUE* and optical photometric study of II Peg (K2 IV + ?) on 1–7 October 1981. Their results for this 6.72 day orbital period system are shown in Figure 6. The asymmetric visual light curve with an amplitude of about 0.25 magnitudes (Rodonò *et al.* 1986) requires at least two large spots, and the coincidence of photometric minimum at reddest (V–I) color requires that the spots be cooler than the unspotted photosphere. Since the inclination of the rotational axis with respect to the line of sight is unknown, spot models were computed for several assumed inclination angles (see Fig. 12 in Rodonò *et al.* 1986). The visibilities of the spots for inclination angles of 35° and 90° are shown in Figure 6. Also shown are emission line fluxes as a function of phase obtained with IUE. The enhancement of fluxes in all spectral lines between phases 0.45 and 0.95 indicates the presence of a compact bright source located near the equator and crossing the meridian near phase 0.70. Since the bright source was observed at the beginning and end of the 7 day observing run, the authors called it a plage rather than a flare. Also, for an assumed inclination angle $i = 90°$, spot No. 2 located on the equator appeared and disappeared view in phase with the ultraviolet emission lines. Thus a plausible explanation is that a very bright plage was located above spot No. 2 but that no unusually bright plage was located above the other spots. Ultraviolet spectra of the plage and quiescent regions of II Peg are shown in Fig. 2 in Rodonò *et al.* (1987), and models for the plage and quiescent regions are derived by Byrne *et al.* (1987).

Other observing campaigns have searched for evidence of spatial correlations between spots and plages. For example, Baliunas and Dupree (1982) found that at light minimum for λ And (G7-G8 IV-III + ?) in December 1978 both the Ca II K line flux and the emission lines observed by *IUE* were larger than at light maximum, consistent with a spatial correlation of spots and active regions. During the October 1981 observing campaign previously described, Rodonò *et al.* (1987) found weak but not conclusive evidence for a correlation of spots with plages for V711 Tau (K1 IV + G5 V) and AR Lac (G2 IV–V + KO IV). V471 Tau (K2 V+DA) has characteristics in common with the RS CVn systems, because tidal forces produce very rapid rotaion synchronous with the 12.5 hour orbital period. For the K2 V star there is evidence that large loops containing 10^5 K gas overlie starspots (Guinan *et al.* 1986) and that flares tend to occur when the largest number of spots are on the disk.

(2) What is the energy balance as a function of time during flares, and in what energy bands is the flare radiation emitted? Enhanced microwave observations of flares on RS CVn-type systems can persist for days at flux levels as large 1 Jy. These observations have been summarized by Gibson (1981), Feldman (1983), and Mullan (1985). Microwave brightness temperatures of a few 10^{10} K inferred from VLBI angular diameter measurements (Mutel *et al.* 1984, 1985; Lestrade *et al.* 1984), together with weak circular polarization and the absence of very rapid

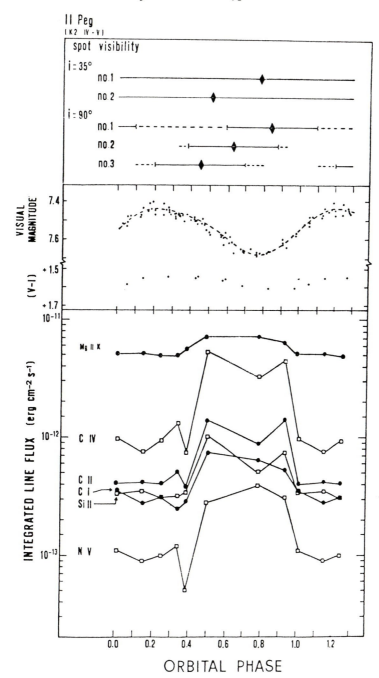

Figure 6 Comparison of visual photometry and V–I color (middle panel) with *IUE* integrated emission line fluxes (bottom panel) of the RS CVn system II Peg (K2 IV+?) from Rodonò *et al.* (1987). The top panel shows the location of the starspots (diamonds) and their visibility (solid lines) for two assumed values of the inclination of the rotational pole to the line of sight. It appears likely that the active region with the bright emission lines lies above spot No. 2 assuming $i = 90°$.

variability, have led most authors to conclude that the flare microwave emission process is incoherent gyrosynchroton emission from a power law distribution of relativistic electrons in a magnetic field of a few Gauss (Mullan 1985). Exploratory VLBI observations indicate that the size of the radio emitting region at flare onset can be smaller than either star.

Unlike flares on M dwarfs, RS CVn flares are rarely detected in broadband optical photometry (Patkos 1981), presumably because of the bright photospheric continuum of the G and K-type stars against which they must be measured. On the other hand, enhanced Hα emission is a good indicator of flares in RS CVn systems (Bopp 1983; Fraquelli 1982, 1984), and the variation of the Hα equivalent width during flares is well correlated in time with the microwave flux.

Flares have been detected with *Copernicus* and with *IUE* as enhanced ultraviolet emission line fluxes in several systems including V711 Tau (Weiler *et al.* 1978), UX Ari (Simon *et al.* 1980), λ And (Baliunas *et al.* 1984), IM Peg (Buzasi *et al.* 1987), and AY Cet (Simon and Sonneborn 1987). For both the 24 September 1976 flare on V711 Tau and the 1 January 1979 flare on UX Ari, a coincident enhancement of the radio flux was detected. In all cases, the *IUE* flare spectra show that the high-temperature TR lines (e.g. C IV and Si IV) strengthen by larger factors than the low-temperature chromospheric lines (e.g. C I and Mg II). Nevertheless, Baliunas *et al.* (1984) found that the majority of the radiative output of the λ And flare at ultraviolet wavelengths is concentrated in the Lα and Mg II h and k lines. They also reported enhanced flux in line-free continuum bands near 1500 Å and 1770 Å. The detected flare emission continued for more than 5 hours, and the estimated time-integrated ultraviolet flare luminosity exceeded 10^{35} erg, several million times that of the moderate size 5 September 1973 solar flare (Doyle and Raymond 1984). Table 5 summarizes the flare peak luminosities in the important ultraviolet emission lines during flares on several systems.

Charles (1983) summarized data on X-ray flares detected from six RS CVn systems. These were extremely energetic events, with peak soft X-ray luminosities of 6×10^{30} to 10^{33} erg s^{-1}. Weiler *et al.* (1978) and White *et al.* (1978) reported on the 24 September 1976 flare on V711 Tau detected with the ultraviolet and X-ray instruments on *Copernicus* and at radio wavelengths. During this flare, the peak X-ray luminosity was 2.2×10^{31} erg s^{-1}, whereas the observed Lα luminosity was 1×10^{30} erg s^{-1} and the peak Mg II luminosity was 3.8×10^{30} erg s^{-1}. If this behavior is typical, then the peak X-ray luminosity of RS CVn flares is several times that of the total ultraviolet emission. Table 6 summarizes peak luminosities in different wavelength bands for different flares.

The nature of the mass motions during flares can be inferred from the observed asymmetric or very broad line profiles. Bopp (1983) summarized the diverse shapes of Hα profiles observed during flares, which range from simple increases in brightness with no additional broadening (Fraquelli 1984), to symmetric broad profiles with total width of 400 km s^{-1} (Furenlid and Young 1978), to the very broad Hα

Table 5 Comparison of Flare Peak Enhancement Luminosities ΔL_{flare}

Spectral feature	V711 Tau[a] 3 OCT 1981	UX Ari[b] 1 JAN 1979	λ And[c] 6 NOV 1982	AY Cet 6 DEC 1983	Sun[d] 5 SEP 1973
		(10^{30} erg s^{-1})			(10^{24} erg s^{-1})
N V 1240 Å	0.23	0.66	0.041	0.10	0.49
C IV 1550 Å	0.87	1.9[e]	0.31	0.80	6.6
Si IV 1400 Å	0.23	0.74	0.10	0.29	3.8
Al III 1860 Å	0.05	0.15[e]	0.021	—	0.12
C III 1175 Å	0.24	−.65	0.14	—	1.2
Si III 1892 Å	0.08	0.23	0.028	—	—
O III 1666 Å	0.04	—	0.021	—	—
C II 1335 Å	0.41	1.3[e]	0.010	0.19	4.7
Si II 1812 Å	0.13	0.61[e]	0.062	0.12	0.12
O I 1304 Å	0.23	0.60	0.12	0.15	0.57
C I 1657 Å	0.01	—	0.021	0.03	0.60
Mg II h and k	2.41	13	1.34	3.0	2.3[f]
He II 1640 Å	0.12	1.0[e]	0.062	0.14	0.19
H I 1216 Å	8.1	43.3[f,g]	2.13	3.0	7.5[g]
SUM	13.2	64.1	4.5	7.82	28.2

[a] Assuming a distance of 31 pc and no correction for interstellar absorption.
[b] Assuming a distance of 50 pc and no correction for interstellar absorption.
[c] Assuming a distance of 24 pc and no correction for interstellar absorption.
[d] From Canfield *et al.* (1980).
[e] Assuming a flare enhancement factor of 5.5 for TR lines and 2.5 for chromospheric lines.
[f] Assuming a flare flux enhancement ratio Lα /Mg II=3.33.
[g] Corrected for interstellar Lα absorption.

profile observed during the August–September 1978 flaring episode of SZ Psc that is suggestive of a mass transfer event (Bopp 1981). Weiler *et al.* (1978) observed emission components in the wings of the Mg II lines shifted by up to ±250 km s^{-1} relative to the K1 IV star during the 24–29 September 1976 flaring episode on V711 Tau and during the 2 October 1976 flare on UX Ari. The authors interpreted the displaced emission features as high-velocity flows, perhaps along magnetic field lines. High-resolution Mg II emission line profiles (Simon *et al.* 1980) observed during the 1 January 1979 flare on UX Ari also exhibited broad wings extending 475 km s^{-1} to the red but no enhancement of the blue wings. The authors attributed the prominent red-shifted line wings to material impacting the G star at the free-fall velocity, and they speculated that large magnetic structures attached to both stars might occasionally interact and interconnect (see Uchida and Sakurai 1983). These interconnections could provide the temporary path by which plasma can stream from one star to the other, and the resulting field annihilation could provide energy for the flare and for accelerating the high-energy particles

Table 6 Peak Luminosities and Microwave Fluxes for Some RS CVn Flares

System	Date	L_x	L_{TR}	L_{MgII}	$L\alpha$	S_{6cm}	Reference
		\(10^{30} erg s^{-1}\)				(mJy)	
V711 Tau	24 Sep 1976	22[a]		3.8	1[b]		Weiler *et al.* (1978) White *et al.* (1978)
V711 Tau	3 Oct 1981		2.3	2.4	8.1	180	Linsky *et al.* (1988)
HD 8357	13 Jan 1978	∼1000					Garcia *et al.* (1980)
HD101379	1978	∼40					Garcia *et al.* (1980)
II Peg	1 Jan 1978	∼140					Schwarz *et al.* (1981)
UX Ari	1 Jan 1979		6.6	13	43		Simon *et al.* (1980)
λ And	6 Nov 1982		0.83	1.3	2.1		Baliunas *et al.* (1984)
AR Lac	5 Oct 1983		∼0.1	0.25		> 50	Walter *et al.* (1987)
AY Cet	6 Dec 1983		4.5	3.0	3.0		Simon and Sonneborn (1987)
IM Peg	5 Jul 1985		70				Buzasi *et al.* (1987)

[a] The peak 2.5–7.5 keV X-ray flux occurred near 0600 UT, whereas the peak Lα flux was recorded at 1200 UT. High levels of 2695 MHz and 8085 MHz emission were detected at 0500, but no data are available for later during the flare.

[b] Not corrected for interstellar absorption.

responsible for the microwave emission.

As an example of the power of multiwavelength studies, I summarize the results of a coordinated optical, ultraviolet, and radio study of the 3 October 1981 flare on V711 Tau analyzed by Linsky *et al.* (1988). This was a unique event in that both short- and long-wavelength high-resolution ultraviolet spectra were obtained by *IUE* during the flaring episode, together with coordinated optical photometry and 6424 MHz observations. The radiative luminosity at flare peak in the temperature range of $4.0 < \log T_e < 5.3$ was 1.2×10^{32} erg s^{-1}, and the total radiative flare energy in this temperature range during the flare was about 2.4×10^{36} erg s^{-1}. The Lα and Mg II emission lines together account for about seven times as much energy as all the TR lines combined. The electron density of the flaring plasma at 6×10^4 K was about 2×10^{11} cm^{-3}, 20 times higher than quiescent, and the radiating volume was about 4×10^{29} cm^3, 200 times smaller than quiescent. A constrained multigaussian fit to the Mg II k line profile (see Fig-

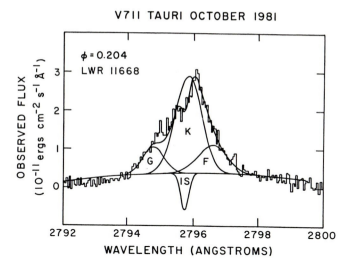

Figure 7 The observed flare peak Mg II k line profile and a four gaussian fit for which the strength and width of three of the gaussians are constrained to be the mean quiescent values for the G star, K star, and interstellar medium, and the radial velocities are those preducted for this phase. The parameters of the fourth gaussian (F) are unconstrained and can be ascribed to the flare. It is centered at 90 ± 30 km s^{-1} relative to the K star. (From Linsky *et al.* 1988).

ure 7) shows that the flare component profile was very broad (66 km s^{-1} FWHM), indicating significant turbulence, and redshifted by 90 ± 30 km s^{-1} relative to the center of mass of the K1 IV star. The authors suggest that the redshift was due to downflowing material, probably above a large starspot known from optical photometry and Doppler imaging to be near disk center of the K1 IV star at that epoch. The flux of kinetic energy at flare peak due to the downflow and turbulence was 2×10^{32} erg s^{-1}, far exceeding the flare radiative luminosity. The microwave emission probably was gyrosynchroton emission from mildly relativistic electrons (typical energy 1.7 MeV) trapped in magnetic flux tubes emerging from the whole area of the spot. The authors developed a working model for RS CVn flares, consistent with all of the data for the V711 Tau event, in which the flare occurs in magnetic flux tubes covering the whole area of the large starspot near disk center with most of the volume filled with relativistic electrons and a small fraction filled with 10^5 K plasma; the latter presumably represents gas ablated by partical beams from the dense chromosphere at the footpoints of the loops.

M Dwarf Stars

In Section 3 I described some of the active phenomena detected on M dwarf stars at a variety of wavelengths and the theoretical tools available for the interpretation of these data. M dwarfs are commonly classified as either dMe or dM depending upon whether the Hα line is in emission or absorption. The dMe stars are the most active, but as emphasized by Cram and Mullan (1979), chromospheres and their attendant heating processes must be present in the dM stars in order to populate the n= 2 level of hydrogen to produce Hα absorption. Although I concentrate on the M dwarfs, many active phenomena readily observed on the M dwarfs are also present in the G and K dwarfs but are more difficult to observe because of the lower contrast with the bright quiescent optical, ultraviolet, and X-ray emission always present. In addition to the review papers previously cited, I call attention to Pettersen's (1983a) review of the physical parameters and Giampapa's (1987) review of the chromospheres and coronae of these stars.

(1) *What is the geometrical relationship of spots (or spot groups) and active regions in the chromosphere and corona?* Contrary to the situation for the RS CVn systems, there are only two relevant data sets for the dMe stars and the answer to the question is not yet firm. Butler *et al.* (1987) observed with IUE two spotted dMe stars, BY Dra, and AU Mic. They were able to observe the binary BY Dra (M0 Ve + M0 Ve) at 9 phases throughout its 3.836 day rotational period and to observe the single star AU Mic (M2 Ve) at 19 phases distributed over most of its 4.865 day rotational period. After two flares are removed from its data set, BY Dra shows a small but probably real variation in the brightest TR lines (C IV 1549 Å and C II 1335 Å) , with maximum ultraviolet line brightness at the time of minimum optical brightness. This is consistent with the hypothesis that chromospheric active regions overlie starspots, but the situation is complicated by the small amplitude of the optical variations (0.10 mag) and the presence of a large polar spot that was always in view and a smaller equatorial spot that disappeared from view at some phases for this low inclination (i = 30°) system (Rodonò *et al.* 1986). On the other hand, the ultraviolet emission line fluxes for AU Mic showed no systematic modulation with rotational phase, but rather sporadic flare-like fluctuations. This star also had a large polar spot (always in view) and a small equatorial spot (Rodonò *et al.* 1986), which together produced only a small amplitude optical light curve. This small asymmetry in the spot distribution with longitude and the numerous flares made it impossible to identify any geometrical relationship of plages with spots. A related data set is the Radick *et al.* (1987) contemporaneous observations of five rapidly rotating Hyades F8 V–G2 V stars that show maximum Ca II H and K line flux at photometric minimum, as expected when plages overlie spots. Contrast is a major difficulty in these data, because the amplitude of the optical variations is only ±0.01 magnitudes and of the Ca II flux is only ±3%. Additional coordinated rotational modulation studies are urgently needed.

(2) *What are the relative times and time scales for emission at different wavelengths during flares?* There is an extensive literature describing flares on M dwarf stars detected only in one wavelength band or a single spectral feature. Here I concentrate on the much smaller number of flares detected simultaneously in sev-

Table 7 Multiwavelength Observations of Flares on M Dwarf Stars

Star	Spectral Type	Date	U (mag)	Balmer Lines	UV Cont.	Mg II Lines	TR Lines	X-ray	Radio	Reference
AU Mic	M2Ve	5 Aug 1980	0.68					No		Butler *et al.* (1987)
YZ CMi	dM4.5e	25 Oct 1979	Yes	Hβ,Hγ				IPC	408MHz	Kahler *et al.* (1982)
YZ CMi	dM4.5e	3 Feb 1983	3.83				x3		6cm	Rodonò *et al.* (1984)
Prox Cen	dM5e	20 Aug 1980					x3	IPC		Haisch *et al.* (1983)
V1005 Ori	dM0.5e	5 Oct 1983					x4		2cm	Rodonò *et al.* (1984)
AD Leo	dM4e	28 Mar 1984	2.1	Hα –H9	x10	x7.5				Rodonò *et al.* (1984)
EV Lac	dM4.5e	3 Sep 1981	>2				x1.5			Bromage *et al.* (1983)
UV Cet	dM6e	9 Jan 1980	3.02	Hα ,Hβ						Pettersen (1983b)
UV Cet	dM6e	17 Sep 1980	2.4					No		Bromage *et al.* (1983)
V577 Mon	dM4.5e +dM4.5e	9 Jan 1980	2.70	Hα ,Hβ						Pettersen (1983b)
EQ Peg AB	dM4e +dM5.5e	7 Dec 1984				x1.2		LE,ME		Haisch *et al.* (1987)
EQ Peg B	dM5.5e	2 Sep 1981		Hα ,Hβ	Yes	x1.05	x3.6			Baliunas and Raymond (1984)

eral wavelength bands. An admittedly incomplete summary of such observations is given in Table 7, with emphasis on those flares detected in the ultraviolet by IUE and in the X-ray region by the Imaging Proportional Counter (IPC) on *Einstein* or the Low Energy (LE) and Medium Energy (ME) detectors on *EXOSAT*. Included in the table are U-band enhancements, the particular Balmer lines and radio bands that brightened, and enhancement factors for the ultraviolet features.

It is important to keep in mind the systematic biases that often are present. For example, instrumental sensitivity limits can prevent detections where flare emission should be present. This is especially true of the *IUE* and the radio observations. A second problem is a lack of time resolution. Again, this is a particularly serious limitation for IUE, where typical integration times of one hour proevents the proper recording of short-duration flares in the far-ultraviolet region. Perhaps this is why Butler *et al.* (1987) failed to detect flares from BY Dra in the ultraviolet when optical flares as large as $\triangle U = 0.68$ mag were recorded in optical photometry.

The first evidence of a flare generally is a rapid increase in the U-band flux manifested as one or more spikes. Simultaneous with the impulsive event, which might be due to particle streams impacting the photosphere, is the onset of X-

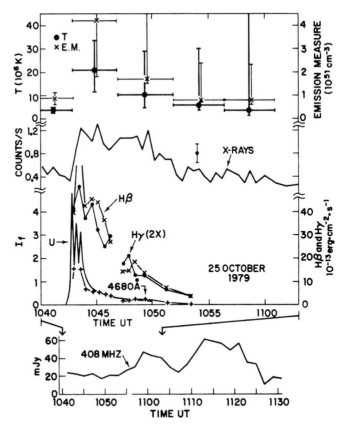

Figure 8 X-ray, optical, and radio fluxes during a flare on YZ CMi on 25 October 1979. X-ray temperatures and emission measures (with 90% confidence error bars) are shown along with 30 s averages of the *Einstein* X-ray counting rates. (From Kahler *et al.* 1982).

ray emission as, for example, was observed during the 25 October 1979 event on YZ CMi (Kahler *et al.* 1982). As illustrated in Figure 8, the soft X-rays detected by *Einstein* reached maximum flux several minutes after the impulsive U-band spike, and the inferred plasma temperature decreased with time (see also Haisch *et al.* 1983, 1987), indicating that the X-rays originated in a cooling thermal plasma. The Balmer line emission also began to increase with the optical flare (Figure 8 and Pettersen 1983b), but the peak emission was 7 minutes after the peak in the U-band flux for the flares observed by Pettersen. This indicates that the Balmer line emission also forms primarily in the cooling plasma during the flare decay phase.

The radio emission often is more complex. For the 3 February 1983 flare on YZ CMi, the 6-cm flux began to rise coincident with the U-band spike and reached peak flux about 7 minutes later (Rodonò *et al.* 1984). On the other hand, the 408 MHz (74 cm) flux during the 25 October 1979 flare on YZ CMi (Figure 8) did not

Table 8 Flare Peak Luminosities and Total Radiated Energies

Star	Date of Flare	Peak Luminosity 10^{28} erg s^{-1}			Total Radiated Energy 10^{30} erg			
		U-band	Mg II	Soft X-rays	U-band	Mg II	TR+Lα	Soft X-rays
YZ CMi	25 Oct 1979	19	–	8	12	–	–	45
Prox Cen	20 Aug 1980	–	–	1.4	–	–	2.7	35
EQ Peg	7 Dec 1984	–	~0.45	10	–	20	–	500

begin to show significant enhancement until about 10 minutes after the initial U-band spike. Solar radio flares at this frequency commonly show similar behavior, which might correspond to the time interval needed for gyrosynchrotron emission from an expanding shock wave to become visible. The coarse time resolution of the *IUE* data prevent any conclusions concerning the relative onset time of ultraviolet flare emission compared to the U-band spikes or concerning the decay time scales.

(3) *What is the energy balance during flares on M dwarf stars?* Table 8 lists the luminosities at flare peak and the total radiated energy integrated over the flare duration for well-studied flares on YZ CMi, Prox Cen, and EQ Peg. Included are luminosities and energies in the U-band, Mg II lines, the TR lines (primarily C IV), and the Lα values (inferred from typical ratios of Lα to other lines during solar flares), and the soft X-ray emission (only the LE data from *EXOSAT*). These quantities do not include all the emission from the photosphere, chromosphere, TR, and corona, but they are representative of all the emissions from these layers. The data are sparse but they do show that the radiation from the hot coronal plasma during the flare is larger than that from the cooler layers except at the very onset of a flare. Thus cooling of the hot flare plasma by thermal conduction, shock waves, or particle streams directed downward appears to be relatively unimportant in the total energy balance, but the data provide no information concerning cooling by expansion or by hard X-rays. Future coordinated multiwavelength observing campaigns are needed to provide further insight concerning the energy balance as a function of time during flares.

5.   Concluding Thoughts

In my very incomplete summary of stellar activity, I have emphasized two essential themes. The first is that magnetic fields control the energetics and dynamics of active phenomena. Since these fields thread the full range of a stellar atmosphere, from the photosphere to the corona, active phenomena at all heights and temperatures are coupled. The second theme is that coordinated or simultaneous observations in very different wavelength regions are needed to understand active pheonomena, because they encompass plasma with a wide range of temperatures and include a nonthermal high-energy component. The need for coordinated multiwavelength observations places heavy demands upon the infrastructure of contemporary astronomy and upon the psychology of astronomers. One solution would be to build new instruments with multiwavelength capability, such as a future satellite with X-ray, ultraviolet, and optical instruments. A second solution would be to establish a mechanism whereby many different observatories will often schedule coordinated observations. Astronomical research will benefit enormously from progress in both directions.

This work is supported in part by NASA grants NGL 06-003-057 and NAG5-82 to the University of Colorado. The author is very grateful for the assistance of Dr. J.E. Neff in preparing this manuscript and to him and Dr. T. Ayres for critiquing its contents. The author also thanks Dr. G. Dulk, for permission to quote most of Table 4, and Drs. A. Brown, S. Kahler, M. Rodonò, S. Vogt, and F. Walter for permission to include figures from their publications.

References

Ambruster, C. W., Sciortino, S., and Golub, L. 1987, *Ap. J. Suppl.*, **65**, 273.

Ayres, T. R., Marstad, N. C., and Linsky, J. L. 1981, *Ap. J.*, **247**, 545.

Ayres, T. R., Stencel, R. E., Linsky, J. L., Simon, T., Jordan, C., Brown, A., and Engvold, O. 1983, *Ap. J.*, **274**, 801.

Ayres, T. R., Jensen, E., and Engvold, O. 1988, *Ap. J. Suppl.*, in press.

Baliunas, S. L., and Dupree, A. K. 1982, *Ap. J.*, **252**, 668

Baliunas, S. L., Guinan, E. F., and Dupree, A. K. 1984, *Ap. J.*, **282**, 733.

Baliunas, S. L., Hartmann, L., Vaughan, A. H., and Dupree, A. K. 1981, *Ap. J.*, **246**, 473.

Baliunas, S. L., and Raymond, J. C. 1984, *Ap. J.*, **282**, 728.

Baliunas, S. L., and Vaughan, A. H. 1985, *Ann. Rev. Astron. Astrophys.*, **23**, 379.

Baliunas, S. L. *et al.* 1983, *Ap. J.*, **275**, 752.

Basri, G. S., Laurent, R., and Walter, F. M. 1985, *Ap. J.*, **298**, 761.

Basri, G. S. 1987, *Ap. J.*, **316**, 377.

Bastian, T. S., and Bookbinder, J. A. 1987, *Nature*, **326**, 678.

Bennett, J. O., Ayres, T. R., and Rottman, G. J. 1984, in *Future of Ultraviolet Astronomy Based on Six Years of IUE Research*, NASA CP-2349, p. 437.

Bidelman, W. P. 1954, *Ap. J. Suppl.*, **1**, 175.

Bohn, U. H. 1984, *Astr. Ap.*, **136**, 338.

Bopp, B. W. 1981, *A. J.*, **86**, 771.

Bopp, B. W. 1983, in *Activity in Red Dwarf Stars*, eds. P. B. Byrne and M. Rodonò (Dordrecht:Reidel), p. 363.

Bopp, B. W., and Moffett, T. J. 1973, *Ap. J.*, **185**, 239.

Bouvier, J., Bertout, C., and Bouchet, P. 1986, *Astr. Ap.*, **158**, 149.

Bromage, G. E., Patchett, B. E., Phillips, K. J. H., Dufton, P. L., and Kingston, A. E. 1983, in *Activity in Red-Dwarf Stars*, eds. P. B. Byrne and M. Rodonò (Dordrecht:Reidel), p. 245.

Brown, A., Jordan, C., Stencel, R. E., Linsky, J. L., and Ayres, T. R. 1984, *Ap. J.*, **283**, 731.

Butler, C. J., Doyle, J. G., Andrews, A. D., Byrne, P. B., Linsky, J. L., Bornmann, P. L., Rodonò, M., Pazzani, V., and Simon, T. 1987, *Astr. Ap.*, **174**, 139.

Butler, C. J., Rodonò, M., Foing, B. H., and Haisch, B. M. 1986, *Nature*, **321**, 679.

Buzasi, D. L., Ramsey, L. W., Huenemoerder, D. P. 1987, *Ap. J.*, **322**, 353.

Byrne, P. B. 1983, in *Activity in Red-Dwarf Stars*, eds. P. B. Byrne and M. Rodonò (Dordrecht:Reidel), p. 157.

Byrne, P. B. 1987, in *Cool Stars, Stellar Systems, and the Sun*, eds. J. L. Linsky and R. E. Stencel (Berlin:Springer-Verlag), p. 491.

Byrne, P. B., Doyle, J. G., Brown, A., Linsky, J. L., and Rodonò, M. 1987, *Astr. Ap.*, **180**, 172.

Caillault, J.-P., and Helfand, D. J. 1985, *Ap. J.*, **289**, 279.

Canfield, R. C., *et al.* 1980, in *Solar Flares*, ed. P. A. Sturrock, (Boulder:Colorado Associated University Press), p. 451.

Catalano, S. 1983, in *Activity in Red Dwarf Stars*, eds. P. B. Byrne and M. Rodonò (Dordrecht:Reidel), p. 343.

Chambliss, C. R. *et al.* 1978, *A. J.*, **83**, 1514.

Charles, P. A. 1983, in *Activity in Red Dwarf Stars*, eds. P. B. Byrne and M. Rodonò (Dordrecht:Reidel), p. 415.

Cram, L. E., and Mullan, D. J. 1979, *Ap. J.*, **234**, 579.

Dere, K. 1982, *Solar Phys.*, **77**, 77.

Dorren, J. D., Sick, M. J., Guinan, E. F., and McCook, G. P. 1981, *A. J.*, **86**, 572.

Doyle, J. G., and Raymond, J. C. 1984, *Solar Phys.*, **90**, 97.

Drake, S. A., Simon, T., and Linsky, J. L. 1988, *A. J.*, submitted.

Drake, S. A., and Linsky, J. L. 1983, *Ap. J.*, **273**, 299.

Drake, S. A., and Linsky, J. L. 1986, *A. J.*, **91**, 602.

Dulk, G. A. 1985, *Ann. Rev. Astron. Astrophys.*, **23**, 169.

Dulk, G. A. 1987, in *Cool Stars, Stellar Systems, and the Sun*, eds. J. L. Linsky and R. E. Stencel (Berlin:Springer-Verlag), p. 72.

Dupree, A. K., and Reimers, D. 1987, in *Exploring the Universe with the IUE Satellite*, eds. Y. Kondo *et al.* (Dordrecht:Reidel), p. 321.

Durney, B. R. and Latour, J. 1978, *Geophys. Astrophys. Fluid Dyn.*, **9**, 241.

Feldman, P. A. 1983, in *Activity in Red-Dwarf Stars*, eds. P. B. Byrne and M. Rodonò (Dordrecht:Reidel), p. 429.

Felli, M., Lang, K. R., and Willson, R. F. 1981, *Ap. J.*, **247**, 325.

Fraquelli, D. 1982, *Ap. J. (Letters)*, **254**, L41.

Fraquelli, D. 1984, *Ap. J.*, **276**, 243.

Furenlid, I., and Young, A. 1978, *A. J.*, **83**, 1527.

Garcia, M. *et al.* 1980, *Ap. J. (Letters)*, **240**, L107.

Gary, D. E., and Linsky, J. L. 1981, *Ap. J.*, **250**, 284.

Gary, D. E. 1985, in *Radio Stars*, eds. R. M. Hjellming and D. M. Gibson (Dordrecht:Reidel), p. 185.

Giampapa, M. S. 1983, in *Activity in Red Dwarf Stars*, eds. P. B. Byrne and M. Rodonò (Dordrecht:Reidel), p. 223.

Giampapa, M. S. 1987, in *Cool Stars, Stellar Systems, and the Sun*, eds. J. L. Linsky and R. E. Stencel (Berlin:Springer-Verlag), p. 236.

Gibson, D. M. 1981, in *Solar Phenomena in Stars and Stellar Systems*, eds. R. M. Bonnet and A. K. Dupree, (Dordrecht:Reidel), p. 545.

Golub, L. 1983, in *Activity in Red Dwarf Stars*, eds. P. B. Byrne and M. Rodonò (Dordrecht:Reidel), p. 83.

Grinin, V. P. 1976, *Izu. Krymskoi Astrofiz. Obs.*, **55**, 179.

Guinan, E. F., Wacker, S. W., Baliunas, S. L., Loeser, J. G., and Raymond, J. C. 1986, in *New Insights in Astrophysics*, ESA SP–263, p. 197.

Haisch, B. M. 1983, in *Activity in Red-Dwarf Stars*, eds. P. B. Byrne and M. Rodonò (Dordrecht:Reidel), p. 255.

Haisch, B. M., Butler, C. J., Doyle, J. G., and Rodonò, M. 1987, *Astr. Ap.*, **181**, 96.

Haisch, B. M., Linsky, J. L., Bornmann, P. L., Stencel, R. E., Antiochos, S. K., Golub, L., and Vaiana, G. S. 1983, *Ap. J.*, **267**, 280.

Hale, G. E., and Ellerman, F. 1903, *Publ. Yerkes Obs.*, **3**, 3.

Hale, G. E., and Ellerman, F. 1904, *Ap. J.*, **19**, 41.

Hall, D. S. 1976, in *Multiple Periodic Variable Stars*, ed. W. S. Fitch (Dordrecht:Reidel), p. 287.

Hall, D. W. 1981, in *Solar Phenomena in Stars and Stellar Systems*, eds. R. M. Bonnet, A. K. Dupree, (Dordrecht:Reidel), p. 431.

Hartmann, L., Baliunas, S. L., Duncan, D. K., and Noyes, R. W. 1984, *Ap. J.*, **316**, 377.

Hartmann, L., Bopp, B. W., Dussault, M., Noah, P. V., and Klimke, A. 1981, *Ap. J.*, **249**, 602.

Herbst, W. *et al.* 1986, *Ap. J. (Letters)*, **310**, L71.

Hubeny, I. 1985, in *Progress in Stellar Spectral Line Formation Theory*, eds. J. E. Beckman and L. Crivellari (Dordrecht:Reidel), p. 27.

Imhoff, C. L., and Appenzeller, I. 1987, in *Exploring the Universe with the* IUE *Satellite*, eds. Y. Kondo *et al.* (Dordrecht:Reidel), p. 295.

Johnson, H. M. 1983, in *Activity in Red Dwarf Stars*, eds. P. B. Byrne and M. Rodonò (Dordrecht:Reidel), p. 109.

Jordan, C., and Brown, A. 1981, in *Solar Phenomena in Stars and Stellar Systems*, eds. R. M. Bonnet and A. K. Dupree (Dordrecht:Reidel), p. 199.

Jordan, C., and Linsky, J. L. 1987, in *Exploring the Universe with the* IUE *Satellite*, eds. Y. Kondo *et al.* (Dordrecht:Reidel), p. 259.

Jordan, C., and Wilson, R. 1971, in *Physics of the Solar Corona*, ed. C. J. Macris (Dordrecht: Reidel), p. 211.

Jordan, C., Ayres, T. R., Brown, A., Linsky, J. L., and Simon, T. 1987, *M.N.R.A.S.*, **225**, 903.

Joy, A. H. 1949, *Pub. Astr. Soc. Pacific*, **105**, 96.

Kahler, S. *et al.* 1982, *Ap. J.*, **252**, 239.

Kelch, W. L., Linsky, J. L., and Worden, S. P. 1979, *Ap. J.*, **229**, 700.

Kron, G. E. 1947, *Pub. Astr. Soc. Pacific*, **59**, 261.

Kunkel, W. E. 1973, *Ap. J. Suppl.*, **25**, 1.

Kuipers, J. 1985, in *Radio Stars*, eds. R. M. Hjellming and D. M Gibson (Dordrecht:Reidel), p. 185.

Kuperus, M., Ionson, J. A., and Spicer, D. S. 1981, *Ann. Rev. Astron. Ap.*, **19**, 7.

Lang, K. R. 1980, *Astrophysical Formulae, Second Edition* (Berlin:Springer-Verlag), p. 510.

Lang, K. R., Bookbinder, J., Golub, L., and Davis, M. M. 1983, *Ap. J. (Letters)*, **272**, L15.

Lestrade, J.-F., Mutel, R. L., Phillips, R. B., Webber, J. C., Niell, A. E., and Preston, R. A. 1984, *Ap. J. (Letters)*, **282**, L23.

Linsky, J. L. 1977, in *The Solar Output and its Variations*, ed. O. R. White (Boulder:Colorado Associated University Press), p. 477.

Linsky, J. L. 1980, *Ann. Rev. Astron. Astrophys.*, **18**, 439.

Linsky, J. L. 1981, in *X-ray Astronomy in the 1980's*, NASA TM-83848, p. 13.

Linsky, J. L. 1983, in *Activity in Red Dwarf Stars*, eds. P. B. Byrne and M. Rodonò (Dordrecht:Reidel), p. 39.

Linsky, J. L. 1984, in *Cool Stars, Stellar Systems and the Sun*, eds. S. L. Baliunas and L. Hartmann (Berlin:Springer-Verlag), p. 244.

Linsky, J. L. 1985, in *Progress in Stellar Spectral Line Formation Theory*, eds. J. E. Beckman and L. Crivellari (Dordrecht:Reidel), p. 1.

Linsky, J. L. 1987, *Astro. Lett. and Comm.*, **26**, 21.

Linsky, J. L., and Avrett, E. H. 1970, *Pub. Astr. Soc. Pacific*, **82**, 485.

Linsky, J. L., and Gary, D. E. 1983, *Ap. J.*, **274**, 776.

Linsky, J. L., and Saar, S. H. 1987, in *Cool Stars, Stellar Systems and the Sun*, eds. J. L. Linsky and R. E. Stencel (Berlin:Springer-Verlag), p. 44.

Linsky, J. L. *et al.* 1988, *Astr. Ap.*, submitted.

Mangeney, A., and Praderie, F. 1984, *Astr. Ap.*, **130**, 143.

Melrose, D. B. 1987, in *Cool Stars, Stellar Systems, and the Sun*, eds. J. L. Linsky and R. E. Stencel (Berlin:Springer-Verlag), p. 83.

Mewe, R. and Gronenschild, E.H.B.M. 1981, *Astr. Ap. Suppl.*, **45**, 11.

Mewe, R. *et al.* 1982, *Ap. J.*, **260**, 233.

Mullan, D. J. 1985, in *Radio Stars*, eds. R. M. Hjellming and D. M. Gibson (Dordrecht:Reidel), p. 185.

Mutel, R. L., Doiron, D. J., Lestrade, J.-F., and Phillips, R. B. 1984, *Ap. J.*, **278**, 220.

Mutel, R. L., Lestrade, J.-F., Preston, R. A., and Phillips, R. B. 1985, *Ap. J.*, **289**, 262.

Neff, J. E., and Neff, D. H. 1987, in *Cool Stars, Stellar Systems and the Sun*, eds. J. L. Linsky and R. E. Stencel (Dordrecht:Reidel), p. 531.

Noyes, R. W., Hartmann, L. W., Baliunas, S. L. Duncan, D. K., and Vaughan, A. H. 1984, *Ap. J.*, **279**, 778.

Oranje, B. J., Zwaan, C., and Middlekoop, F. 1982, *Astr. Ap.*, **110**, 30.

Pallavicini, R., Golub, L., Rosner, R., Vaiana, G. S., Ayres, T. R., and Linsky, J. L. 1981, *Ap. J.*, **248**, 279.

Patkos, L. 1981, *Astrophys. Letters*, **22**, 1.

Pettersen, B. R. 1983a, in *Activity in Red-Dwarf Stars*, eds. P. B. Byrne and M. Rodonò (Dordrecht:Reidel), p. 17.

Pettersen, B. R. 1983b, in *Activity in Red-Dwarf Stars*, eds. P. B. Byrne and M. Rodonò (Dordrecht:Reidel), p. 239.

Popper, D. M., and Ulrich, R. K. 1977, *Ap. J. (Letters)*, **212**, L131.

Pottasch, S. R. 1964, *Space Science Reviews*, **3**, 816.

Radick, R. R., Thompson, D. T., Lockwood, G. W., Duncan, D. K., and Baggett, W. E. 1987, *Ap. J.*, **321**, 459.

Ramsey, L. W., and Nations, H. L. 1980, *Ap. J. (Letters)*, **239**, L121.

Robinson, R. D., Worden, S. P., and Harvey, J. W. 1980, *Ap. J.*, **236**, L155.

Rodonò, M. 1983, *Adv. Space Res. 1*, **No. 9**, 225.

Rodonò, M. 1986, in *Cool Stars, Stellar Systems and the Sun*, eds. M. Zeilik and D. M. Gibson (Berlin:Springer-Verlag), p. 475.

Rodonò, M. *et al.* 1984, in Proc. Fourth *IUE* Conference, ESA SP-218, p. 247.

Rodonò, M. *et al.* 1986, *Astr. Ap.*, **165**, 135.

Rodonò, M. *et al.* 1987, *Astr. Ap.*, **176**, 267.

Rosner, R., Golub, L., and Vaiana, G. S. 1985, *Ann. Rev. Astron. Astrophys.*, **23**, 413.

Rust, D. M. 1986, in *Proceedings of the NSO/SMM 1985 Summer Meeting on "The Lower Atmosphere in Solar Flares"*, ed. D. F. Neidig, National Solar Observatory.

Rutten, R.G.M. 1984, *Astr. Ap.*, **130**, 353.

Saar, S. H. 1987, in *Cool Stars, Stellar Systems and the Sun*, eds. J. L. Linsky and R.E. Stencel (Berlin:Springer-Verlag), p. 10.

Saar, S. H. 1988, *Ap. J.*, in press.

Saar, S. H., and Linsky, J. L. 1985, *Ap. J. (Letters)*, **299**, L47.

Saar, S. H., Linsky, J. L., and Beckers, J. M. 1986, *Ap. J.*, **302**, 777.

Schmitt, J.H.M.M., Golub, L., Harnden, F. R. Jr., Maxson, C. W., Rosner R., and Vaiana, G. S. 1985, *Ap. J.*, **290**, 307.

Schrijver, C. J. 1987a, *Astr. Ap.*, **172**, 111.

Schrijver, C. J. 1987b, in *Cool Stars, Stellar Systems and the Sun*, eds. J. L. Linsky and R. E. Stencel (Berlin:Springer-Verlag), p. 135.

Schrijver, C. J. and Mewe, R. 1986, in *Cool Stars, Stellar Systems, and the Sun*, eds. M. Zeilik and D. M. Gibson (Berlin:Springer-Verlag), p. 300.

Schwartz, D. A. *et al.* 1981, *M.N.R.A.S.*, **196**, 95.

Schwarzschild, K., and Eberhard, G. 1913, *Ap. J.*, **38**, 292.

Shine, R. A., Milkey, R. W., and Mihalas, D. 1975, *Ap. J.*, **199**, 724.

Shine, R. A., and Linsky, J. L. 1974, *Solar Phys.*, **39**, 49.

Simon T., Herbig, G., and Boesgaard, A. M. 1985, *Ap. J.*, **293**, 551.

Simon, T., Linsky, J. L., and Schiffer, F. H. III 1980, *Ap. J.*, **239**, 911.

Simon, T., and Sonneborn, G. 1987, *A. J.*, **94**, 1657.

Skumanich, A. 1972, *Ap. J.*, **214**, L35.

Skumanich, A., and MacGregor, K. 1986, *Adv. in Space Res.*, **6**, No. 8, 151.

Spicer, D. S., Mariska, J. T., and Boris, J. P. 1986, in *Physics of the Sun*, eds. P. A. Sturrock *et al.* (Dordrecht:Reidel), Vol 2, Ch 12.

Stencel, R. E., Linsky, J. L., Ayres, T. R., Jordan, C., and Brown, A. 1982, in *Advances in UV Astronomy: Four Years of* IUE *Research*, NASA CP-2238, p. 259.

Swank, J. H., White, N. E., Holt, S. S., and Becker, R. H. 1981, *Ap. J.*, **246**, 208.

Swank, J. H., and Johnson, H. M. 1982, *Ap. J. (Letters)*, **259**, L69.

Uchida, Y., and Sakurai, T. 1983, in *Activity in Red Dwarf Stars*, eds. P. B. Byrne and M. Rodonò (Dordrecht:Reidel), p. 411.

Vaiana, G. S., and Rosner, R. 1978, *Ann. Rev. Astron. Astrophys.*, **16**, 393.

Van Leeuwen, F., and Alpenhaar, P. 1982, **ESO Messenger**, No. 28, p. 15.

Vaughan, A. H. *et al.* 1981, *Ap. J.*, **250**, 276.

Vilhu, O., and Rucinski, S. M. 1983, *Astr. Ap.*, **127**, 5.

Vogt, S. S. 1983, in *Activity in Red-Dwarf Stars*, eds. P. B. Byrne and M. Rodonò (Dordrecht: Reidel), p. 137.

Vogt, S. S., and Penrod, G. D. 1983, *Pub. Astr. Soc. Pacific*, **95**, 565.

Vogt. S. S., Penrod, G. D., and Hatzes, A.P. 1987, *Ap. J.*, **321**, 496.

Walter, F. M. 1981, *Ap. J.*, **245**, 677.

Walter, F. M. 1982, *Ap. J.*, **253**, 745.

Walter, F. M., Neff, J. E., Gibson, D. M., Linsky, J. L., Rodonò, M., Gary, D. E., and Butler, C. J. 1987, *Astr. Ap.*, **186**, 241.

Walter, F. M., and Bowyer, S. 1981, *Ap. J.*, **245**, 671.

Weiler, E. J., *et al.* 1978, *Ap. J.*, **226**, 919.

White, N. E., Sanford, P. W., Weiler, E. J. 1978, *Nature*, **274**, 569.

Wilson, O. C. 1978, *Ap. J.*, **226**, 379.

Worden, S. P. 1983, in *Activity in Red Dwarf Stars*, eds. P. B. Byrne and M. Rodonò (Dordrecht:Reidel), p. 207.

Young, A., Klimke, A., Africano, J. L., Quigley, R., Radick, R. R, and Van Buren, D. 1983, *Ap. J.*, **267**, 655.

Hot Stars

Theodore P. Snow

Center for Astrophysics and Space Astronomy

University of Colorado

1   Introduction

Hot stars emit energy over a broad portion of the electromagnetic spectrum, and are therefore appropriate targets for multiwavelength observations and analysis. Fig. 1 shows the full flux distribution for a ζ Puppis, an O supergiant. By far the major portion of the emitted energy is thermal radiation from the photosphere, which for a hot star typically peaks in the ultraviolet (or extreme ultraviolet, for very hot objects such as some white dwarfs or central stars of planetary nebulae). The visible emission for most hot stars is also thermal, corresponding to the low-frequency portion of the blackbody curve. Thus the energy budget of most hot stars is governed primarily by thermal processes.

For many O and B stars there are significant contributions in other wavelength regimes as well. Most have extended atmospheres or stellar winds, which give rise to both infrared and x-ray emission. As seen in the figure, the magnitude of this emission lies far below that of the thermal continuum; typically the x-ray luminosity is about 10^{-7} of the total luminosity (Harnden *et al.* 1979; Vaiana *et al.* 1981), and the infrared flux is lower than the optical/UV emission by a similar factor. Radio emission also occurs in OB stars, but contributes very little (about 10^{-10}) to the overall energy output. Despite the relatively minor portion of the total luminosity of OB stars from processes other than the normal thermal emission by the photosphere, a great deal of very important information on the nature of the stars and of their extended atmospheres can be derived from multispectral observations.

Much, or perhaps all, of the emission arising from sources other than the photospheres of hot stars is time variable, so the most useful multiwavelength observations are those made simultaneously. Since the variations to a large extent can be viewed as fluctuations about some mean or average behavior, there is some benefit to be derived from multiwavelength observations that are not simultaneous, but when attempts are made to understand the fundamental nature of extended atmospheres and stellar winds, the properties of circumstellar envelopes, the relationship of photospheric phenomena to stellar wind behavior, or the propagation of disturbances through the atmosphere, simultaneous observations are required. Because few such observations have been possible so far, this review will frustrate

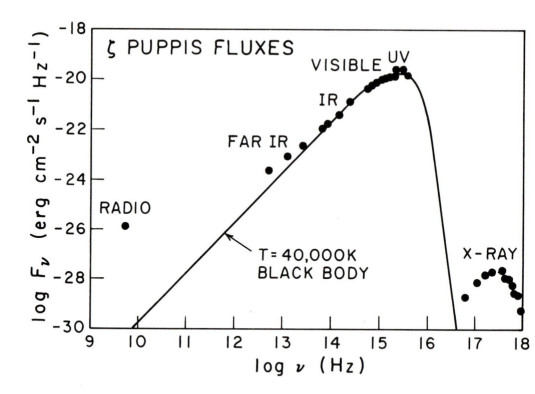

Figure 1 Observed fluxes for ζ Puppis. The dots represent measured values for the radio (Abbott *et al.* 1980), infrared (compilation of Gezari, Schmitz, and Mead 1987); visible (compilation of Hoffleit and Jaschek 1982); ultraviolet (Thompson *et al.* 1978) and far ultraviolet (Longo *et al.* 1988); and x-ray (Long and White 1980) portions of the spectrum. The curve represents a black-body flux distribution for T = 40,000 K, normalized to fit the observed ultraviolet and visible data. This figure is believed to represent the broadest spectral coverage ever achieved for a single astrophysical object; it is unfortunate that the data were not all obtained at the same time.

those seeking to know the results of such studies. (It is noteworthy that observing success in several hot star programs involving *Voyager, IUE,* and ground-based data has been achieved very recently, but the results are not yet available. These programs have been carried out by a number of groups, instigated in many cases by G.J. Peters to whom the reader is referred for the most recent analyses.) Instead we will report on the conclusions that can be drawn from non-coordinated multiwavelength observations, and will discuss the things we would like to learn from coordinated or simultaneous observations.

The next section summarizes the known physical processes in hot stars that can be studied through observations in the various wavelength regimes separately. The following section (section III) is devoted to what is learned from combining observations from different spectral bands. This will include subsections on OB star winds, on Be stars, on hot stellar remnants, and on OB stars in binary systems.

2. Physical Processes

In this section we review the emission mechanisms responsible for observed fluxes from hot stars in each spectral band, beginning with long wavelengths.

Radio: Many O stars have been detected at radio wavelengths, primarily with the VLA (e.g. Abbott *et al.* 1980; Abbott, Bieging, and Churchwell 1981; Abbott *et al.* 1986). There is extended radio continuum emission in many of these stars due to free-free emission in the ionized stellar winds. The ultimate source of the emitted energy is the ionizing photospheric radiation from the star, which injects into the free electrons the kinetic energy that they give up upon emitting radio photons. Observations of the free-free emission yield information on the wind density, which leads to estimated mass-loss rates when combined with other data on the wind terminal velocity.

The free-free continuum emission is readily detected only in stars with substantial mass-loss rates; it is rare, for example, for this emission to be detected from B main sequence stars or Be stars, even though these stars have high-speed stellar winds, because the mass-loss rates are down from those of OB supergiants by several orders of magnitude (e.g. Taylor *et al.* 1988). One difficulty in the interpretation of the emission when it is seen is that the geometry may not be spherically symmetric; that is, the emitting gas may be concentrated in an equatorial disk or some other nonspherical distribution, so that mass-loss rates derived from the assumption of spherical symmetry are overestimated. This problem has arisen especially for Be stars, for which many lines of evidence suggest equatorial concentration of the circumstellar material (discussed below, in section IV).

Nonthermal radio emission has been detected in some hot stars, including OB supergiants and Wolf-Rayet stars (Abbott, Bieging, and Churchwell 1981, 1984; Abbott *et al.* 1986; White and Becker 1983). This indicates that relativistic electrons are present in these winds, and are moving in a strong magnetic field (White 1985). There have been suggestions (Pollock 1987) that the relativistic electrons create x-ray and gamma-ray emission through Compton scattering and bremsstrahlung, but as we will discuss, there are other, more widely-accepted x-ray emission mechanisms, and it is controversial whether hot stars are gamma-ray sources.

Infrared: Most OB stars and Be stars, as well as Wolf-Rayet stars, have excess infrared emission, with respect to blackbody radiation. There are in general two mechanisms for producing this excess: free-free emission, as discussed above; and thermal emission from circumstellar dust grains.

The free-free emission arises through the same mechanism and has the same ultimate source of energy as just described for the radio continuum emission from hot stars. The infrared contribution to the free-free excess emission arises closer to the star, where densities are higher, and thus samples a physically different region than does the radio emission. Infrared free-free excesses are detected in a wider variety of stars than is radio emission; for example, most Be stars are detected (Gehrz, Hackwell, and Jones 1974), and many OB supergiants (e.g. Barlow and Cohen 1977; Leitherer *et al.* 1982; Abbott, Telesco, and Wolff 1984; Castor and

Simon 1983). Wolf- Rayet stars also commonly display excess infrared emission, but in some cases at least there is a component due to dust rather than free-free emission by electrons (Williams, van der Hucht, and Thé, 1987).

In similar manner as is done with radio data, the infrared free-free excesses from hot stars can be used to interpret mass- loss rates, when combined with information on the wind velocity law. Barlow and Cohen (1977), Leitherer *et al.* (1982), and Castor and Simon (1983) surveyed a number of OB supergiants in this fashion. The infrared flux distributions of Be stars have been interpreted in similar fashion by several workers (Waters 1986; Waters, Coté, and Lamers 1987; Lamers and Waters 1987). For the Be stars it is likely that the infrared- emitting gas is not spherically symmetric, because the mass-loss rates derived from the infrared are far higher in most cases than those derived from ultraviolet line profiles. The discrepant rates can be reconciled only if it is assumed that the infrared emission and the ultraviolet absorption arise in physically distinct regions. Indeed there is ample evidence from many lines of study supporting the view that Be stars have equatorially concentrated circumstellar material (see, for example, many articles in Slettebak and Snow 1987; also, Waters, Coté, and Lamers 1987).

Most "normal" O and B stars do not show evidence of circumstellar dust emission, but this is seen in many classes of peculiar stars, including the so-called peculiar Be stars (e.g. Swings, 1976; Coté 1987), the Herbig Ae and Be stars, which are pre- main sequence objects (e.g. Finkenzeller and Mundt 1984), and Wolf-Rayet stars (Williams, van der Hucht, and Thé, 1987). The dust emission is recognizable because of spectral features such as the 11-micron silicate emission, or because it far exceeds the flux attributable to free-free emission, based on other information about the density of the circumstellar gas. As in the case of free-free emission from electrons, the ultimate source of the dust emission is photospheric radiation from the star. In this case the grains absorb over a very broad range from the ultraviolet through the visible, whereas the flux that ionizes hydrogen to produce free electrons comes from very short ultraviolet wavelengths only.

The emission from circumstellar dust can be interpreted to provide information on the size distribution and composition of the grains, which is useful for studies of the interstellar medium, but not essential to understanding the nature of the OB and Be stars where grains are formed (furthermore, hot stars are not generally thought to be the most important sources of interstellar dust; red giants and supergiants are thought to inject far more dust into the interstellar medium).

Visible: The dominant source of optical emission from hot stars is the photospheric continuum, which is essentially thermal radiation. Measurements of the flux distribution can therefore lead to determination of photospheric properties such as effective temperature, and integration of the continuum over all wavelengths provides information on the luminosity.

Of course, a great wealth of information is also provided by spectral lines. The absorption spectrum can be analyzed in various classical ways to derive composition, rotational broadening (although we will have more to say about that in the next section), surface gravity, and so on.

Emission lines arise in many stars that have extended atmospheres or winds.

The emission features invariably indicate the presence of gas above the photosphere (e.g. Mihalas 1970), either in the form of a stationary shell or disk, or in the form of a high-speed wind (or both). The profiles can be interpreted to provide information on the density and velocity law in the winds (see review by Cassinelli 1979), which lead in turn to estimates for the mass-loss rates. For Be stars the emission lines of hydrogen are often self-reversed (the absorption component often being referred to as a shell absorption line), and correlations of the presence of the absorption reversals with projected rotational velocity lead to the widespread belief that the circumstellar material is equatorially concentrated (Struve 1931; Poeckert and Marlborough 1978; Dachs 1987 and references cited therein). Both Be and OB stars have emission lines in the infrared (usually high hydrogen series such as the Brackett, Paschen, and Pfund series), which can be interpreted in the same manner as the visible emission lines, but which refer to a different region in the wind (e.g. Andrillat and Houziaux 1967; Andrillat 1987; Hamann and Simon 1987).

Returning to photospheric absorption lines, recently a great deal of interest has developed in the evidence that they have been providing on nonradial pulsations in hot stars. This is worth special mention here because of the growing body of evidence that the nonradial pulsations might be intimately linked with the presence and the behavior of the winds in at least some hot stars (principally Be stars). The pulsations are detected through very high-quality, high-resolution observations of features that change with time. Extensive observational work is being done by Smith (1977, 1987 and references cited therein), Vogt and Penrod (1983), Bolton and co-workers (Bolton *et al.* 1987), and Baade (1987 and references cited therein), most of it concentrated on main sequence B or Be stars where the lines are clean enough to reveal the pulsations. These also happen to be the stars whose winds require some trigger mechanism other than radiation pressure (Abbott 1982), so there is interest in the possibility that the pulsations might somehow be the trigger (Willson 1986; Smith 1987).

Ultraviolet: The peak of the photospheric continuum energy distribution for most hot stars falls in the ultraviolet, so everything said in the preceding discussion about the interpretation of the optical continuum applies to the ultraviolet as well. The ultraviolet has many advantages linked to the fact that most of the energy comes out in this spectral region; it is expected, for example, that more accurate effective temperatures are derived from the ultraviolet flux distribution, and of course ultraviolet data are required for any direct measurement of bolometric magnitudes or luminosities.

Similarly, the ultraviolet line spectrum is rich in information on stellar composition and surface physical conditions, although because ultraviolet spectra are generally more difficult to obtain than optical spectra, most of this work is still based on visible-wavelength observations.

The ultraviolet spectrum becomes uniquely important in studies of stellar winds, however, because the low-density gas typical of these winds gives rise to ultraviolet absorption but often not to visible absorption lines. From the first ultraviolet rocket spectra (Morton 1967) to the *Copernicus* satellite (1972-1980) to *IUE*

(1978-present), the study of stellar winds has been an important task. The winds give rise to P Cygni profiles, which are composites of emission and absorption or, in lower-density cases, to extended or asymmetric absorption. These profiles yield information on the wind terminal velocity and, with appropriate modelling (e.g Castor and Lamers 1979), on the wind velocity law and mass-loss rate (e.g. Garmany *et al.* 1981; Gathier, Lamers, and Snow 1981; Garmany and Conti 1985; and many others, most of them cited in the review by de Jager, Nieuwenhuijzen, and van der Hucht 1987). The ultraviolet wind features are the best indicators of time variability in stellar winds, a potentially important topic (to be discussed later in this review).

X rays: It was found, primarily from the observations made with the *Einstein Observatory*, that many OB stars are sources of soft x-ray emission (Vaiana *et al.* 1981). The emission scales well with optical luminosity, usually in a ratio of about 10^{-7}. There has been considerable discussion of its origin, the initial possibilities being either a hot corona at the base of the wind (Hearn 1976; Cassinelli 1979) or shock-induced emission farther out in the wind (Lucy and White 1980). The spectrum of the x-ray emission appears now to have ruled out the coronal models, leaving many to believe that the emission arises at high levels in the wind, presumably due to shocks (Cassinelli *et al.* 1981). Recent upper limits to the FeXIV and FeXV emission lines in the optical have given further evidence that hot stars do not have hot coronae at the bases of their winds (Baade and Lucy 1987). As mentioned earlier, there is one suggestion that the x-ray emission, at least in some Wolf-Rayet stars, is produced by Compton scattering due to relativistic electrons (Pollock 1987), but this does not appear to be the general mechanism for OB stars. It is not yet known whether Be stars have soft x-ray emission similar to that of the OB stars, because current x-ray instruments are too insensitive to detect x-rays at 10^{-7} of the optical luminosity of these stars.

Some OB stars are sources of hard x rays, but these are probably binary systems with compact companions. The interpretation of such systems is beyond the scope of this review.

3. Multispectral Research on Hot Stars

Having reviewed what is learned from studies of hot stars in the separate spectral regimes, we concentrate now on studies in which data from different spectral bands are combined to produce information that could not be derived as well from any individual band alone. Specific studies will be cited here to illustrate the application of techniques. These references are meant to be representative, rather than constituting a complete listing. The chapter in this volume by Córdova summarizes the current status of efforts to carry out multiwavelength observing campaigns for various classes of objects, including hot stars; the reader seeking a more complete listing should look there.

Basic stellar properties

We have already alluded to determinations of stellar effective temperatures based on either ultraviolet or optical continuum measurements. Here we make the simple point that these determinations are much better done if both spectral bands are observed and the data are combined into a continuous spectral energy distribution covering both. There are numerous examples in the literature, usually involving IUE data, where this has been done successfully (e.g. de Loore *et al.* 1984; Fitzpatrick, Savage, and Sitko 1982; Fitzpatrick 1987; Remie and Lamers 1982). The value of extending such flux distribution measurements to the infrared lies not so much in establishing the basic stellar properties but in determining the presence and properties of circumstellar material (dust in this case).

Multispectral continuum energy distributions are needed in order to make direct measurements of stellar luminosities, and this has been done in some cases. For example Code *et al.* (1975) used data from the *OAO-2* Wisconsin experiment to obtain luminosities for several stars, then used existing interferometric angular diameters to determine effective temperatures from the Stefan-Boltzmann law.

One shortcoming of attempts to combine multiband data in order to determine luminosities of hot stars is that for many of them, particularly the O stars, a significant portion of the emitted flux lies below the 1200A short-wavelength cutoff of most present or past ultraviolet instruments. Only *Copernicus* and the *Voyager* broad-band spectrometer extended well below this cutoff, but the sensitivity and data acquisition rate of *Copernicus* were both limited (and the instrument was not well calibrated absolutely), so that it was not practical to carry out many extensive flux distribution measurements. Some work on basic stellar properties has been done with the *Voyager* instrument (e.g. Stalio, Polidan, and Peters 1987), which extends well into the extreme ultraviolet.

In certain circumstances, multiwavelength observations can be used to derive stellar radii. When hot stars with extended atmospheres are members of eclipsing binaries, measurements of the light curve in different spectral bands can resolve ambiguities regarding the extent of the star as opposed to the extent of its circumstellar envelope. The best example of this to date is probably the study of the binary Wolf-Rayet system V444 Cygni, for which Cherepashchuk, Eaton, and Khalliullin (1984) were able to obtain light curves in six bands ranging from 2460A to 3.5 microns. From the detailed analysis of the eclipse durations in the separate bands, the authors were able to deduce unambiguously the radius (hence the effective temperature) of the WR star, as well as many details of its envelope, including the height dependence of temperature and density and the wind velocity law.

Another basic parameter that can in principle be derived from multiwavelength observations is the inclination angle of a star, which is very important because knowing i can lead to firm knowledge of the rotational velocity, which in turn allows the possibility of determining the effect of rotation on other stellar properties such as the winds. Knowledge of the inclination angle can also help resolve ambiguities in the geometry of the circumstellar material. The method for determining i, effective only for rapid rotators, is based on the observation that photospheric line widths are often observed to be narrower in the ultraviolet than in the visible (Morton 1967; Heap 1976; Hutchings 1976; Hutchings and Stoeckly 1977). This is interpreted as due to equatorial darkening caused by rapid rotation, so that the ultraviolet flux arises primarily in the polar regions, whereas the visible emission comes from the entire disk. The observed ratio of visible to ultraviolet line widths, combined with theoretical models that include the effects of equatorial darkening (e.g. Collins and Sonneborn 1977), can provide information on the inclination angle and the true rotational velocity. Unfortunately, in practice this is very difficult to accomplish unambiguously, because the models have many undetermined parameters, and because the required data quality in observing the line profiles is difficult to achieve. Nevertheless, estimates of i have been made for several stars (Hutchings and Stoeckly 1977; Collins and Sonneborn 1977).

OB star winds

Most stars in the upper left-hand corner of the H-R diagram undergo mass loss through rapid winds (e.g. Snow and Morton 1976; de Jager, Nieuwenhuijzen, and van der Hucht 1987 and references cited therein). The wind velocities may be in excess of 2000 km s^{-1}, and the mass-loss rates are typically of order 10^{-6} solar masses per year (e.g. Garmany *et al.* 1981; references cited in de Jager, Nieuwenhuijzen, and van der Hucht 1987). The ultraviolet line profiles are often, perhaps always, characterized by narrow absorption components superposed on the broad absorption (Lamers, Gathier, and Snow 1982; Prinja and Howarth 1986), and these narrow components may vary in strength but little or not at all in velocity.

Multispectral observations of OB stars have helped in the analysis of the winds in several ways. Comparison of mass-loss rates derived from the ultraviolet absorption profiles and from the infrared excess emission place constraints on the wind geometry, indicating that the winds are generally spherically symmetric (this kind of analysis is described in more detail in the next section). The derivation of mass-loss rates from observations of the free-free radio or infrared emission (e.g. Abbott, Bieging, and Churchwell 1982; Abbott *et al.* 1986; Barlow and Cohen 1977; Castor and Simon 1983; Leitherer *et al.* 1982) relies on wind terminal velocities measured from the ultraviolet line profiles. Further information on the wind velocity law may be forthcoming from the analysis of soft x-ray emission, which is thought to be induced by shocks (Lucy and White 1980), raising the possibility that the complex velocity structure caused by the shocks may be inferred from analysis of the x-ray data in conjunction with the ultraviolet line profiles (work on this is in progress; Owocki 1987).

Perhaps the most important multispectral studies of OB star winds are yet to be made. It seems probable that simultaneous observations, over several timescales (ranging from hours to weeks) would provide new information on the fundamental causes of the winds and on their geometry. For example, if visible, infrared, and ultraviolet line profiles could be observed simultaneously at a time when changes in the lines were occurring, it might be possible to determine whether the changes are caused by the propagation of a density enhancement outward through the wind, because these lines are all formed at different heights. It has been suggested on the basis of variations in the ultraviolet lines alone (Lamers *et al.* 1988) that such density enhancements or "puffs" do frequently form and move outward. Further information on the vertical structure of the wind, as well as the cause of the variations, would be forthcoming if soft x-ray observations could be made at the same time as the line profile observations, to determine whether the x-ray flux varies as the alleged puffs flow outward (to date the only data on x- ray variations in OB stars failed to answer this question because of the difficulty in arranging simultaneous *IUE* and *Einstein* observations; Snow, Cash, and Grady 1981). It might also be useful to obtain simultaneous infrared and radio continuum data, although it seems unlikely that short-term changes would occur in the free-free emission, which arises over such a large volume around the star that small-scale variations observed in the ultraviolet and visible line profiles would probably have little effect, except possibly over long times.

B-emission stars

Be stars have been defined in various ways, but here we intend to describe primarily the so-called "classical" Be stars, which may be defined (Collins 1987) as non-supergiant B stars having Balmer emission lines. These stars tend to lie on or near the main sequence, and are often rapid rotators. In addition to the Balmer emission lines, Be stars are often characterized by high-velocity winds that create displaced or asymmetric ultraviolet absorption lines (Snow and Marlborough 1976) with derived mass-loss rates of order 10^{-11} to 10^{-9} solar masses per year (Snow 1981; several references in Jaschek and Groth 1982); they sometimes have shell absorption lines in the visible and the ultraviolet (Snow, Peters, and Mathieu 1979); they commonly display linear polarization in the visible (e.g. Poeckert and Marlborough 1978 and references cited therein); and above all, they are almost universally variable in all these phenomena.

The cause of the extended atmosphere or circumstellar envelopes that gives rise to the Be phenomena is not known, and there may be different causes for different stars. Among the candidates are rotationally-induced stellar winds, mass exchange in binaries, and the relatively new suggestion that nonradial pulsations may trigger mass loss in the equatorial region (for recent reviews, see the conference proceedings edited by Jaschek and Groth 1982 and by Slettebak and Snow 1987).

Multiwavelength observations are particularly useful in determining the properties of the winds and the geometry of the circumstellar material in Be stars. Mass-loss rates can be derived from analysis of the ultraviolet absorption line profiles (e.g. Snow 1981) or from modelling of the infrared excess free- free emission (e.g. Waters 1986; Waters, Coté, and Lamers 1987). The infrared data invariably imply much higher mass-loss rates than do the ultraviolet absorption lines, leading to the consensus that the infrared excesses arise in a physically distinct region from the ultraviolet emission. This, in turn, supports the view that there must be a dense, cool equatorial disk where the Balmer emission and the infrared excess are formed, and outside of that, a high-velocity, low-density wind which forms the ultraviolet absorption lines (e.g. Lamers and waters 1987; Lamers 1987 and references cited therein). This picture was recently given an important boost by radio observations of the Be star ψ Persei, in which radio free-free emission was detected (Waters *et al.* 1987a). There is no way to reconcile the radio, infrared, and ultraviolet data for this star unless the free-free emission is attributed to an equatorially confined disk which is distinct from the ultraviolet line-forming region.

Interestingly, x-ray data on a binary system containing a compact source and a Be star have recently provided a wholly new line of evidence for equatorial confinement of the circumstellar envelopes of Be stars. Waters *et al.* (1987b) found variations in the x- ray emission from the compact source, which he attributed to variations in the ambient density of the circumstellar material of the Be star that occur as the compact companion follows its moderately inclined orbit. From this analysis, Waters was able to show that the circumstellar material around the Be star is confined to a thin equatorial disk.

Perhaps the most important aspect of multiwavelength studies of Be stars is coordinated or simultaneous observations of time- variable behavior of the circumstellar phenomena. Changes may occur on timescales of decades, months, weeks, or days. Long- term multiband observing campaigns (e.g. Doazan *et al.* 1985) have suggested that these variations are related, but it is not clear how. It is not known whether the short-term variations are triggered by long-term changes, or the long-term variations are caused by changes in the short-term activity level, or whether both are induced by some deep-seated changes that occur in the stars. The long-term changes, occurring over decades, usually involve the onset of a new shell or emission phase, followed by an extended period of continued activity, and then a return to a quiescent phase that may not even be characterized by Balmer emission. Multiwavelength observations show that these variations in visible-wavelength phenomena are accompanied by changes in the degree of ultraviolet activity (described in the next paragraph), but as yet little is known about other wavelength bands. Of particular interest would be long-term infrared data, to determine whether the free-free excess varies along with the visible-wavelength manifestations of the circumstellar material.

The short-term variations can involve changes in the Balmer emission profiles (as well as infrared hydrogen emission line profiles; Hamann and Simon 1987) and are especially characterized by variations in the ultraviolet line profiles (Doazan *et al.* 1985). The ultraviolet resonance lines, particularly CIV (1550A), often display narrow displaced components, and these components can be highly variable, appearing and disappearing, or shifting in velocity, in times of hours or days. Detailed analyses (e.g. Henrichs *et al.* 1983) show that the variations are consistent with the ejection of puffs or shells by the star, although others have argued for complex velocity structure, perhaps induced by shocks in the winds (Barker 1987). Of course it is possible that different mechanisms are responsible for the variations in different stars.

Apart from the apparent linkage between the long-term changes observed primarily in the visible and the level of activity seen in the ultraviolet lines, little else is known about the possible relationship of variations in different spectral bands.

What is needed are truly simultaneous observations over many different timescales, and this has been difficult or impossible to achieve. If such multiwavelength observations could be carried out, it might be possible not only to see how the various kinds of activity are related, but also to determine the fundamental causes of the Be phenomenon. For example, if monitoring of the circumstellar activity could be coordinated with observations of nonradial pulsations, it might be possible to establish whether changes in the mode of pulsations trigger the changes in the Be phenomena, as some have suggested (e.g. Smith 1987; Baade 1987; Willson 1986).

Another multiwavelength approach would be to obtain sensitive soft x-ray observations of Be stars, to determine whether these stars are x-ray sources, in analogy with the OB stars, and if so, to see how changes in the x-ray emission might be related to changes in other manifestations of the winds. The only survey instruments available to date, the *Einstein Observatory* and *EXOSAT*, have not had sufficient sensitivity to detect Be stars, if their soft x-ray fluxes scale with optical luminosity in the same way as for OB stars. *ROSAT* might, however, be sufficiently sensitive, and it is gratifying to know that coordinated multispectral observations will be specifically supported by *ROSAT*.

Compact stellar cores

There are several classes of hot stars that are thought to be the remnant cores of stars that have lost their outer layers due to winds or pulsational instabilities. These include hot subdwarfs, central stars of planetary nebulae, white dwarfs, and Wolf-Rayet stars.

The hot subdwarfs are thought to be the cores of low-mass stars that have insufficient envelope mass to ignite nuclear shell sources and go through red giant and planetary nebula stages before becoming white dwarfs. They have a wide range of effective temperatures (25,000 to 100,000 K), but all are hot, so that the bulk of the continuum is emitted in the ultraviolet (for a review of *IUE* observations, see Vauclair and Liebert 1987).

Multiwavelength studies of hot subdwarfs have been important in establishing the fundamental surface properties of the stars, such as effective temperatures and surface gravities. Typically low-resolution *IUE* spectra are combined with optical photometry and, in some cases, far-ultraviolet Voyager data to obtain these results (e.g. many references cited in Vauclair and Liebert 1987). In the case of one very hot subdwarf, the temperature (120,000 K) was established by combining x-ray data from *Einstein* and *EXOSAT* with ultraviolet fluxes measured by *Voyager* and *IUE* and optical data (Holberg 1986).

Central stars of planetary nebulae share many characteristics with the hot subdwarfs, and are similarly studied with multispectral techniques. These stars are so hot in general that most of their fluxes are emitted shortward of the *IUE* cutoff, so effective temperatures are often estimated from the analysis of nebular emission lines, which are used to infer the intensity of the extreme ultraviolet ionizing flux from the central star. The combination of this optical technique with measured ultraviolet flux distributions can improve estimates of effective temperatures and other atmospheric properties.

Both the hot subdwarfs and the central stars of planetary nebulae become white dwarfs in time, and the latter objects are also the subject of multispectral analyses. As in the other cases, improved effective temperatures are derived from the combination of optical and ultraviolet flux distributions (e.g. Holberg *et al.* 1986; Koester *et al.* 1985; Liebert *et al.* 1984). X-ray data, in combination with optical and ultraviolet observations, allow accurate determinations of the He/H ratios in some white dwarfs, an important test of gravitational diffusion theory in these stars (Kahn *et al.* 1984; Petre, Shipman, and Canizares 1986; Heise 1985). Knowledge of the atmospheric composition of white dwarfs has been greatly enhanced in general by the combination of line identifications and analyses in the optical with those obtained in the ultraviolet, where trace elements show up more readily (many references cited by Vauclair and Liebert 1987).

The hot subdwarfs, central stars of planetary nebulae, and white dwarfs are all low-mass remnants of stars whose initial masses were either low or intermediate. By contrast, the Wolf-Rayet stars are thought to be the cores of massive stars whose outer layers have been stripped away by high-velocity stellar winds, exposing layers rich in either CNO products (the nitrogen- rich WN stars) or in helium-burning products (the carbon-rich WC stars). Both types of Wolf-Rayet stars have masses of 15 or more solar masses and dense, high-velocity winds producing mass-loss rates of order 10^{-5} solar masses per year.

The determination of basic parameters of Wolf-Rayet stars has been aided substantially by multispectral observations, so that optical and ultraviolet energy distributions could be combined (e.g. Nussbaumer *et al.* 1982; Underhill 1980, 1981; Barlow, Smith, and Willis 1981; Fitzpatrick, Savage, and Sitko 1982). Because the winds are so dense, it is often difficult to distinguish photospheric fluxes from these emitted at high levels in the wind, so that determinations of stellar parameters from flux distributions can be ambiguous. This problem was neatly avoided for one Wolf-Rayet star in an eclipsing binary, where light curves obtained at different wavelengths ranging from the ultraviolet to the infrared revealed the true dimensions and basic properties of the core star (Cherepaschuk, Eaton, and Khalliullin 1984). Studies of the winds from Wolf-Rayet stars have been accomplished using multispectral data, particularly through the analysis of infrared free-free emission in the context of wind terminal velocities derived from ultraviolet line profiles (e.g. Barlow, Smith, and Willis 1981). Finally, abundance analyses in Wolf-Rayet stars have combined optical and ultraviolet data in order to improve the range of elements and ionization stages that could be observed (e.g. Garmany and Conti 1982; Smith and Willis 1982a,b; Hillier 1986).

Hot stars in binary systems

The stellar winds characteristic of hot stars provide for interactive effects in binary systems containing compact companions. The general nature of these systems and their multiwavelength study is included in another chapter (By K. Mason), but here we refer briefly to situations where fundamental properties of the hot primary are derived from multiwavelength data.

It is sometimes possible to infer parameters characterizing the winds from the x-ray emission due to the accretion of wind material onto the compact companion. The x-ray emission in this situation is very sensitive to the mass influx from the wind, so the wind density at the location of the compact companion can be inferred from the x-ray flux. This information, combined with the wind terminal velocity inferred from the ultraviolet, can provide unique information on the velocity and density structure of a stellar wind. This technique is especially useful when the companion has an eccentric or inclined orbit, so that it probes different regimes in the wind at different times. Waters *et al.* (1987b) were recently able to determine the geometry of the wind in a binary containing a Be star and a compact x-ray source, showing that the wind is highly confined to the equatorial plane of the Be star.

The combination of ultraviolet and x-ray data have allowed the study of x-ray ionization effects in the winds from OB stars, since the orbiting x-ray source is expected to ionize the wind material in phase with the binary period (Hatchett and McCray 1977). As yet observations have not perfectly matched theory in these systems, but in principle new information on the nature of the winds can be inferred (see the review by Córdova and Howarth 1987).

4. Summary and Conclusions

The benefits to be derived from multiwavelength observations of hot stars have only begun to be accrued. Even so, important and unique information on the basic properties of the stars has been acquired, and we have seen that multispectral data are absolutely essential for many studies of the extended atmospheres of these stars.

The practical difficulties of obtaining simultaneous multispectral observations are enormous, with the result that for every success story reported here, there have been several failures, either because it proved impossible to arrange the observations, or the weather was poor for the ground-based work, or, in some cases, the target object failed to perform during the brief interval when the coordinated observations were planned.

The future of multispectral research on hot stars is uncertain. At present we have, in addition to ground-based radio, optical, and infrared observatories, only a single ultraviolet telescope in space, yet the ultraviolet is crucial to nearly all multispectral studies of hot stars. When the *IUE* ceases operations, the *Hubble Space Telescope* will be in orbit, but it is not likely that *HST* will be available for extensive coordinated observations of stellar targets. Some data on far-UV and EUV fluxes are available from the *Voyager* spacecraft, but scheduling restrictions make coordinated observations difficult. The upcoming launches of *ROSAT* and the *Extreme Ultraviolet Explorer* promise to improve the chances of success in multispectral programs requiring high-energy data.

It appears that the kind of coordinated multiwavelength observing campaigns that have so much promise for revealing the fundamental secrets of hot stars and their winds will not be possible until there is either a dedicated space mission containing a package of instruments for different spectral bands, or a vastly different system for scheduling observations on existing instruments in which priority is given to coordinated observations.

The author is grateful for helpful comments from several colleagues, particularly H. Lamers, C. D. Garmany, P. S. Conti, and F. A. Córdova. R.S. Polidan was kind enough to provide far-ultraviolet *Voyager* data on ζ Puppis in advance of publication. The work presented here was sponsored by NASA grant NSG-5300 to the University of Colorado.

References

Abbott, D. C. 1982, *Ap.J.*, **259**, 282.

Abbott, D. C., Bieging, J. H., and Churchwell, E. 1981, *Ap.J.*, **250**, 645.

Abbott, D. C., Bieging, J. H., and Churchwell, E. 1984, *Ap.J.*, **280**, 671.

Abbott, D. C., Bieging, J. H., Churchwell, E., and Cassinelli, J. P. 1980, *Ap. J.*, **238**, 196.

Abbott, D. C., Bieging, J. H., Churchwell, E., and Torres, A. V. 1986, *Ap. J.*, **303**, 239.

Abbott, D. C., Telesco, C. M., and Wolff, S. C. 1984, *Ap.J.*, **279**, 225.

Andrillat, Y. 1987, *Physics of Be Stars*, A. Slettebak and T.P. Snow, eds. (Cambridge:Cambridge University Press), p. 237.

Andrillat, Y. and Houziaux, L. 1967, *J. Obs.*, **50**, 107.

Baade, D. 1987, Physics of Be Stars, A. Slettebak and T. P. Snow, eds. (Cambridge:Cambridge University Press), p. 361.

Baade, D. and Lucy, L. 1987, *Astr. Ap.*, in press.

Barlow, M. J. and Cohen, M. 1977, *Ap.J.*, **213**, 737.

Barlow, M. J., Smith, L. J., and Willis, A. J. 1981, *M.N.R.A.S.*, **196**, 101.

Barker, P. K. 1987, *Physics of Be Stars*, A. Slettebak and T. P. Snow, eds. (Cambridge:Cambridge University Press), p. 431.

Bolton, C. T., Fullerton, A. W., Bohlender, D., Landstreet, J. D., and Gies, D. R. 1987, *Physics of Be Stars*, A. Slettebak and T. P. Snow, eds. (Cambridge:Cambridge University Press), p. 82.

Cassinelli, J. P. 1979, *Ann. Rev. Astr. Ap.*, **17**, 275.

Cassinelli, J. P., Waldron, W. L., Sanders, W. T., Harnden, F. R., Rosner, R., and Vaiana, G. S. 1981, *Ap.J.*, **250**, 677.

Castor, J. I. and Lamers, H. J. G. L. M. 1979, *Ap. J. Suppl.*, **39**, 481.

Castor, J. I. and Simon, T. 1983, *Ap.J.*, **265**, 304.

Cherepaschuk, A. M., Eaton, J. A., and Khalliullin, K. F. 1984, *Ap.J.*, **281**, 774.

Code, A. D., Davis, J., Bless, R. C., and Hanbury-Brown, R. 1976, *Ap. J.*, **203**, 417.

Collins, G. W. 1987, *Physics of Be Stars*, A. Slettebak and T. P. Snow, eds. (Cambridge:Cambridge University Press), p. 3.

Collins, G. W. and Sonneborn, G. H. 1977, *Ap. J. Suppl.*, **34**, 41.

Córdova, F. A. and Howarth, I. D. 1987, *Scientific Accomplishments of the IUE*, Y. Kondo, ed. (Dordrecht:Reidel), p. 395.

Coté, J. 1987, *Astr. Ap.*, **181**, 77.

Dachs, J. 1987, *Physics of Be Stars*, A. Slettebak and T. P. Snow, eds. (Cambridge:Cambridge University Press), p. 149.

Doazan, V., Grady, C. A., Snow, T. P., Peters, G. J., Marlborough, J. M., Barker, P. K., Bolton, C. T., Bourdonneau, B., Kuhi, L. V., Lyons, R. W., Polidan, R. S., Stalio, R., and Thomas, R. N. 1985, *Astr. Ap.*, **152**, 182.

Finkenzeller, U. and Mundt, R. 1984, *Astr. Ap. Suppl.*, **55**, 109.

Fitzpatrick, E. D. 1987, preprint.

Fitzpatrick, E. D., Savage, B. D., and Sitko, M. L. 1982, *Ap.J.*, **256**, 578.

Garmany, C. D. and Conti, P. S. 1984, *Ap. J.*, **284**, 705.

Garmany, C. D. and Conti, P. S. 1985, *Ap. J.*, **293**, 407.

Garmany, C. D., Olson, G. L., Conti, P. S., and Van Steenberg, M. 1981, *Ap. J.*, **250**, 660.

Gathier, R., Lamers, H.J.G.L.M., and Snow, T.P. 1981, *Ap. J.*, **247**, 173.

Gehrz, R. D., Hackwell, J. A., and Jones, T. W. 1974, *Ap. J.*, **191**, 675.

Gizari, D. Y., Schmitz, M., and Mead, J. M. 1987, *Catalog of Infrared Observations*; (*NASA Ref. Pub. 1196*), (2nd Ed.), (Washington:NASA).

Harnden, F. R., Branduardi, G., Elvis, M., Gorenstein, P., Grindlay, J., Pye, J. P., Rosner, R., Topka, K., and Vaiana, G. S. 1979, *Ap. J. (Lett.)*, **234**, L51.

Hatchett, S. and McCray, R. 1977, *Ap. J.*, **211**, 552.

Heap, S. R. 1976, *Be and Shell Stars*, A. Slettebak, ed. (Dordrecht:Reidel), p. 165.

Hearn, A. C. 1975, *Astr. Ap.*, **40**, 277.

Heise, J. 1985, in *X-ray Astronomy in the EXOSAT Era*, ed. A. Peacock (Dordrecht:Reidel), p. 79 (also in *Space Sci. Rev. 40*).

Henrichs, H. P., Hammerschlage-Hensberge, G., Howarth, I. D., and Barr, P. 1983, *Ap. J.*, **268**, 807.

Hillier, D. 1986, *Wolf-Rayet Stars (IAU Symp. 116)*, C. de Loore, A. J. Willis, and P. Laskarides, eds. (Dordrecht:Reidel), p. 261.

Hoffleit, D. and Jaschek, C. 1982, *The Bright Star Catalogue* (New Haven:Yale University Observatory).

Holberg, J. 1986 (reported by Vauclair and Liebert 1987).

Holberg, J., Wesemael, F., and Basile, J. 1986, *Ap. J.*, **306**, 629.

Hutchings, J. B. 1976, *P.A.S.P.*, **88**, 5.

Hutchings, J. B. and Stoeckly, T.R. 1977, *P.A.S.P.*, **89**, 19.

de Jager, C., Nieuwenhuijzen, H., and van der Hucht, K. A. 1987, *Astr. Ap.*, in press.

Jaschek, M. and Groth, H.-G. (eds.) 1982, *Be Stars (IAU Symp. 98)*, (Dordrecht:Reidel).

Kahn, S. M., Wesemael, F., Liebert, J., Raymond, J. C., Steiner, J. E., and Shipman, H. L. 1984, *Ap. J.*, **278**, 255.

Koester, D., Vauclair, G., Dolez, N., Oke, J. B., Greenstein, J. L., and Weidemann, V. 1985, *Astr. Ap.*, **142**, L5.

Lamers, H. J. G. L. M. 1987, *Physics of Be Stars*, A. Slettebak and T. P. Snow, eds. (Cambridge:Cambridge University Press), p. 219.

Lamers, H. J. G. L. M., Gathier, R., and Snow, T. P. 1982, *Ap. J.*, **258**, 186.

Lamers, H. J. G. L. M., Snow, T. P., de Jager, C., and Langerwert, A. 1988, *Ap. J.*, in press.

Lamers, H. J. G. L. M. and Waters, L. B. F. M. 1987, *Astr. Ap.*, **182**, 80.

Leitherer, C., Hefele, H., Stahl, O., and Wolf, B. 1982, *Astr. Ap.*, **108**, 102.

Liebert, J., Wesemael, F., Sion, E. M., and Wegner, G. 1984, *Ap. J.*, **277**, 692.

Longo, R., Stalio, R., Polidan, R. S., and Rossi, L. 1988, *Ap. J.*, in press.

de Loore, C., Giovannelli, F., van Dessel, E.L., Bartolini, C., Burger, M., Fervari-Toniolo, M., Giangrande, A., Guarnieri, A., Hellings, P., Hensberge, H., Persi, P., Picciori, A., and Van Diest, H. 1984, *Astr. Ap.*, **141**, 279.

Lucy, L. B. and White, R. L. 1980, *Ap. J.*, **241**, 300.

Mihalas, D. 1970, *Stellar Atmospheres* (San Francisco:Freeman).

Morton, D. C. 1967, *Ap. J.*, **147**, 1017.

Morton, D. C., Jenkins, E. B., Matilsky, T. A., and York, D. G. 1972, *Ap. J.*, **177**, 219.

Nussbaumer, H., Schmutz, W., Smith, L. J., and Willis, A. J. 1982, *Astr. Ap. Suppl.*, **47**, 257.

Owocki, S. 1987, unpublished.

Petre, R., Shipman, H. L, and Canizares, C. R. 1986, *Ap. J.*, **304**, 356.

Poeckert, R. and Marlborough, J. M. 1978, *Ap. J.*, **220**, 940.

Pollock, A. M. T. 1987, *Astr. Ap.*, **171**, 135.

Prinja, R. K. and Howarth, I. D. 1986, *Ap. J. Suppl.*, **61**, 357.

Remie, H. and Lamers, H. J. G. L. M. 1982, *Astr. Ap.*, **105**, 85.

Slettebak, A. and Snow, T. P. (eds.) 1987, *Physics of Be Stars* (Cambridge:Cambridge University Press).

Smith, L. J. and Willis, A. J. 1982a, *Wolf-Rayet Stars (IAU Symp. 99)*, C. de Loore and A. J. Willis (eds.), (Dordrecht:Reidel), p. 113.

Smith, L. J. and Willis, A. J. 1982b, *M.N.R.A.S.*, **201**, 451.

Smith, M. A. 1977 *Ap. J.*, **215**, 927.

Smith, M. A. 1987, preprint.

Snow, T. P. 1981, *Ap. J.*, **251**, 139.

Snow, T.P., Cash, W.C., and Grady, C.A. 1981, *Ap. J. (Lett.)*, **244**, L19.

Snow, T. P. and Marlborough, J. M. 1976, *Ap. J. (Lett.)*, **203**, L87.

Snow, T. P. and Morton, D. C. 1976, *Ap. J. Suppl.*, **32**, 429.

Snow, T. P., Peters, G. J., and Mathieu, R. 1979, *Ap. J. Suppl.)*, **39**, 359.

Stalio, R., Polidan, R.S., and Peters, G.J. 1987, *Ap. J.*, in press.

Struve, O. 1931, *Ap. J.*, **73**, 94.

Swings, J.P. 1976, *Be and Shell Stars*, ed. A. Slettebak (Dordrecht:Reidel), p. 219.

Taylor, A. R., Waters, L. B. F. M., Lamers, H. J. G. L. M., Persi, P., and Bjorkman, K. S. 1988, *M. N. R. A. S.*, in press.

Thompson, G. I., Nandy, K., Jamar, C., Monfils, A., Houziaux, L., Carnochan, D. J., and Wilson, R. 1978, *Catalogue of Stellar Ultraviolet Fluxes*, (London:UK Science Research Council).

Underhill, A. B. 1980, *Ap. J.*, **239**, 220.

Underhill, A. B. 1981, *Ap. J.*, **244**, 963.

Vauclair, G. and Liebert, J. 1987, *Scientific Accomplishments of the IUE*, Y. Kondo, ed. (Dordrecht:Reidel), p. 355.

Vaiana, G.S., Casinelli, J.P., Fabbiano, G., Giacconi, R., Golub, L., Gorenstein, P., Haisch, B.M., Harnden, F.R., Johnson, H.M., Linsky, J.L., Maxson, C.W., Mewe, R., Rosner, R., Seward, F., Topka, K., and Zwaan, C. 1981, *Ap. J.*, **245**, 163.

Vogt, S. S. and Penrod, G. D. 1983, *Ap. J.*, **275**, 661.

Waters, L. B. F. M. 1986, *Astr. Ap.*, **162**, 121.

Waters, L. B. F. M., Coté, J., and Lamers, H. J. G. L. M. 1987, *Astr. Ap.*, in press.

Waters, L. B. F. M., Lamers, H. J. G. L. M., and Coté 1987, *Physics of Be Stars*, A. Slettebak and T. P. Snow, eds. (Cambridge:Cambridge University Press), p. 245.

Waters, L. B. F. M., Taylor, A. R., van den heuvel, E. P. J., Habets, G. M. H. J., and Persi, P. 1987, *Astr. Ap.*, in press.

Waters, L. B. F. M., Taylor, A. R., van den heuvel, E. P. J., Habets, G. M. H. J., and Persi, P. 1987b, *Astr. Ap.*, in press.

White, R. L. 1985, *Ap. J.*, **289**, 698.

White, R. L. and Becker, R. H. 1983, *Ap. J. (Lett.)*, **272**, L19.

Williams, P.M., van der Hucht, K.A., and Thé, P.S. 1987, *Astr. Ap.*, **182**, 91.

Willson, L. A. 1986, unpublished.

Cataclysmic Variables

Paula Szkody

Department of Astronomy, University of Washington

Mark Cropper

Mullard Space Science Laboratory, University College, London

1. Introduction

In recent years, there have been a variety of general reviews on cataclysmic variables (CVs) covering a wide range of wavelengths e.g. X-ray (Mason 1985; Osborne 1986), ultraviolet (Cordova and Howarth 1987; Verbunt 1987), optical (Wade and Ward 1987); infrared (Berriman, Szkody and Capps 1985) and radio (Chanmugam 1987). Although there is much to learn from looking at the members of this class of variables at one specific wavelength, there is much more to be gained in understanding the physical processes at work in these systems through a multiwavelength approach. A few past reviews have touched on some aspects of multiwavelength data (Cordova and Mason 1982; Pringle and Verbunt 1984; Szkody 1985a, b). In this review, we will concentrate on the major astrophysical results on CVs that have emerged due to a concentrated effort to obtain and apply multiwavelength data to specific problems.

We will use the commonly accepted definition of a CV as a close binary with a white dwarf primary and late spectral type Roche Lobe-filling secondary. This means we will bypass the work on symbiotics and neutron star low mass X-ray binaries and also novae as they are covered in the following two chapters. Since our purpose is to concentrate on a few success stories from which important results have emerged, we will not attempt to review every multiwavelength campaign in the field, although the literature will attest to the importance of each program to the total picture.

2. Data Base

In virtually every case of existing multiwavelength observations, the optical region of the spectrum is combined with satellite data (Einstein or EXOSAT for the X-ray, ANS or IUE and/or Voyager for the ultraviolet) or ground-based infrared or radio. The optical is the cornerstone of the determination of the "state" of the system i.e. quiescence, outburst, high or low accretion rate. It is also the easiest wavelength in which to obtain data for bright CVs due to a) the continuous coverage of many bright systems by variable star organizations such as the AAVSO and the RASNZ without any special requests and b) the optical Fine Error Sensor (FES) on board the IUE satellite which is normally used for centering the correct targets gives a visual magnitude which is good to 10%. While the optical magnitude is thus easily available, the coordination of several satellites (e.g. EXOSAT and IUE and Voyager) has presented a monumental scheduling problem, especially when coverage of an unpredicted event like the rise of a dwarf nova outburst is desired. The fact that several data sets of this type do exist attests to the perseverance of the observers and the willingness of satellite directors to override normal observations to accomplish target of opportunity programs.

3. Results

CVs can be classified according to their dominant source of light. This depends principally on the strength of the magnetic field. In high magnetic field systems (over a million gauss), the emission from the system can be dominated by one or more accretion columns (the AM Hers) while the lower field systems are dominated by the light of accretion disks (the dwarf novae and novalike systems) or a combination of disk and accretion column (the DQ Hers). Thus, the major research areas within the field of CVs center on the geometry and energetics of the accretion column and the mass transfer in the disk and its role in the outburst mechanism. In both groups, there is a major ongoing effort to detect the underlying white dwarfs and secondary stars which are normally swamped by the light of the column or the disk. The detection and spectral typing of the secondary enables the determination of the system distance, while the detection and temperature determination of the white dwarf presents the opportunity to study the effects of accretion heating and subsequent cooling in comparison to single white dwarfs. We will concentrate on the advances made in these particular areas.

Disk CVs

The long standing debate among astronomers working on disk CVs centers on the cause of the outbursts which occur in dwarf novae but not in novalike systems. While there is a general agreement that the outbursts are related to an increase of accretion through the disk and onto the white dwarf, it is not totally clear whether the accretion is modulated by the transfer rate from the secondary (recently reviewed in Bath 1985) or by an instability in the disk (recently reviewed in Cannizzo and Kenyon 1987). A successful model must predict the correct flux distribution that is actually observed for the disk, as well as how the flux changes over the course of an outburst, and it must also predict the correct timescales for the observed changes.

A. The Flux Distribution of the Disk

A multiwavelength approach to this problem is mandated by the large temperature range of the disk, which results in prominent emission in the EUV (hot inner zones) through the infrared (cool outer zones). Limited success in reproducing the observed flux distributions from 1000-5500Å (the IUE to FES range) has been obtained for the high mass accretion systems which appear to have optically thick steady state disks. A large data set exists in this regard, including the novalike systems V3885 Sgr, LSI+558, KQ Mon and UX UMa (Guinan and Sion 1982a,b; Sion and Guinan 1982; King *et al.* 1983) and more than 20 dwarf novae at outburst (summarized in Szkody 1985a and Pringle, Verbunt and Wade 1986). A few examples are shown in Figure 1.

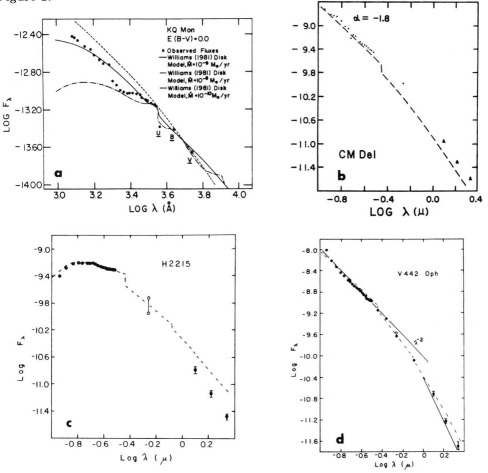

Figure 1 Examples of flux distributions:

a. The novalike system KQ Mon from Sion and Guinan (1982).

b. The dwarf nova CM Del near outburst from Szkody (1985c). The dashed line is a Williams and Ferguson (1982) model with $\dot{M} = 10^8 M \odot$ /yr.

c. The DQ Her novalike system H2215–084 (FO Aqr) from Szkody 1985b. The dashed line is a Williams and Ferguson $10^{-10} M \odot$ /yr model.

d. The novalike system V442 Oph from Szkody and Shafter 1983. Courtesy of the Publications of the Astronomical Society of the Pacific. The fluxes have been corrected for a reddening of E(B–V) = 0.25. The dashed line is the Williams and Ferguson $10^{-8} M \odot$ /yr model.

Although these are the easiest cases to model, the fits between models and observations are not perfect, especially as concerns the strength of the Balmer jump and the ratio of UV to optical flux. The standard accretion disk models (reviewed in Pringle 1981) generally use black bodies or a combination of stellar atmospheres to calculate the resultant disk flux distribution. Wade (1984, 1987) has discussed the differences in these two approaches as compared to the observations. In general, the stellar atmosphere models give Balmer jumps larger than observed and too much optical flux but good fits to the UV fluxes (Figure 2), while black body models give no Balmer jump at all. So far, there are no major attempts to fit the large range of quiescent flux distributions that have been observed (shown in Verbunt 1987). Quiescence is a difficult time to model, since the disks can be highly unstable on short timescales and large areas may not be optically thick. In addition, there may be large contributions to the disk light from a hot spot and/or stellar sources which must be correctly removed before attempting to fit the observations to a model. Thus, the wealth of observational data must await a better treatment of the radiative transfer physics in the disk from a theoretical standpoint in order to realize improvement in the fitting of disk flux distributions. The models of Kriz and Hubeny (1986) are a step in the right direction, since they take into account the effects of radiation interacting with a finite thickness disk. These effects are different from the assumptions of the previous classic stellar atmosphere models and the resulting computed flux distributions provide better fits to the observations.

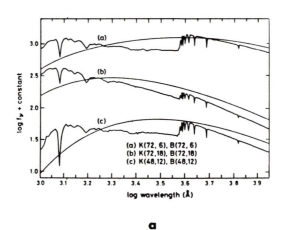

a

Figure 2 The differences between various theories and observations:

a. Calculated disk spectra for stellar atmosphere and black body models. From Wade (1984).

b. Flux distributions computed with stellar atmosphere models for different types of instability models vs. mass transfer models as compared to the UV–optical observations of VW Hyi and CN Ori (from Verbunt 1986).

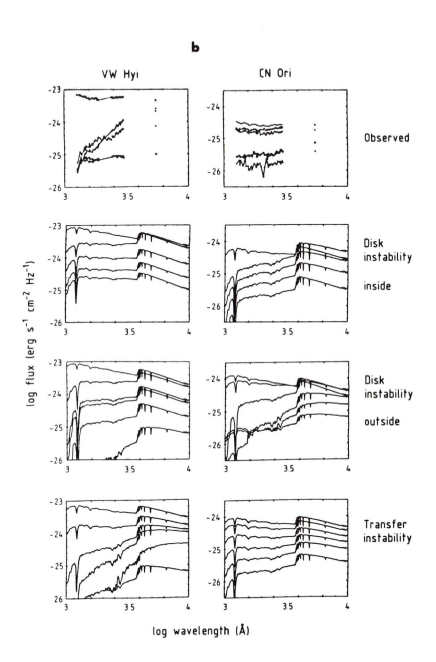

b

log wavelength (Å)

B. Changes During Dwarf Nova Outbursts

Simultaneous multiwavelength data over a large wavelength range throughout the course of an entire outburst are crucial to the understanding of what happens to a disk during an outburst and which model provides the best explanation of the changes. Good recent reviews of the mass transfer and disk instability models, with complete references, are contained in Bath (1985) and Cannizzo and Kenyon (1987). Among the differences between the models is the location of the start of the outburst – whether it starts from the outside of the disk (as in the mass transfer models and in the asymmetrical outbursts of the disk instability models), or from the inside (as in the symmetrical outbursts of the disk instability models). Other differences include the timescale for the rise to outburst and the disk luminosity during the interval between outbursts.

In the mass transfer models, the disk is fed by an increased rate of mass transfer from the secondary at the time of outburst. Since the additional material must interact with a large part of the disk before the temperature goes up, the rise to outburst usually results in a considerable delay between the optical and ultraviolet light. After the outburst, the disk cools down to some equilibrium value.

In the disk instability models, the mass transfer rate from the secondary is constant, but the change in the temperature dependence of the opacity in the hydrogen and helium ionization zones leads to an instability in the disk which makes the disk change from a cool mass-storing state to a hot mass-dumping state. There is an S shaped relationship between the effective temperature and the surface density of the disk (Smak 1984). Between eruptions, the disk stores up the mass transferred from the secondary and the luminosity gradually increases as the disk surface density reaches the values at the bottom branch of the instability curve. Conditions in the disk then change rapidly to those of the hot branch of the instability curve (resulting in short time lags between various wavelengths). The timescale varies depending on whether the eruption starts in the inner disk (for small mass transfer rate systems) or near the outer edge (for large mass transfer rate systems) (Smak 1984; Cannizzo and Kenyon 1987). After the high state is completed, the disk returns to the cool quiescent state as a small fraction of the stored material accretes onto the white dwarf.

As well as investigating the inter-outburst quiescent behavior, the multiwavelength campaigns have tried to determine the timescale for the rise to outburst in the inner (UV emitting) disk compared to that of the outer (optical emitting) disk, in order to test the model predictions. Although several IUE-FES measurements exist on the rise to outburst (summarized in Pringle and Verbunt 1984; Szkody 1985a and Pringle, Verbunt and Wade 1986) the best and most extensive multiwavelength results come from the large coordinated campaigns on VW Hyi (results in Pringle *et al.* 1987), SS Cyg (Watson, King and Heise 1985; Cannizzo, Wheeler and Polidan 1986) and OY Car (Naylor *et al.* 1987a,b). These campaigns are summarized in Table 1. Figure 3 shows the VW Hyi data from the Pringle *et al.* paper.

Table 1 Extensive Multiwavelength Outburst Campaigns

Object	Outburst Coverage	Satellites and Wavelengths[a]
VW Hyi	Outburst, Superoutburst, Quiescence	EXOSAT, Voyager, IUE, Optical
SS Cyg	Asymmetrical Wide and Narrow Outbursts	EXOSAT, Voyager, IUE, Optical
OY Car	Superoutburst	EXOSAT, IUE, Optical, IR

[a]The wavelengths for each of the above regions involve: EXOSAT (0.05-6 keV), Voyager (950-1150Å), IUE (1150-3000Å), Optical from ground-based and FES measurements, IR (1.25-2.2 microns)

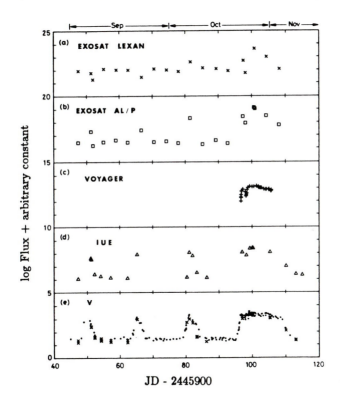

Figure 3 The VW Hyi Multiwavelength Observations from Pringle *et al.* 1987.

The major results from these monumental efforts are discussed below.

1) The UV and soft X-ray lag the optical rise to outburst by about half a day for the ordinary asymmetrical outbursts of VW Hyi and SS Cyg. This confirms results seen in shorter campaigns on VW Hyi (Hassall *et al.* 1983; Schwarzenberg-Czerny *et al.* 1985), WX Hyi (Hassall, Pringle and Verbunt 1985) and RX And (Pringle and Verbunt 1984). Efforts to model these delays have led to some support for the mass transfer model, which can more easily account for the length of the delay than the disk instability models (Pringle, Verbunt and Wade 1986). However, the disk instability modelling was based on the use of a combination of Kurucz stellar atmosphere models to make up the UV and optical flux distribution. Cannizzo and Kenyon (1987) have confirmed that shorter delays are obtained with these atmosphere models compared to models with combinations of black bodies for outbursts that start at a large disk radius and move inward (the asymmetrical outbursts). Since Kurucz models cannot reproduce the correct flux distributions, it is not clear that this is a better treatment of the radiative transfer than the black body approach. Recent disk instability models by Meyer-Hofmeister (1987) have shown that using lower temperatures and appropriate opacities for the disk can give lag times up to one day, which is larger than needed. Models by Mineshige (1987), which concentrate on the energetics of the outer disk, produce stagnation stages in the disk instability model which can also account for lags up to one day. Only once a model can reproduce a flux distribution in detail at outburst can we have some confidence that the appropriate physics is being applied and only then can we proceed from that point to evolve that disk through the outburst instability to determine if the correct lag time can be obtained. Until we reach this position, we are limited in our understanding of the outburst mechanism.

Further simultaneous multiwavelength data on symmetrical outbursts are still needed to determine the timelag differences between symmetrical and asymmetrical outbursts. IUE data on SS Cyg at the same point on the rise of a symmetrical outburst as an asymmetrical one shows much stronger UV continuum levels and line emission, indicating that the UV lag time is probably small (Polidan 1987). Unfortunately, the symmetrical outbursts are rare in SS Cyg and so it is difficult to obtain adequate data. IUE archival data on a slow rise of CN Ori (Pringle, Verbunt and Wade 1986) also show the simultaneous rise at UV and optical wavelengths. More complete model fitting of the observed flux distributions during the rise of the symmetrical vs asymmetrical outbursts will determine if they are due to different mechanisms (transfer vs disk instability) or to different sites for the start of the outburst (outside or inside of the disk).

2) At outburst, there is a flux deficiency at short UV (less than 1200Å) wavelengths in comparison to model atmospheres (Polidan and Holberg 1987). This effect is also evident in O stars but not in hot white dwarfs. A better understanding of the radiation emitted by O and B main sequence stars at short wavelengths is needed. Since high gravity stars are not affected, it may be that either an outflowing wind (common in O and B stars and in disks at outburst) or the effects of lower gravity in the atmosphere of the O stars and outer reaches of the disk as

compared to white dwarfs are causing the discrepancy (Polidan 1987).

3) The observed superoutbursts in VW Hyi and OY Car reveal several interesting points. The superoutburst of VW Hyi looked normal at optical wavelengths but showed a double peaked (i.e. precursor) structure at Voyager and IUE wavelengths (Pringle *et al.* 1987). Since the precursor had similar characteristics to a normal outburst, this observation gives some support to the suggestion that superoutbursts may be triggered by normal outbursts (Bateson 1977). Time resolved orbital coverage of the superoutburst in the eclipsing system OY Car showed that the X-ray flux (which is not eclipsed) must originate from a large (Roche lobe size) optically thin corona (Naylor *et al.* 1987a), the superhump light originates from a large (projected area of secondary star or the accretion disk) optically thick zone with temperature of about 8000K (Naylor *et al.* 1987b) and there is extended vertical structure in the disk which modulates the ultraviolet to infrared continuum light.

All of these characteristics seem to be in accord with a relatively new idea for the cause of the superoutburst and associated superhump phenomena (Whitehurst 1987; Charles 1987; Honey *et al.* 1987). This approach postulates that the superoutbursts are due to increased mass transfer after a normal outburst has started. Due to tidal instabilities existing in systems with mass ratios greater than 4, the disk becomes eccentric and precesses, resulting in the superhump phenomenon. The tidal stresses result in disk flaring and large brightness zones which could account for some of the large scale modulations seen in the OY Car data. This is one area that would benefit from more multiwavelength data (especially any modulations that can be traced to the disk structure at superoutburst) as well as the further development of the tidal theory.

4) The soft X-ray and UV behavior of systems near outburst reveal some clues as to the structure of the boundary layer and corona. Although interstellar absorption permits the detection of a soft X-ray component in only the nearest systems, the novalike system IX Vel (van der Woerd 1987) and the dwarf novae SS Cyg, U Gem (Cordova *et al.* 1980b), VW Hyi (van der Woerd 1987) and SW UMa (Szkody, Osborne and Hassall 1988) (at outburst or superoutburst) have all been found to show a soft X-ray component of 5-30 eV. Models of Pringle and Savonije (1979) and King and Shaviv (1984) postulate that the quiescent X-ray emission is from bremsstrahlung cooling in a corona about the white dwarf, while outburst emission should be from an optically thick boundary layer. If inclination and absorption effects are included, the model can explain most of the observed data. In the low inclination systems VW Hyi, SS Cyg and U Gem, the hard quiescent X-ray spectrum changes to a soft X-ray spectrum at outburst, while the UV shows the typical hot outburst disk with absorption lines. The high inclination system OY Car shows an optically thin coronal component at outburst and a cooler UV continuum with emission lines, consistent with a boundary layer hidden by an extended disk rim due to the high inclination of the system. In the lower inclination systems SW UMa and IX Vel, however, because of the uncertainties in the interstellar absorption, the X-ray emission could arise either from a corona or a boundary layer. More tests of the boundary layer model must await the next satellite capable of soft X-ray

measurements.

Another interesting aspect of the soft X-ray observations at outburst has been the quasi-periodic pulsations evident in SS Cyg, U Gem and VW Hyi with periods near 9, 25 and 14 seconds respectively (Cordova *et al.* 1980a,b; van der Woerd 1987). There are possible connections to the optical oscillations which are observed in these same systems at outburst, although the amplitudes, periods and coherence times in the X-ray and optical are different. Van der Woerd (1987) summarizes some of the possibilities such as the optical being the tail of the spectral distribution of the oscillating X-ray flux, the inner disk being heated by the varying X-ray flux or irregularities in the outer disk being heated by a constant X-ray flux but acknowledges that there is no existing model which can explain the origin and properties of both the X-ray and optical pulsations.

5) The decline from outburst has been observed to start first in X-ray wavelengths for SS Cyg and VW Hyi (Watson, King and Heise 1985; van der Woerd and Heise 1987), while there is generally a similar decline in UV and optical until the optical reaches quiescent values. After this point, the UV and X-ray in VW Hyi continue to decline until the start of the next outburst (Verbunt *et al.* 1987). This UV decline has also been noted in WX Hyi (Hassall *et al.* 1985), U Gem (Szkody and Kiplinger 1985) and WZ Sge (Hassall 1987) but VW Hyi is the only case where the X-ray is linked to the ultraviolet decline. At face value, this can be taken as strong support for the mass transfer model, which predicts a decreasing flux during quiescence (Pringle *et al.* 1987). However, in U Gem, VW Hyi and WZ Sge, the UV is dominated by the white dwarf (Panek and Holm 1984; Mateo and Szkody 1984; Verbunt 1987). Thus, the decreasing UV flux can also be explained by the cooling off of the white dwarf which has been heated by the outburst (discussed in Pringle *et al.* 1987). Further simultaneous X-ray, ultraviolet and optical observations throughout the quiescent interval of long outburst period systems (with and without prominent white dwarfs) are needed to separate the effects of a decreasing accretion rate from the cooling of the white dwarf.

Magnetic CVs

Magnetic CVs radiate energy over a wide range of wavelengths from the X-ray through to the radio, and multiwavelength studies have been essential in making sense of the large body of data available. The central issue is the understanding of the energetics of the accretion column. In view of the fact that these systems can be found in different luminosity states, simultaneous or nearly simultaneous multiwavelength observations are mandatory if the overall energy balance of the accretion process is to be understood.

In the AM Her stars, unlike non-magnetic CVs, there is no storage of accretion energy in a disk because accretion takes place directly at or near one or more magnetic poles. There, material travelling at a velocity of ~4000 km/s is decelerated at a shock front before settling onto the white dwarf surface as it cools. In the process, X-rays are emitted as bremsstrahlung. Approximately half of these will be incident on the white dwarf photosphere and will be reprocessed as soft X-rays. The soft X-ray component may be increased by a large factor as the result of non-radiative effects (flow inhomogeneities, conduction etc. – see below). In addition, cyclotron radiation will be emitted in the optical/infrared region of the spectrum by electrons in the accreting gas which are threaded on the magnetic field lines. The relative importance of these three components depends primarily on the magnetic field, the area over which the accretion takes place and the mass of the white dwarf. The other sources of electro-magnetic radiation from the system are line and continuum emission from the accretion stream between the two stars, synchrotron or maser activity in the radio from a region near the red dwarf, and the photospheres of the two stars. Of all of the above, only the last two are relatively constant in luminosity.

The DQ Her stars have additional emission sources. In these CVs the system dimensions are large enough or the magnetic field small enough for some sort of a disk to form. Whether these disks have the same morphology as those of the non-magnetic systems is still controversial, but it is probable that in most DQ Her systems accretion does not occur directly from the stream onto or near to the magnetic poles. The disk and hotspot (where the stream strikes the disk) are therefore additional emission sources. These structures and the secondary star may also intercept the X-ray beam from the accretion shock at the magnetic poles, causing reprocessing and reemission at longer wavelengths.

A. The Flux Distribution

The only simultaneous multifrequency observations spanning the X-ray to the optical with sufficient energy resolution in the X-rays to constrain the soft X-ray flux in any magnetic CV are those on QQ Vul by Osborne *et al.* (1986), using EXOSAT, IUE and optical observations (see Figure 4). However, compilations of the spectra from X-ray to infrared also exist for AM Her (Lamb 1985) and VV Puppis (Patterson *et al.* 1984). These are all AM Her systems: unfortunately no such compilations exist for DQ Her systems, although Sherrington *et al.* (1980) give energy distributions from 1200Å to the infrared K band. These important investigations have allowed the energy balance to be determined and compared to the calculations of Masters (1978) and Lamb and Masters (1979), which is still the standard theory. This predicts $L_{soft\ X-ray} \simeq L_{hard\ X-ray} + L_{cyclotron}$. The results of the AM Her and VV Pup investigations countered earlier suggestions that the soft to hard X-ray ratio might be as high as 100 – the "soft X-ray problem" (Tuohy *et al.* 1978, Raymond *et al.* 1979, Fabbiano *et al.* 1981, Fabbiano 1982, Rothschild *et al.* 1981 and Tuohy *et al.* 1981) – by concluding that the ratio was approximately unity. The results from QQ Vul, however, suggest that the ratio is at least 4.5. In addition, King and Watson (1987) find that the mass transfer rates and accretion luminosities calculated for AM Her systems are an order of magnitude lower than those expected from evolutionary considerations unless the ratio is considerably larger than unity. These and the earlier results have spawned a number of refinements to the basic theory which include the effects of nuclear burning, in homogeneities in the accretion stream, electron conduction, end losses of the high velocity tail of the Maxwellian electron velocity distribution directly into the photosphere and different temperatures for the ions and electrons in the plasma (for an excellent discussion on the present state of play and a complete set of references see Imamura *et al.* 1987, sections I and II). On the observational side, further progress will depend on satellites with superior energy resolution and sensitivity, particularly in the soft X-ray band.

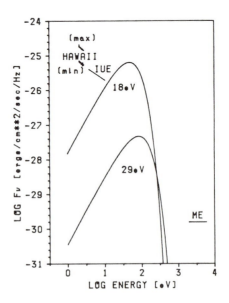

Figure 4 The QQ Vul Overall energy distribution from Osborne *et al.* 1986.

One of the problems with the standard theory discussed above has been that for typical magnetic fields of approximately 20 million G found in the AM Her systems the cyclotron radiation is predicted to peak in the UV or violet. Instead, observations find that the flux peaks in the near infrared at ∼ 1 micron. Some progress in improving the models has been made by paying attention to the shape of the shock front and modifying the mass accretion rate profile across the shock (Stockman and Lubenow 1987; Wickramasinghe and Ferrario 1987). It is clear, after consideration of the manner in which the gas stream threads on the magnetic field, that the shape of the accretion region will not be the idealized circular cap with a uniform accretion rate profile. Ever since the AM Her systems were recognized as a separate class, observational evidence for this fact has been available from the asymmetry (with respect to orbital phase) of the optical and infrared light and polarization curves. Moreover, the earliest theoretical studies (e.g., Fabian, Pringle, and Rees

976) had established that the height of the shock front at any point on the surface
of the white dwarf depended on the accretion rate. Nevertheless, it is only recently
with more extensive X-ray data and optical and infrared polarimetry coverage that
these effects have been recognized as being central to a better understanding of the
accretion region. A number of theoretical investigations have concentrated on the
flux and polarization distribution of the cyclotron component (Wickramasinghe and
Meggitt 1985 and references therein, Barrett and Chanmugam 1984 and references
therein). These have been compared to simultaneous optical/infrared multicolor
photometry and polarimetry available mostly from investigations by Bailey (see
e.g. Bailey *et al.* 1985) to draw conclusions about the dimensions, temperatures
and magnetic fields of the cyclotron emitting regions. The results indicate the cyclotron region is coincident with the shock front, although it covers a larger area on
the white dwarf than the X-ray emitting region (Beuermann, Stella and Patterson
1987).

Compilations of energy distributions for the DQ Her systems AO Psc, FO Aqr
and EX Hya covering the IUE ultraviolet to infrared range have been given in
Mouchet (1983), Szkody (1985b) and Sherrington *et al.* (1980), while distributions
over the shorter IUE to optical range are available for V1223 Sgr, TV Col and
BG CMi (3A0729+103) in the above investigations. These have been fitted to disk
models with some success: Szkody found reasonable agreement between the observations and Williams and Ferguson (1982) disk models and Mouchet was able to
calculate distances and mass transfer rates using standard optically thick disk models described by Bath *et al.* (1980). However, these calculations are bound to be
approximate in view of the many sources of light in these systems. DQ Her systems
are not strong soft (< 0.25 keV) X-ray emitters (see Watson 1986 for a review), but
spectral information in the hard X-ray band is available for a number of systems
e.g. BG CMi (McHardy *et al.* 1987) which suggest the presence of more than one
emitting component.

B. Variable Components

In addition to the questions of overall flux distribution discussed above, multifrequency observations have been crucial to our understanding of magnetic CVs in a
number of ways which make use of the rich temporal behaviour of these objects.
For example, it is important to know whether the pulse fractions and shapes at
hard X-ray wavelengths relative to those at soft X-ray and optical wavelengths are
a result of beaming due to anisotropic optical depths or emission coefficients or
whether they are a result of obscuration, absorption by cold matter or absorption
by hot ionized matter. The variety of applications discussed below attests to the
importance of such observations.

1) In the AM Her systems, knowledge of the inclination and angle of the magnetic axis relative to the rotation axis from optical polarimetry (Brainerd and Lamb
1985; Cropper 1987) has permitted the angle at which the accretion column is
viewed at any orbital phase to be calculated and X-ray and optical light and po-

larization curves to be synthesized (e.g. Imamura 1984, Cropper *et al.* 1986) with
a measure of success for a number of systems. However, the complex and variable
behavior shown in soft X-rays by some systems such as QQ Vul (Osborne *et al.*
1987), when simultaneous observations in the optical show it to be relatively normal, and the phase shifts seen in the soft X-ray light curves of AM Her (Heise *et al.*
1985), remain to be explained. Recent simultaneous optical/infrared polarimetry of
ST LMi and BL Hyi by Bailey *et al.* (1987) indicates that accretion may be taking
place near a second pole in the infrared: these and other multicolor observations by
Piirola *et al.* (1987) find that this second emission region is probably not diametri-
cally opposed to the main emission region. It is possible that some of the soft X-ray
behaviour may be understood when these additional components are included in
the models.

2) The X-ray and optical light curves of DQ Her systems are generally sinusoidal when folded on the white dwarf rotation period and have similarly shaped
profiles (see for example those for EX Hya in Mason, 1985). The wavelength dependence of the amplitude of the modulation is consistent with different photoelectric
absorbing columns perpendicular and parallel to the accretion column, but not if
the absorbing matter is cold (King and Shaviv 1984). These authors also suggest
that the ubiquitous sinusoidal shape of the variations seen over a wide wavelength
range in the DQ Her systems are a result of accretion caps with large extent. However, it has been pointed out (Cropper 1986; Rosen, Mason and Cordova 1987) that
the cap need only be extended in one direction, for instance into a crescent shape
which is the footprint of the magnetic field lines threaded by accreting material
from the disk. The Rosen *et al.* investigation extends this concept. They propose
that accretion takes place as a curtain and find that photoelectric absorption and
a grazing eclipse geometry in such a model is able to account for the details of the
energy dependence of the rotational modulation and eclipse depths in the X-ray
and optical observations of EX Hya.

3) DQ Her systems are characterized by a number of coherent periodicities based
on the orbital and white dwarf rotation frequencies, along with sidebands and amplitude modulations due to the beat interaction of these two frequencies (Warner
1986). The current model (see Warner 1983 for a review), suggests that an X-ray
beam from the rotating white dwarf illuminates a feature fixed in the frame of the
binary system (the secondary or hot spot) to create an optical modulation. Multi-
frequency observations have been crucial in identifying which are the rotation and
which are the reprocessed periods, in finding the phase relationship between them
and in identifying the reprocessing sites (Mateo, Szkody and Hutchings 1985).

4) Sharp dips are visible in the X-ray, optical and infrared light curves of a
number of AM Her and DQ Her systems (see the example of EF Eri from Motch
in Watson, 1986 and TV Col from Schrijver, Brinkman and van der Woerd 1987)
Multiwavelength observations indicate that they are probably caused by absorption
by the accretion stream as it passes in front of the shock region as the binary rotates
(King and Williams 1985).

5) Optical high speed photometry has shown that many AM Her systems show

quasi-periodic oscillations on a timescale of ~ 1 sec (Middleditch 1982). Langer *et al.* (1982) investigated the stability of the accretion shock and found that periods of this order were expected, although subsequent investigations have found that other factors might damp out the oscillations. No oscillations at such short periods have been found in X-ray observations (but count rates are too low for any strong upper limits to be set). This suggests that the oscillations may be a second order effect. However, quasi-periodic oscillations are observed in X-rays at longer periods, for example 400 sec in AM Her (Stella *et al.* 1986). These may or may not be related to the oscillations seen in the non-magnetic systems; again, a comprehensive explanation of their origin has proved elusive because of the complicated physical processes occurring at the inner Lagrangian point and the field threading region.

6) In order to test the standard model in which the soft X-rays are generated by reprocessing of hard X-rays, a number of investigations into the correlation of X-ray fluxes with optical flickering exist for AM Her systems. Watson *et al.* (1987) found a significant degree of correlation between the hard X-ray, soft X-ray and optical bands in EF Eri. The correlation is particularly strong between the hard X-ray and optical. The lags between the various bands were less than 10 sec. Simultaneous optical/infrared high speed photometry by Bailey *et al.* (1983) also found that the flaring was strongly correlated, suggesting that these emissions arose from the same accretion column. These results are in rough agreement with the standard model. On the other hand Szkody *et al.* (1980) found little or no correlation between the soft X-rays and the optical in AM Her and a similar result was found for the soft and hard X-rays by Stella *et al.* (1986). The negative result may be understood if a substantial fraction of the soft X-rays in AM Her are not the result of reprocessing. This is also indicated by the 'reversed X-ray mode' seen in EXOSAT data from this system (Heise *et al.* 1985). Correlations between the X-ray and optical flickering were also found for the suspected DQ Her system TT Ari (Jensen *et al.* 1983), but here a finite lag between the bands was observed, with the hard X-rays being delayed by approximately 1 min. This may be consistent with a model in which the X-rays are produced in a corona above the disk.

The Underlying Stars

There are three chances for detecting the underlying stars in CVs:

1) during eclipse when most of the disk light is blocked by the secondary (or when the accretion column is eclipsed by the body of the white dwarf in the case of some of the AM Her systems)

2) in very low accretion rate (short orbital period) systems and especially in disk systems of high inclination where the disk contributes a minimal amount to the system light

3) during the low states of the novalike systems (VY Scl type) and AM Her stars when the mass transfer is reduced to negligible levels and most of the disk/column disappears.

In each of these cases, observations spanning a large wavelength range are necessary to pick out the white dwarf (ultraviolet) and the cool secondary (infrared).

One of the best examples of an eclipsing low accretion rate multiwavelength study involves the dwarf nova Z Cha. Marsh, Horne and Shipman (1987) used the optical emission lines to study the disk by means of the eclipse and the IUE spectra to isolate the white dwarf after the subtraction of the disk. Wade and Horne (1988) used the near IR (7000-9000Å) to study TiO bands and Na 8183, 8194 lines in order to classify the secondary spectral type and determine the velocity amplitude. The results of the various wavelength solutions for the white dwarf are discrepant and point out the problems of determining reliable solutions for the orbital parameters of CVs from only study of the optical emission lines. Pushing the study of velocities from the optical to the IR has also illuminated some problems in interpreting the behavior of the lines from the secondary in terms of heating effects (Wade and Horne 1988; Martin, Jones and Smith 1987).

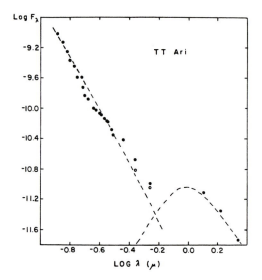

Figure 5 The low state flux distribution of TT Ari from Shafter *et al.* (1985). The dashed lines are a 50,000 K Wesemael *et al.* (1980) white dwarf model and a representation of an M2–M3 main sequence star.

The study of the low state VY Scl and AM Her systems have resulted in some of the best estimates of the white dwarf temperatures and spectral types (hence distances) of the secondaries. Figure 5 shows the multiwavelength flux distribution during one of the well studied low states of TT Ari (Shafter *et al.* 1985). The 50,000-60,000K hot white dwarf is evident in the ultraviolet (IUE regime) from the steep flux distribution in the continuum light and from the width of the Lyman alpha line and also in the optical from the broad Balmer absorption lines. The secondary spectral type is determined from the IR (1.25-2.2 micron) flux distribution. The optical shows a small residual flux from the disk when these two components are subtracted. Since the VY Scl systems occupy the orbital period range of 3-4 hrs (Shafter *et al.* 1985), the mass accretion rates are normally high. To study these systems when the accretion turns off is to look at the white dwarf as it cools down from the accretion heating. In AM Her, Heise and Verbunt (1987) have found that large areas on the white dwarf are heated from the normal temperature of 20000 K to ~30000 K when the system is not in its lowest state. From the relative rates at which the short wavelength IUE flux declined relative to the longer wavelength IUE and optical, they found that the flux from the accretion stream dropped more rapidly than that from the heated caps as the system fell to fainter states. From compilations of the white dwarfs in both low accretion systems and low state systems, it appears that the temperatures of the white dwarfs in systems which have entered low states are generally higher than those of the white dwarfs viewed in low accretion rate systems (Szkody, Downes and Mateo 1988). Theoretical models of the accretion heating of white dwarfs are just beginning to emerge (Sion 1987; Pringle 1987; Shaviv and Starrfield 1987).

When magnetic systems drop to a low state as the accretion flow diminishes, it is possible to measure the magnetic field on the white dwarf by Zeeman spectroscopy in the optical. This important parameter is necessary for our understanding of the overall energy balance in the accretion processes referred to above, and is available for 4 AM Her systems. Shafter *et al.* (1985) were also able to set upper limits on the magnetic field of the DQ Her system TT Ari when it dropped to a low state – the first such direct measurement in these systems. Optical photometry and polarimetry had already provided some upper limits to the field strength in DQ Her systems from spinup rates and the absence of circular polarization (reviewed in Lamb 1985). The confirmation of the presence of this magnetic field has recently come from the detection of circular polarization in the optical and infrared in BG CMi (Penning *et al.* 1986; West, Berriman and Schmidt 1987) and the recent discovery of optical circular polarization at a period slightly shorter than the orbital period in the old Nova V1500 Cyg (Schmidt and Stockman 1987).

Our resulting knowledge of the stellar parameters in CVs from the low accretion rate and low state studies are summarized in Table 2.

Table 2 Stellar Parameters from UV and IR Data

Object	Type	WD Temp	Secondary	D(pc)	Field (MG)	References
TT Ari	Disk	50000	M3	214	< 6	1
V834 Cen	AM Her	26500	M3-8	50-200	–	2
Z Cha	Disk	13000	M5.5	97	–	3,4,5
U Gem	Disk	30000	M5	76	–	6,7
AM Her	AM Her	20000	M4.5	76	20	8,9,10,11
ST LMi	AM Her	13000	M5.5	130	30	12,13,14
MV Lyr	Disk	60000	M5	320	–	15,16,17
IP Peg	Disk	15000	M4.5	130	–	18,19,20
VV Pup	AM Her	9000	M4	150	32	21,22,23
MR Ser	AM Her	20000	M5.5	142	–	13,24
AN UMa	AM Her	20000	not seen	>120	–	24,25
BL Hyi	AM Her	~20000	—	—	30	26

References

1 Shafter *et al.* 1985; 2 Maraschi *et al.* 1984; 3 Wood *et al.* 1986

4 Marsh, Horne and Shipman 1987; 5 Wade and Horne 1988; 6 Wade 1979

7 Panek and Holm 1984; 8 Young Schneider and Shectman 1981

9 Heise and Verbunt 1987; 10 Schmidt, Stockman and Margon 1981

11 Latham, Liebert and Steiner 1981; 12 Szkody, Liebert and Panek 1985

13 Mukai and Charles 1986; 14 Schmidt, Stockman and Grandi 1983

15 Schneider, Young and Shectman 1981; 16 Szkody and Downes 1982

17 Chiappetti *et al.* 1982; 18 Szkody and Mateo 1986

19 Martin, Jones and Smith 1987; 20 Szkody 1987; 21 Liebert *et al.* 1978

22 Szkody, Bailey and Hough 1983; 23 Wickramasinghe and Meggitt 1982

24 Szkody, Downes and Mateo 1987; 25 Liebert *et al.* 1982

26 Wickramasinghe, Visvanathan and Tuohy 1984

4. Summary

We have attempted to point out the significant contributions of multiwavelength studies in the field of cataclysmic variables. For disk CVs, this type of work is very important in discriminating between mass transfer and disk instability models for the outbursts of dwarf novae, since the models predict different timescales for the rise to outburst in different wavelengths and predict different inter-outburst behavior of the fluxes at different wavelengths. The observations at outburst and superoutburst also provide tests of the models of boundary layers and the superoutburst phenomenon. For magnetic AM Her type CVs, the multiwavelength approach is mandatory for the determination of the energy balance of the accretion column and the source of the temporal behaviour ranging from seconds to orbital timescales that is observed at many wavelengths (i.e. optical depth or absorption effects). For the DQ Her systems with combinations of disks and columns, the multiwavelength work has been crucial in identifying the spin and reprocessing periods and the reprocessing sites as well as the geometry of the accretion columns. For all types of CVs, multiwavelength observations during low states, when the mass transfer rate is very low, enables the determination of the secondary spectral type (hence distance), white dwarf temperature and magnetic field strength. Knowledge of these basic stellar parameters puts valuable constraints on the theoretical physics of the accretion flow and the accretion heating of the white dwarf.

However, all of the available data and analysis are far from providing a complete unambiguous picture of the physical structure of disk and magnetic CVs. The correct treatment of viscosity and radiative transfer in accretion disks must be accomplished in order to provide explanations of the observed multiwavelength flux distributions in a variety of systems. Multiwavelength data during the rise of symmetrical outbursts and during the superoutburst state are needed to test the predictions of disk models. Further soft X-ray and UV data during outbursts of dwarf novae are needed to test the characteristics of the boundary layer and inner disk at times of increased mass accretion. X-ray through infrared observations of more AM Her stars are still needed to determine the ratio of the soft X-ray to the hard X-ray and cyclotron fluxes, which is an indication of the reprocessing occuring in the accretion column. Further multiwavelength circular polarization observations are needed to determine the magnetic field strengths in DQ Her systems and the long term accretion column changes in the AM Her systems. Finally, the opportunities to detect the underlying stars presented by low state situations should not be missed by multiwavelength observers, as this is often the only chance to separate out the contribution from the secondary and the white dwarf from the total system light and to study the effects of the mass transfer on the observed characteristics of the primary.

PS gratefully acknowledges discussions with the participants of the 1987 Taos Workshop on Multiwavelength Astrophysics, especially John Cannizzo, Phil Charles, France Cordova, Juan Echevarria, and Ron Polidan. MC acknowledges useful conversations with Keith Mason. Paula Szkody was supported through NSF grant AST 86-10226 and NASA grant NSG 5395. Mark Cropper was supported by an SERC AAT Fellowship.

References

Bailey, J., Hough, J. H., Gatley, I., and Axon, D. J. 1983, *Nature*, **301**, 223.
Bailey, J. *et al.* 1985, *MNRAS*, **215**, 179.
Bailey, J., Peacock, T. I., Wickramasinghe, D. T., Cropper, M. S. 1987, in preparation.

Barrett, P. E., and Chanmugam, G. 1984, *Ap. J.*, **278**, 298.

Bateson, F. M. 1977, *N. Z. Jour. Sci.*, **20**, 73.

Bath, G. T., Pringle, J. E., and Whelan, J. A. 1980, *MNRAS*, **190**, 185.

Bath, G. T. 1985, *Rep. Prog. Phy.*, **48**, 483.

Berriman, G., Szkody, P. and Capps, K. 1985, *MNRAS*, **217**, 327.

Beuermann, K., Stella, L., and Patterson, J. 1987, *Ap. J.*, **316**, 360.

Brainerd, J. J. and Lamb, D. Q. 1985, in *Cataclysmic Variables and Low Mass X-Ray Binaries*, Cambridge, Mass., eds. Lamb, D. Q. and Patterson, J., Reidel, Dordrecht, Holland, p. 247.

Cannizzo, J. K., Wheeler, J. C. and Polidan, R. S. 1986, *Ap. J.*, **301**, 634.

Cannizzo, J. K. and Kenyon, S. 1987, *Ap. J.*, in press.

Chanmugam, G. 1987, *IAU Coll. 93, Cataclysmic Variables. Recent Multi-frequency Observations and Theoretical Development* eds. H. Drechsel, Y. Kondo, and J. Rahe, p. 53.

Charles, P. 1987, Talk presented at Taos meeting on Multiwavelength Astrophysics.

Chiappetti, L., Maraschi, L., Tanzi, E. G. and Treves, A. 1982, *Ap. J.*, **258**, 236.

Cordova, F. A., Chester, T. J., Tuohy, I. R., Garmire, G. P. 1980a, *Ap. J.*, **235**, 163.

Cordova, F. A., Nugent, J. J., Klein, S. R., and Garmire, G. P. 1980b, *MNRAS*, **190**, 87.

Cordova, F. A. and Mason, K. O. 1982, IAU Trans. Vol. XVIIIA ed. P. A. Wayman, D. Reidel, p. 289.

Cordova, F. A. and Howarth, I. D. 1987, in *Exploring the Universe with the IUE Satellite* ed. Y. Kondo *et al.* D. Reidel, p. 395.

Cropper, M. S. 1986, *MNRAS*, **222**, 225.

Cropper, M. S. 1987, *MNRAS*, in press.

Cropper, M. S., Menzies, J. W., and Tapia, S. 1986, *MNRAS*, **218**, 201.

Fabbiano, G. 1982, *Ap. J.*, **262**, 709.

Fabbiano, G. *et al.* 1981, *Ap. J.*, **243**, 911.

Fabian, A. C., Pringle, J. E. and Rees, M. J. 1976, *MNRAS*, **175**, 43.

Guinan, E. F. and Sion, E. M. 1982a, *Ap. J.*, **258**, 217.

Guinan, E. F. and Sion, E. M. 1982b, NASA CP-2238 *Advances in UV Astronomy*, p. 465.

Hassall, B. J. M. *et al.* 1983, *MNRAS*, **203**, 865.

Hassall, B. J. M., Pringle, J. E., and Verbunt, F. 1985, *MNRAS*, **216**, 353.

Hassall, B. J. M. 1987, Poster at 1987 North American Workshop on Cataclysmic Variables, Aspen.

Heise, J. *et al.* 1985, *A. and Ap.*, **148**, L14.

Heise, J. and Verbunt, F. 1987, *A and Ap.*, in press.

Honey, W. B., Charles, P. A., Whitehurst, R., Barrett, P. E. and Smale, A. P. 1987, *MNRAS*, in press.

Imamura, J. N. 1984, *Ap. J.*, **285**, 223.

Imamura, J. N., Durisen, R. H., Lamb, D. Q., and Weast, G. J. 1987, *Ap. J.*, **313**, 298.

Jensen, K. A. *et al.* 1983, *Ap. J.*, **270**, 211.

King, A. R., Frank, J., Jameson, R. F. and Sherrington, M. R. 1983, *MNRAS*, **203**, 677.

King, A. R., and Shaviv, G. 1984, *MNRAS*, **211**, 883.

King, A. R., and Williams, G. 1985, *MNRAS*, **215**, 1P.

King, A. R. and Watson, M. G. 1987, *MNRAS*, **227**, 205.

Kriz, S. and Hubeny, J. 1986, *Bull. Ast. Inst. Czech.*, **37**, 129.

Lamb, D. Q., and Masters, A. R. 1979, *Ap. J.*, **234**, L117.

Lamb, D. Q. 1985, in *Cataclysmic Variables and Low Mass X-Ray Binaries*, Cambridge, Mass., eds. Lamb, D. Q. and Patterson, J., Reidel, Dordrecht, Holland, p. 179.

Langer, S. H., Chanmugam, G., and Shaviv, G. 1982, *Ap. J.*, **258**, 859.

Latham, D., Liebert, J. and Steiner, J. 1981, *Ap. J.*, **246**, 919.

Liebert, J. *et al.* 1978, *Ap. J.*, **225**, 201.

Liebert, J., Tapia, S., Bond, H. E., and Grauer, A. D. 1982, *Ap. J.*, **254**, 232.

Maraschi, L. *et al.* 1984, *Ap. J.*, **285**, 214.

Marsh, T. R., Horne, K. and Shipman, H. L. 1987, *MNRAS*, **225**, 551.

Martin, J. S., Jones, D. H. P., and Smith, R. C. 1987, *MNRAS*, in press.

Mason, K. O. 1985, *Space Sci. Rev.*, **40**, 99.

Masters, A. R. 1978, Ph.D. Thesis, University of Illinois.

Mateo, M. and Szkody, P. 1984, *A. J.*, **89**, 863.

Mateo, M., Szkody, P., and Hutchings, J. 1985, *Ap. J.*, **288**, 292.

McHardy, I. M., Pye, J. P., Fairall, A. P., and Menzies, J. W. 1987, *MNRAS*, **225**, 355.

Meyer-Hofmeister, E. 1987, *IAU Coll. 93, Cataclysmic Variables. Recent Multi-Frequency Observation and Theoretical Development*, eds. H. Drechsel, Y. Kondo, and J. Rahe, D. Reidel, p. 327.

Middleditch, J. 1982, *Ap. J.*, **257**, L71.

Mineshige, S. 1987, *Pub.A.S.* Japan, in press.

Mouchet, M. 1983, *IAU Colloquium 73, Cataclysmic Variables and Related Objects*, eds. Livio, M. and Shaviv, G., Reidel, Dordrecht, p. 173.

Mukai, K. and Charles, P. A. 1986, *MNRAS*, **222**, 1p.

Naylor, T. *et al.* 1987a, *MNRAS*, **229**, 183.

Naylor, T. *et al.* 1987b, *MNRAS*, submitted.

Osborne, J. P. 1986, *IAU Coll. 93, Cataclysmic Variables. Recent Multi-Frequency Observation and Theoretical Development*, eds. H. Drechsel, Y. Kondo, and J. Rahe, D. Reidel, p. 207.

Osborne, J. P. *et al.* 1986, *MNRAS*, **221**, 823.

Osborne, J. P. *et al.* 1987, *Ap. J.*, **315**, L123.

Panek, R. J. and Holm, A. V. 1984, *Ap. J.*, **277**, 700.

Patterson, J., Beuermann, K., Lamb, D. Q., Fabbiano, G., and Raymond, J. C. 1984, *Ap. J.*, **279**, 785.

Penning, W. R., Schmidt, G. D., and Liebert, J. 1986, *Ap. J.*, **301**, 881.

Piirola, V., Reiz, A., and Coyne, S. J. 1987, preprint.

Polidan, R. and Holberg, J. 1987, *MNRAS*, **225**, 131.

Polidan, R. 1987, Talk presented at Taos meeting on Multiwavelength Astrophysics.

Pringle, J. E. 1981, *Ann. Rev. A. Ap.*, **19**, 137.

Pringle, J. E. 1987, preprint.

Pringle, J. E. and Savonije, G. 1979, *MNRAS*, **187**, 777.

Pringle, J. E. and Verbunt, F. 1984, ESA SP-218, 377.

Pringle, J. E., Verbunt, F. and Wade, R. A. 1986, *MNRAS*, **221**, 169.

Pringle, J. E. *et al.* 1987, *MNRAS*, **225**, 73.

Raymond, J. C. *et al.* 1979, *Ap. J.*, **230**, L95.

Rosen, S. R., Mason, K. O., and Cordova, F. A. 1987, *MNRAS*, in press.

Rothschild, R. E. 1981, *Ap. J.*, **250**, 723.

Schmidt, G. D., Stockman, H. S. and Margon, B. 1981, *Ap. J.*, **243**, L157.

Schmidt, G. D., Stockman, H. S. and Grandi, S. A. 1983, *Ap. J.*, **271**, 735.

Schmidt, G. D. and Stockman, H. S. 1987, IAU Circ. 4458.

Schrijver, J., Brinkman, A. C. and van der Woerd, H. 1987, *IAU Coll. 93 Cataclysmic Variables: Recent Multifrequency Observations and Theoretical Developments*, eds. H. Drechsel, Y. Kondo and J. Rahe, D. Reidel, p. 261.

Schwarzenberg-Czerny, H. *et al.* 1985, *MNRAS*, **212**, 645.

Shafter, A. W. *et al.* 1985, *Ap. J.*, **290**, 707.

Shaviv, G. and Starrfield, S. 1987, *Ap. J.*, **321**, L51.

Sherrington, M. R., Lawson, P. A., King, A. R., and Jameson, R. F. 1980, *MNRAS*, **191**, 185.

Sion, E. M. 1987, *IAU Coll. 95, Second Conference on Faint Blue Stars*, eds. A. G. D. Philip, D. S. Hayes, and J. Liebert, L. Davies Press, p. 413.

Sion, E. M. and Guinan, E. F. 1982, NASA CP-2238, p. 460.

Smak, J. 1984, *PASP*, **96**, 5.

Stella, L., Beuermann, K., and Patterson, J. 1986, *Ap. J.*, **306**, 225.

Stockman, H. S. and Lubenow, A. F. 1987, *IAU Coll. 93, Cataclysmic Variables. Recent Multi-Frequency Observation and Theoretical Development*, eds. H. Drechsel, Y. Kondo, and J. Rahe, D. Reidel, p. 607.

Szkody, P., Córdova, F. A., Tuohy, I. R., Stockman, H. S., Angle, J. R. P., and Wisniewski, W. 1980, *Ap. J.*, **241**, 1070.

Szkody, P. 1987, *A. J.*, **94**, 1055.

Szkody, P. 1985a, in *Proc. of ESA Workshop: Recent Results on CVs, ESA SP-236*, p. 39.

Szkody, P. 1985b, in *Cataclysmic Variables and Low Mass X-Ray Binaries*, eds. D. Q. Lamb and J. Patterson, D. Reidel, p. 385.

Szkody, P. 1985c, *A. J.*, **90**, 1837.

Szkody, P. and Downes, R. A. 1982, *PASP*, **94**, 328.

Szkody, P. and Shafter, A. W. 1983, *PASP*, **95**, 509.

Szkody, P., Bailey, J. A. and Hough, J. H. 1983, *MNRAS*, **203**, 749.

Szkody, P. and Kiplinger, A. 1985, *BAAS*, **17**, 839.

Szkody, P., Liebert, J. and Panek, R. 1985, *Ap. J.*, **293**, 321.

Szkody, P. and Mateo, M. 1986, *A. J.*, **92**, 483.

Szkody, P., Downes, R. A., and Mateo, M. 1988, *PASP*, in press.

Szkody, P., Osborne, J. and Hassall, B. J. M., 1988, *Ap. J.*, in press.

Tuohy, I. R., Lamb, F. K., Garmire, G. P., and Mason, K. O. 1978, *Ap. J.*, **226**, L17.

Tuohy, I. R., Mason, K. O., Garmire, G. P., Lamb, F. K. 1981, *Ap. J.*, **245**, 183.

van der Woerd, H. 1987, Ph.D. Thesis, Amsterdam.

van der Woerd, H. and Heise, J. 1987, *MNRAS*, in press.

Verbunt, F. 1986, in *Physics of Accretion onto Compact Objects*, eds. Mason, K. O., Watson, M. G. and White, N. E., Springer-Verlag, p. 59.

Verbunt, F. 1987, *Ap. J. Suppl.*, in press.

Verbunt, F., Hassall, B. J. M., Pringle, J. E., Warner, B. and Marang, F. 1987, *MNRAS*, **225**, 113.

Wade, R. A. 1979, *A. J.*, **84**, 562.

Wade, R. A. 1984, *MNRAS*, **208**, 381.

Wade, R. A. and Ward, M. J. 1985, in *Interacting Binary Stars*. eds. J. E. Pringle and R. A. Wade, Camb. U. Press, p. 85.

Wade, R. A. 1987, *IAU Coll. 95, Second Conference on Faint Blue Stars*, eds. A. G. D. Philip, D. S. Hayes, and J. Liebert, L. Davies Press, p. 685.

Wade, R. A. and Horne, K. 1988, *Ap. J.*, in press.

Warner, B. 1983, *IAU Colloquium 73, Cataclysmic Variables and Related Objects*, eds. Livio, M. and Shaviv, G., Reidel, Dordrecht, p. 155.

Warner, B. 1986, *MNRAS*, **219**, 347.

Watson, M. G., King, A. R. and Heise, J. 1985, *Space Sci. Rev.*, **40**, 127.

Watson, M. G. 1986, *In Physics of Accretion onto Compact Objects*, eds. Mason, K. O., Watson, M. G., and White, N. E., Springer-Verlag, p. 97.

Watson, M. G., King, A. R., and Williams, G. A. 1987, *MNRAS*, **226**, 867.

Wesemael, F., Auer, L. H., Van Horn, H. M. and Savedoff, M. P. 1980, *Ap. J. Suppl.*, **43**, 159.

West, C. W., Berriman, G. and Schmidt, G. D. 1987, *Ap. J.*, **322**, L35.

Whitehurst, R. 1987, presentation at 1987 North American Workshop on Cataclysmic Variables, Aspen.

Wickramasinghe, D. T. and Meggitt, S. M. A. 1982, *MNRAS*, **198**, 975.

Wickramasinghe, D. T., Visvanathan, N. and Tuohy, I. R. 1984, *Ap. J.*, **286**, 328.

Wickramasinghe, D. T. and Meggitt, S. M. A. 1985, *MNRAS*, **214**, 605.

Wickramasinghe, D. T. and Ferrario, L. 1987, *MNRAS*, in press.

Williams, R. E. and Ferguson, D. 1982, *Ap. J.*, **257**, 672.

Wood, J. H. *et al.* 1986, *MNRAS*, **219**, 629.

Young, P., Schneider, D. P. and Shectman, S. A. 1981, *Ap. J.*, **245**, 1043.

X-ray Binaries

Keith O. Mason

Mullard Space Science Laboratory

University College London

Holmbury St. Mary, Dorking, Surrey, U.K.

1. Introduction

In few branches of astronomy are multiwavelength observations so tightly woven into the fabric of the field as in the study of accreting binary X-ray sources. This follows naturally from the fact that these objects emit radiation over a large expanse of the electromagnetic spectrum. It also stems from historical necessity. Cosmic X-ray sources were discovered in the 1960s when proportional counter detectors were first carried above the Earth's atmosphere by sounding rockets. These early rocket-borne X-ray detectors, and their immediate satellite-borne successors of the early 1970s, gave only crude estimates of the position, spectrum and temporal characteristics of sources. As a result, complementary observations in the optical, and to a lesser extent the radio, bands of the spectrum have played an crucial role in developing our understanding of binary X-ray sources, beginning with the optical identification of Sco X-1 in 1966. The importance of such a multiwavelength approach continues to the present day, and indeed is becoming increasingly vital to progress in the field as experience is gained in piecing together clues found in different parts of the electromagnetic spectrum. In this chapter I use examples taken from the literature to illustrate the impact of multiwavelength observations on X-ray binary research. It should be recognized at the outset that the full potential of such observations has yet to be realized, largely because they are so difficult to execute. Thus it is hoped that the subject matter contained herein, will also, by inference, serve as a guide to likely areas of expansion for the future.

2. General Background

A striking example of the early 'multiwavelength' approach in X-ray astronomy is the sequence of observations that secured the identification of the original, and arguably still the best, black-hole candidate, Cygnus X-1. A variable radio source was found to be positionally coincident with the $9^{th.}$ magnitude optical

star HDE 226868. However, the association of this star with the X-ray source
Cygnus X-1 was based only on its location within the relatively large positional
'error box' of the X-ray source. The crucial link came when it was noticed that
the radio source had risen above the limits of detectability at the same time as a
transition of the X-ray flux from a 'high' to a 'low' state (Hjellming 1973; Tananbaum *et al.* 1972). This provided a direct association between the X-ray source
and HDE 226868, which optical observations showed to be a 5.6 day period, single-
lined, spectroscopic binary with a massive unseen companion (Webster and Murdin
1972; Bolton 1972).

Cygnus X-1 was the first X-ray source to be associated with a binary system, and gave impetus to the idea that the prodigious energy output of many
cosmic X-ray sources is derived from the gravitational potential energy released
when matter is transferred from one component of a binary system onto a highly
compact companion (neutron star or black hole). This model was supported by
the discovery that the source Cen X-3 was pulsed with a period of 4.8 seconds
(Giacconi *et al.* 1971) and that the period of the pulse varied in a cyclical fashion consistent with motion in a 2.1-day binary orbit (Shreier *et al.* 1972). This,
together with the observation that the X-ray source was also eclipsed every two
days, led to the conclusion that Cen X-3 was a spinning, magnetized neutron star
in orbit about a large companion. The companion star was eventually identified
in the optical band by Krzeminski (1974) and found to be an O-type supergiant.

Other examples were rapidly found of X-ray sources, many pulsing, some
eclipsing, which were associated with massive, early-type companions. However,
it became increasingly clear that not all variable X-ray sources were of this type.
Sco X-1, the brightest steady X-ray source in the sky, had been identified with a
$12^{th.}$ magnitude blue star some years before the work on Cygnus X-1 and Cen X-3
(Sandage *et al.* 1966; Johnson and Stephenson 1966). This was an emission line
object whose spectrum was quite unlike that of a normal star. Murdin *et al.* (1974)
and Davidsen *et al.* (1974) independently associated the X-ray source 3U 0614+09
with an $18^{th.}$ magnitude blue star on the basis of an accurate X-ray position measured with the *Copernicus* satellite (Willmore *et al.* 1974). They noted that the
ratio of X-ray to optical flux for this source was similar to that of Sco X-1, and
much greater than that of the supergiant X-ray binaries.

As better X-ray positions became routinely available in the latter half of the
1970's, largely as a result of measurements made with modulation collimators on
the SAS-3 and HEAO-1 satellites, many more such 'faint blue' X-ray source optical
counterparts were discovered. Though much more difficult to study than the X-
ray pulsar/supergiant systems, it is becoming increasingly clear that the 'Sco X-
1' type of X-ray source is also a binary, but that the mass-donating companion

is a low-mass, late-type star whose intrinsic optical luminosity is in many cases insignificant compared to that generated by reprocessing of X-rays in the system. The separation of X-ray binaries into two distinct populations is most clearly seen when they are sorted according to their X-ray to optical flux ratio; but the two kinds of system also have a different distribution on the sky (e.g. van Paradijs 1983), the 'high-mass' systems being more concentrated towards the galactic plane than the 'low-mass' systems. Thus there are at least two major populations of accreting binary in the Galaxy, with very different properties and evolutionary history.

The high-mass and low-mass groups can be further subdivided. Among the high-mass systems are those with early-type supergiant companions and those associated with B_e stars. The former usually have orbital periods of a few days, while the orbital periods of the B_e sub-class may extend to hundreds of days. Both types of system can include an X-ray pulsar. The low-mass X-ray binaries, too, may be a grouping which encompasses a number of subclasses with different evolutionary histories. In this regard, LMXRB with orbital periods of a few hours, which have dwarf companions, can be contrasted with the systems which have orbital periods of several days and where the low-mass companion star must be evolved. A further example of diversity among LMXRB is that a number of them are found in Globular Clusters. The positions of these sources in the cluster potential well indicates that they are low-mass binaries, yet they were probably formed under much different conditions to the low-mass systems outside of clusters.

3. Orbital Variations in High-Mass X-ray Binaries

It was noted in the previous section that the accreting object in many high-mass X-ray binaries (HMXRB) is a magnetized neutron star. The magnetic field channels the accreting matter onto only part of the neutron star's surface. The consequence is that the X-ray emission from the neutron star is beamed, and the distant observer records an X-ray flux that is modulated with the spin period of the neutron star. The spin periods of the known 'X-ray pulsars' range from 0.069 seconds (A0538–66; Skinner *et al.* 1982) to 835 seconds (4U0352+30 = X Per; White *et al.* 1976).

The presence of an X-ray pulsar in a system provides an accurate clock whose apparent variation, due to Doppler motion, can be used to investigate the orbital motion of the neutron star. When such X-ray pulse timing studies are combined with radial velocity measurements of lines in the optical spectrum of the companion star, the orbital parameters of the system can be derived in an analogous way to those of classical double-lined spectroscopic binaries. Particularly when coupled with measurements of the X-ray eclipse duration, when present, such information can be used to estimate accurately the masses of the component stars of the binary, and the size, evolutionary state and structure of the mass-donating companion.

The results of such studies are reviewed by Rappaport and Joss (1983). Among the principal conclusions are that the masses of the neutron stars in HMXRB are all consistent with the value 1.4 ± 0.2 M$_\odot$, similar to the Chandrasekhar limit for the mass of a degenerate dwarf. Thus the mass estimates are in line with the expectation that the neutron stars in these binaries were formed as a result of collapse of either the degenerate core of a highly evolved star, or an accreting degenerate dwarf. The evidence also suggests that the companion stars in the short period HMXRB, for example Cen X-3, Vela X-1, SMC X-1, 4U1538–52, and LMC X-4, are undermassive for their luminosity (Rappaport and Joss 1983). A number of the longer period HMXRB have significant orbital eccentricity, for example Vela X-1 (e = 0.092), 4U0115+63 (e = 0.34), and GX 301–2 (e $\sim$ 0.45).

The material that is accreted onto the compact X-ray emitting star is probably captured from the stellar wind of the companion star in most HMXRB, particularly those in which the companions are of the earliest spectral type. Roche lobe overflow may also be significant in some systems, whereas in the long-period Be star binaries matter ejected in a dense equatorial ring is probably important in powering the X-ray source. By studying the X-ray light curves of these objects it is possible to map out the distribution of matter around the companion star, using the degree of obscuration of the X-ray source as a measure of the column density of material along each line of sight. This can be combined with ultraviolet data on the profiles of resonance lines formed in the stellar wind of the companion to study the density and velocity profile of the wind as a function of radius, and the effects of the strong X-ray source on the wind.

A measurement that is directly related to the density and extent of the envelope surrounding the companion star is the duration of X-ray eclipse in highly inclined systems. The length of the X-ray eclipse in the HMXRB 3U1700–37, a 3.4 day binary containing an O6.5f companion (HD153919), has been observed to vary between 1.1 days and 0.75 days (Branduardi, Mason and Sandford 1978; Mason, Branduardi and Sanford 1976; Jones *et al.* 1973), implying considerable variations in the companion star's envelope. Observed changes in the mean X-ray flux density and absorption of the source may be a direct consequence of the changes in the extent and structure of the companion's stellar wind. The binary phase dependence of the low-energy turnover in the spectrum of the X-ray emission due to photoelectric absorption can also be measured during the ingress and egress from eclipse to determine the density profile of the companion star's atmosphere. In this way, for example, Sato *et al.* (1986) measure a scale height of $\sim 3 \times 10^{11}$ cm for the atmosphere of the star HD 77581, the B0.5Ib companion of the X-ray source

Vela X-1. This is approximately 15% of the star's radius.

The stellar winds in early type supergiants are believed to be radiatively accelerated. The effect of embedding a strong ionizing X-ray source in this wind is to destroy the ions responsible for coupling the wind to the star's radiation field (Hatchett and McCray 1977). As a result the velocity profile of the stellar wind is expected to be distorted in the vicinity of the X-ray source, and this should have a noticeable effect on the profile of P-Cygni lines formed in the wind and observed in the UV part of the spectrum. Such effects are indeed seen, particularly in *IUE* high resolution spectra of Vela X-1/HD 77581 (Dupree *et al.* 1980) and in 3U1700–371 (Howarth, Hammerschlag-Hensberge and Kallman 1986). Evidence that material around the X-ray source is not distributed uniformly, and particularly that there is an enhancement in the density of material trailing the X-ray source in its orbit, is also found from an examination of the X-ray light curves of these two sources (Mason, Branduardi and Sandford 1976; Branduardi, Mason and Sandford 1978; Charles *et al.* 1978). Explanations for this concentration of material trailing the X-ray source include a supersonic accretion wake (e.g. Eadie *et al.* 1975; Charles *et al.* 1978), material collecting as the result of Roche lobe overflow from the companion (White, Kallman and Swank 1983; Prendergast and Taam 1974), and the effects of collision between the radiatively driven wind and the bubble of wind material photoionized by the X-ray source (Fransson and Fabian 1980).

Evidence for inhomogeneities in the envelopes of X-ray source companion stars comes from the observations of discrete absorption features in the X-ray light curves of a number of HMXRB, particularly near superior conjunction where the line of sight passes close to the companion star. Such absorption events are seen, for example, in Cyg X-1 (Mason *et al.* 1974; Li and Clark 1974; Pravdo *et al.* 1980), SMC X-1 (Marshall, White and Becker 1983) and Vela X-1 (Charles *et al.* 1978). These 'blobs' of absorbing material may be representative of density enhancements within the normal stellar wind of the companion, or they may be associated with separate organized flows, for example a gas stream due to Roche lobe overflow. Both Vela X-1 and particularly 3U1700–37 exhibit considerable flux variability on time scales of hours, suggesting corresponding variations in the rate of accretion onto the X-ray emitting star. White, Kallman and Swank (1983) have argued in the case of 3U1700–37 that these variations in accretion rate reflect density inhomogeneities in the stellar wind of the companion. They find changes in the amount of absorption of the X-ray source on a similar time scale to the flares, but not correlated with them. White, Kallman and Swank suggest that the variable absorption is also due to inhomogeneities in the wind, and that the high density structure is elongated along directions of constant radius from the supergiant companion, forming radially expanding 'sheets' of high density matter in the stellar wind. These inhomogeneities may be the same as those invoked by Lucy and White (1980) to form shocks in the stellar winds of isolated OB stars in order to account for their (weak) X-ray emission.

4. Orbital Variations in Low-Mass X-ray Binaries

The nature of the Low-Mass X-ray Binaries (LMXRB), also known as 'Population II' X-ray sources, or X-ray bulge sources, because of their concentration towards the Galactic Bulge, was a mystery for many years. Despite the expectation, formed by analogy with the high-mass systems, that the Population II objects were accreting binary stars, most steadfastly refused to exhibit evidence to confirm their binary nature. Particularly hard to understand was the failure to detect X-ray eclipses in all but one of the class (Her X-1), despite the fact that many of the Population II sources are among the brightest in the sky at X-ray wavelengths. Because of their relative faintness in the optical range, it was recognized early on that the companion star in a Population II source, assuming they were binaries, would have to be a low-luminosity late-type star. Joss and Rappaport (1979) calculated that if a population of semi-detatched binaries containing a late type dwarf were distributed with random inclinations, approximately one in five should be eclipsing.

The above situation has changed drastically in recent years, thanks largely to the application of sensitive observing techniques in both the X-ray and the optical band. At the time of writing the orbital periods of over 20 LMXRB have been measured with reasonable confidence. The periods measured range from 0.19 hours (11 minutes!) for the globular cluster X-ray source XB1820–303 (Stella, Priedhorsky and White 1987) to 235 hours (9.8 days) for the source Cyg X-2 (Cowley, Crampton and Hutchings 1979). The techniques brought to bear on the problem of period determination in LMXRB include X-ray and optical photometry, optical spectroscopy and X-ray and optical pulse timing measurements. The 'honors' for discovering orbital periods in LMXRB divide approximately equally between X-ray and optical observers.

Determining the orbital periods of LMXRB is of fundamental importance to understanding their origin and structure. The orbital period itself severely constrains the nature of the secondary, assuming Roche geometry. Thus the companion in XB1820–303, which has a 11-minute orbital period, must be degenerate; whereas the companion's Roche lobe in the 9.8 day binary Cyg X-2 can only be filled by a giant star. The majority of LMXRB measured have periods of a few hours, indicative of a late-type dwarf companion. It should be noted, though, that the large concentration of sources at a period of a few hours might be partly a selection effect because these are the periods most easily observed.

Why was evidence of orbital motion initially so difficult to find in LMXRB? The answer is essentially the idea put forward by Milgrom (1978). He pointed out that if LMXRB had thick accretion disks (due for example to the effects of X-ray heating) then the mass-donating companion star could lie in the X-ray shadow of the thick disk. As a result there would be a severe selection against detecting highly inclined LMXRB – the view of the X-ray source would be blocked by the disk – and consequently against detecting eclipsing systems. The situation in practice turns out to be somewhat more complicated than this, as described below. However, the essence is that it is easier to detect orbital signatures in systems with fainter apparent X-ray brightness than in the very bright systems that were (naturally) studied first.

In many LMXRB there is evidence that the thickness of the accretion disk in the frame of reference of the binary changes with angle around the disk (see for example Mason 1986). This can cause various unforseen orbital phenomena, an example of which is irregular dips in the X-ray flux that recur with the binary period. These are observed in systems where the orbital inclination is such that the X-ray source is obscured by disk material for only part of the orbital cycle. In some systems there is also evidence for significant amounts of (hot) scattering material above and below the disk plane (an accretion disk corona, or ADC). Scattered X-rays can be seen from some such sources even though the direct line of sight to the X-ray emitter is blocked by the disk. It is characteristic of these 'ADC' sources that they have lower than average apparent L_x/L_{opt} ratios, and that the apparent size of the X-ray emitter is large, so that eclipses by the companion are partial, with gradual ingress and egress. Disk structure of the kind that causes X-ray dips in some objects can lead to quasi-sinusoidal orbital variations in the X-ray flux of ADC sources. The asymmetry of the disk can, further, give rise to orbital modulation of an LMXRB's optical light, as the solid angle subtended by the X-ray heated disk changes. In a similar way X-ray heating of one side of the companion star can also yield an orbital modulation of the optical light. An advantage of searching for such modulations in the optical band is that they are observable over a much wider range of orbital inclinations than the various X-ray orbital signatures, which rely on the line of sight to the X-ray source passing through disk material, or through the companion star, at certain orbital phases.

Multiwavelength observations have played an important role in determining our current understanding of LMXRB disks. One example of such observations concerns the line of sight in the binary frame along which X-ray dips occur. Most X-ray dipping sources do not exhibit X-ray eclipses, and their optical companions are too faint for phase resolved spectroscopy to be practical yet. Thus relating the phase of the X-ray dips to the binary frame is not straightforward. This problem has been solved for two dipping sources, XB1254–690 and X1755–338, by obtaining simultaneous X-ray and optical photometry (Motch *et al.* 1987; Mason, Parmar and White 1985). The optical counterparts of both sources have V magnitude close to 19, and have orbital periods of 3.9 hours and 4.4 hours respectively. Both stars exhibit a quasi-sinusoidal optical flux modulation, with a full amplitude of approximately 0.5 magnitudes, with a period identical to the X-ray dip recurrence period. The optical modulation is probably caused primarily by the changing aspect of the companion star, one side of which is heated by X-rays. In both XB1254–690 and X1755–338 the X-ray dips occur about one-fifth of an orbital cycle before the optical minimum. This locates the thickest part of the accretion disk in the general vicinity of the impact point of the gas stream from the companion.

The classic 'ADC' source is X1822–371 (= V691 CrA). The 5.57 hour orbital period of this system, and indications of its peculiar nature, were first discovered as a result of optical photometry (Mason *et al.* 1980). The optical light curve has a minimum about 1 magnitude deep which lasts for about an hour and is preceeded by a gradual decline in light level lasting for about a third of the orbital cycle. In the X-ray band, the light curve is resolved into a narrow, partial dip, and a quasi-sinusoidal modulation whose minimum occurs about an hour before the narrow dip (White *et al.* 1981). These features are presently understood as occultations of an extensive accretion disk corona by the companion and structure on the accretion disk respectively. Mason and Córdova (1982), building on modeling of the X-ray light curve performed by White and Holt (1982), have combined X-ray, ultraviolet, optical and infrared photometry, as well as optical spectroscopy, of X1822–371 to study the structure of the source. They are able to construct a self consistent model which explains the changes in the shape of the light curve between wavelength bands. In this model the far ultraviolet emission is dominated by the hot central regions of the disk, producing a deeply modulated light curve, whereas in the optical and infrared bands emission from the outside of the disk is also important. The optical line emission probably arises in the X-ray shadow of the disk. Strangely, unlike the similar system X2129+470 (=V1727 Cyg; McClintock, Remillard and Margon 1981), heating of the companion appears to contribute only a small fraction of the optical modulation in X1822–371.

A final note on orbital modulation in low-mass X-ray binaries concerns those systems that contain an X-ray pulsar. In contrast to the high mass systems, where the pulsar phenomenon is commonplace, only two known LMXRB exhibit X-ray pulsations, Her X-1 (P_{spin}=1.24 s) and X1626–673 (P_{spin}=7.68 s). Pulse timing studies are a sensitive probe of the orbital dynamics of these two systems. Her X-1 is a complex, eclipsing system which has a 2-day orbital period and also a long-term X-ray on/off cycle which has a period of 35 days. No reliable optical spectroscopic orbit exists for the system because of X-ray heating effects which distort the spectral lines as a function of orbital phase. However, Middleditch and Nelson (1976) have detected optical pulsations from the binary at certain times in the orbital and 35-day cycles which are explained as reprocessed, pulsed X-ray emission. Optical pulsations are detected both from the vicinity of the X-ray emitting star and from the side of the companion star facing the X-ray source. Pulses from these two sites are shifted in frequency with respect to one another because of their relative orbital motion, and the velocity curve of the pulsations from the heated face of the companion can be used to recover the orbital velocity of that star. By combining this information with the velocity curve of the X-ray pulsar and the measured X-ray eclipse duration, Middleditch and Nelson (1976) have been able to deduce accurate estimates of the component masses and inclination of the binary ($M_x = 1.30 \pm 0.14\ M_\odot$, $M_c = 2.18 \pm 0.11\ M_\odot$, $i = 87° \pm 3°$).

Similar work has been done on X1626–673. In this case *no* evidence of Doppler delay effects are seen in the X-ray pulse, which put severe constraints on the nature of the mass donating secondary (Levine *et al.* 1988). However high speed optical photometry by Middleditch *et al.* (1981) reveals not only pulsations at the X-ray frequency but also weak pulsations at a slightly lower frequency than the X-ray pulse. If these downshifted pulsations are interpreted as originating on the surface of the companion star, as in Her X-1, then the magnitude of the frequency shift is the orbital frequency, which yields an orbital period of ~2485 seconds, or about 40 minutes. Middleditch *et al.*'s analysis of the observations, including a direct estimate of the light travel time across the binary orbit derived by comparing the phase difference between the direct and reprocessed pulse, suggests that X1626–673 is a low inclination system with dimensions of about 1 light second. When coupled with limits on the orbital modulation of the X-ray pulse, it is found that the mass of the companion star in this system, most probably a hydrogen-depleted dwarf, can be no more than $0.1\ M_\odot$ (Levine *et al.* 1988).

5. Exotic Radio Objects: Cyg X-3 and SS 433

A number of galactic X-ray binaries emit radio waves, as summarized by Hjellming and Johnston (1986). In many instances detection of the radio flux has been instrumental in securing the optical identification of the star. The case of Cyg X-1 has already been mentioned; other notable examples include Cyg X-3 (Braes and Miley 1972), SS 433 (Clark and Murdin, 1978), Cir X-1 (Whelan *et al.* 1977) and LSI+61°303 (Gregory and Taylor 1978).

Despite this, the study of the radio emission of X-ray binaries has, by and large, yet to contribute substantially to our understanding of the physics of these objects. There are, though, two major exceptions to this statement; Cyg X-3 and SS 433. They are two very different X-ray binary systems: Cyg X-3, with an orbital period of 4.8 hours, is probably related to the low-mass subclass, whereas SS 433, with a 13 day orbital period, is a high mass system. Yet they share the common property that they produce resolvable radio jets. The study of these jets, combined with data in other wavebands, has been instrumental in furthering our understanding of these sources.

Cyg X-3

In late August 1972, a variable radio emitter which was catalogued by Braes and Miley (1972) within the region of positional uncertainty of Cyg X-3 suddenly brightened by approximately two to three orders of magnitude, becoming for a time the brightest radio source in the sky at some wavelengths (Gregory *et al.* 1972). First recorded during a rocket flight in 1966 (Giacconi *et al.* 1967), Cyg X-3 is among the brightest 'steady' X-ray sources. The discovery of the giant radio outburst prompted careful analysis of X-ray data acquired with the *Uhuru* and *Copernicus* satellites, and these revealed that the X-ray flux of Cyg X-3 was modulated in a quasi-sinusoidal manner with a period of 4.8 hours (Parsignault *et al.* 1972; Sanford and Hawkins 1972). The stability of this period (the period derivative measured over 13 years is 7.8 $\times 10^{-10}$) argues strongly that the underlying 4.8 hour clock is orbital motion. Cyg X-3 lies in a heavily obscured region of the galactic plane, and is about 11 kpc distant. Consequently no optical counterpart is visible. The source is, though, detected in the infrared (Becklin *et al.* 1972). A simultaneous X-ray/infrared observation of Cygnus X-3 was made by Becklin *et al.* (1973) using the X-ray instruments on the *Copernicus* satellite, and the 200 inch Mt. Palomar telescope. This observation demonstrated that the infrared source shared the same 4.8 hour modulation observed in the X-ray band, establishing beyond doubt the connection between Cyg X-3 and the flaring radio source because of the high degree of coincidence between the radio and infrared positions.

Further observations of Cyg X-3 have provided evidence of even more remarkable properties. Foremost, and most controversial, are reports that Cyg X-3 is a strong source of ultra-high energy gamma rays (10^{12}–10^{16} eV), and that it may be a dominant source of high energy cosmic rays in the Galaxy (Dowthwaite *et al.* 1983; Danaher *et al.* 1981; Lamb *et al.* 1981; Samorski and Stamm 1983; Lloyd-Evans *et al.* 1983; Hillas 1984; Porter 1984). A remarkable feature of the reported detections is that Cyg X-3 emits a comparable amount of energy in the γ-ray region as in the combined interval between radio and X-rays (Figure 1). The high energy γ-ray emission appears to be confined to one or both of two narrow phase regions near 4.8 h phases 0.2 and 0.6 measured with respect to the X-ray minimum. Chadwick *et al.* (1985) have reported evidence that the 10^{12} eV γ-rays from Cyg X-3 are pulsed with a 12.59 ms period. If confirmed, this would be strong evidence that Cyg X-3 contains a young, rapidly rotating neutron star. Even more controversial is the reported detection of muons from the direction of, and with the 4.8 hour period of, Cyg X-3 (Marshak *et al.* 1985a,b; Battistoni *et al.* 1985). In order to explain these detections in underground proton decay experiments, it may be necessary to invoke an entirely new particle. However, the reality of the detections has yet to be demonstrated to universal satisfaction (e.g. Molnar 1986).

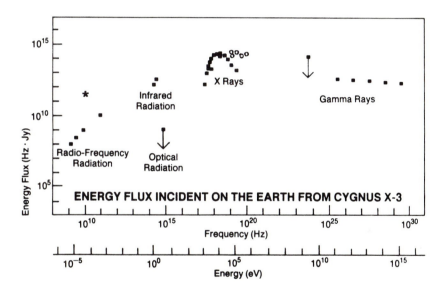

Figure 1 The spectrum of Cyg X-3 from the radio band to ultra-high energy γ-rays plotted as energy flux versus frequency and energy on a logarithmic scale. Plotted in this way equal areas on the diagram represent equal amounts of radiated energy, so that the energy output of the source in the various bands can be assessed. The black squares show a typical flux distribution for this variable source during a relatively quiet state. The asterisk marks the maximum 10.5 GHz flux observed during the first observed radio outburst in 1972. There appears to be some discrepancy in the high-energy X-ray data (cf. Reppin et al. 1979 and Meegan, Fishman, and Haymes 1979), which could be due to source variability. This diagram is taken from Córdova (1986).

A problem peculiar to all observations in the high energy γ-ray part of the spectrum is that the number of source events is low and the background is high. The case for detection of such emission from Cyg X-3 rests heavily on the fact that events recur at special phases in the 4.8 hour cycle of the source, whose ephemeris is established from X-ray observations. The present uncertainties in the ephemeris of the 4.8 hour modulation of Cyg X-3, though small by X-ray standards, nevertheless impact the significance of the detections of the reported sharp γ-ray features in Cyg X-3, which in many cases are based on folding data from many weeks of observations. Thus here is a clear example where simultaneous observations in the high energy γ-ray and X-ray parts of the spectrum could improve the credibility of the γ-ray results. Similar problems exist in the search for ultra-high energy γ-ray emission from other binary X-ray sources (e.g. Cawley 1987).

Turning to the study of Cyg X-3 at lower frequencies, it has been found that the amplitude of the 4.8 hour modulation increases towards lower energies in the X-ray band once the effects of an X-ray halo around the source on the sky, caused by scattering off interstellar material, is removed (Molnar and Mauche 1986). The form of the energy dependence suggests that photoelectric absorption plays a role in producing the 4.8 hour modulation. One possible model is that the X-rays from the compact star are scattered and absorbed in an extensive wind that is being 'boiled' off the companion as a result of X-ray heating (Davidsen and Ostriker 1974; Pringle 1974; Willingale, King and Pounds 1985). The X-ray modulation arises in this model because the wind is centered on the companion, and the column density of material along the line of sight to the X-ray source is therefore a function of binary phase. An alternative model is that Cyg X-3 is an extreme form of ADC source (White and Holt 1982). In this case the 4.8 hour modulation is due to obscuration of the ADC by disk structure.

Several giant radio flares have been observed from Cyg X-3 since the original observation in 1972 (e.g. Johnston *et al.* 1986). These show a spectral evolution characteristic of an expanding, relativistic, synchrotron-emitting cloud (cf. van der Laan 1966). Even during the 1972 outburst there was evidence that the Cyg X-3 radio source was measurably extended, when VLBI observations by Hinteregger *et al.* (1972) failed to detect it. Recent high spatial resolution observations demonstrate that Cyg X-3 becomes substantially elongated at radio frequencies in the weeks following a giant flare (Geldzahler *et al.* 1983; Spencer *et al.* 1986). This result is confirmed by Molnar (1986) who, monitoring weaker, less long-lived flares, measures an expansion rate of the source of about 0.4 mas per hour. This corresponds to an expansion velocity of 0.27 the speed of light assuming that the source is a double-sided jet 10 kpc distant. There is also evidence that the low-level flares characteristic of Cyg X-3 in its 'quiet' state recur periodically, with a period that is a few percent longer than the 4.8 hour orbital period (Molnar, Reid and Grindlay 1984; Molnar 1986). The mechanism behind this radio modulation, and particularly an explanation for why its period is different form the orbital period, is still unknown, though apsidal motion due to the influence of a third body has been considered (Molnar 1986). Flares are also seen in the infrared portion of the spectrum (Mason, Córdova and White 1986; Becklin *et al.* 1974). These have a much shorter time scale (a few minutes) than the radio flares and may represent the initial injection of material into the jet (Mason, Córdova and White 1986).

Cyg X-3 is clearly a bizarre object, and we are a long way from understanding the reasons for its peculiar nature and prodigious energy output. It is an object whose output spans the whole range of the observable electromagnetic spectrum and multiwavelength observations are clearly essential to study it. Several simultaneous multiwavelength campaigns on Cyg X-3 have been conducted already; examples of those that have been reported in the literature include Becklin *et al.* (1974), Mason *et al.* (1976), Molnar (1986) and Mason, Córdova and White (1986). We can look forward to further extensive monitoring of Cyg X-3 over as large a wavelength range as possible as perhaps the best way to probe this enduring cosmic mystery.

SS 433

Clark and Murdin (1978) were searching the error box of a point-like radio source within the radio supernova remnant W50 when they came across a peculiar emission line star. They did not know at the time that the object appeared in Stephenson and Sanduleak's (1977) catalogue of emission line stars (No. 433); indeed Seaquist was searching for radio sources among objects in the SS catalogue, and independently found that SS 433 was a radio emitter (Seaquist *et al.* 1979). The spectrum that Clark and Murdin (1978) obtained of SS 433 contained a number of strong emission lines, but not all had wavelengths corresponding to known spectral features. Moreover, these mystery lines had disappeared when they re-observed the star after an interval of two weeks.

More extensive observations by Margon *et al.* (1979a) showed that the wavelength of the mystery lines changed substantially from night to night, and it quickly became established (Margon *et al.* 1979b) that the line shifts were periodic, with a period of 162 days. In fact there are two distinct sets of 'moving' lines in the spectrum of SS 433, travelling back and forth in anti-phase. The amplitude of the periodic wavelength shift is enormous, amounting to $\pm 40,000$ km s^{-1}. To explain these unprecedented line shifts, a model has been developed in which the moving spectral lines originate in two opposing jets which are being ejected from a binary system with a velocity of 0.26 c. This model has been established by combining information from the X-ray, optical and radio bands.

Clark and Murdin (1978) made the suggestion that SS 433 was the optical counterpart of the weak, variable X-ray source A1909+04 (Seward *et al.* 1976). Simultaneous radio and X-ray flares have been reported (Seaquist *et al.* 1982). However, the most significant finding concerning the radio emission of SS 433 comes from high resolution observations which show extended radio lobes either side of the central source (Spencer 1979; Hjellming and Johnston 1981a,b; Gilmore *et al.* 1981; Hjellming and Johnston 1986). The radio lobes are aligned with the major axis of W50, and also with weak X-ray lobes recorded on high resolution X-ray images taken with the *Einstein Observatory* (Seward *et al.* 1980; Watson *et al.* 1983). Repeated observations with the VLA demonstrate that the morphology of the radio jets is changing with time. These changes can be explained by a model in which the radio emission is produced in a bi-polar jet which is precessing with the 164-day period of the moving optical spectral lines. Remarkably, the parameters obtained from the analysis of the radio jets agree extremely well with those derived from the optical data (Hjellming and Johnston 1986), confirming the expanding jet model as the only viable explanation for the bizarre behavior of SS 433. The cone angle of the precessing jet is found to be 19°.

The optical spectrum of SS 433 contains 'stationary' spectral lines, in addition to the high velocity lines from the jet. Analysis of the 'stationary' He II 4686Å line by Crampton and Hutchings (1981) indicates that the underlying system from which the jets originate in an interacting binary system, similar to the high-mass X-ray binaries. The orbital period is 13 days. X-ray observations with the EXOSAT observatory, recently confirmed by data from the Japanese *Ginga* satellite, demonstrate that the X-ray flux of SS 433 suffers a broad eclipse (Stewart *et al.* 1987; Kawai, Stewart, 1987 private communication) coincident in phase with a similar eclipse of the optical light. Stewart *et al.* (1987) also find a strong iron emission line in the X-ray spectrum that moves back and forth in accordance with the moving optical emission lines. Unlike the optical spectrum, however, only one X-ray line component is ever visible, corresponding in velocity to the most blue-shifted set of optical emission lines. This suggests that X-ray line emission is only visible from the jet that points towards us. The most likely explanation of this is that the X-ray flux originates from the innermost regions of the bi-polar jet and that X-rays from the receding jet are obscured by a thick accretion disk. The available X-ray and optical data, coupled with a consideration of the total energetics of the system, suggest that the compact component of the SS 433 binary system may well be more massive than the maximum allowed mass for a neutron star, and is therefore a black hole (Stewart *et al.* 1987). In any case it seems likely that this compact star is surrounded by a thick accretion disk and is the source of the precessing jets. Why SS 433 has these jets of relativistic particles, and why they precess, however, remains a mystery.

6. Flaring and Bursts

X-ray binaries as a class are noted for their variability on virtually all time scales studied. In many systems, periodic variations betray the presence of an underlying clock such as rotational, orbital or precessional motion. Irregular flaring and transitions between 'active' and 'quiescent' states are common, while 'bursts', caused by a thermonuclear explosion on the surface of a neutron star, are found in a subclass of low-mass X-ray binaries. To study irregularly variable phenomena successfully over a wide range of frequencies it is obviously desirable, and usually necessary, for the observations in the various bands to be simultaneous. In this section I discuss simultaneous observations of two kinds of irregular flux variability: the flaring behavior of Sco X-1, and the thermonuclear 'bursts' that occur in many LMXRB.

Flares in Sco X-1

Sco X-1 is the brightest of the 'permanent' cosmic X-ray sources and, as noted in the introduction, it was the first to be optically identified. It was also the first X-ray binary source from which radio emission was detected (Andrew and Purton 1968)†. Irregular brightness variations were found in the optical counterpart of Sco X-1 (Sandage *et al.* 1966; Johnson and Stephenson 1966; Mook 1967), and also in the radio emission of the star (Ables 1969). It was natural to seek similar variations in X-ray band, and to determine whether the flux variations in the various bands were correlated.

The brightness of Sco X-1 made such investigations tractable with the earliest satellite-borne X-ray detectors (e.g. *OSO-3*, Hudson *et al.* 1970; *Vela 5A, 5B, 6A* and *6B*, Evans *et al.* 1970; Mook *et al.* 1975), and even with instruments on sounding rockets (Chodil *et al.* 1968; Mark *et al.* 1969). The sounding rocket experiments on Sco X-1 indeed mark the beginning of coordinated multiwavelength observations of X-ray binaries, the launching of the X-ray instrument being determined in some cases by the results of simultaneous optical monitoring so as to catch a particular state of the source!

A number of extensive multiwavelength observing campaigns on Sco X-1 have been conducted since then (e.g. Canizares *et al.* 1973; Pelling 1973; Mook *et al.* 1975; Bradt *et al.* 1975; Canizares *et al.* 1975; Ilovaisky *et al.* 1980; Petro *et al.* 1981). Sco X-1 exhibits two distinct modes of X-ray behavior; an active or flaring state, and quiescent intervals where the flux level is relatively constant. During the active state there is generally a good correlation between the X-ray and optical

---

† In fact the radio emission of Sco X-1 has three spatially separate components, one variable and coincident with the X-ray/optical source, and two other components on nearly opposite sides, ∼1.2 arc minutes away; Fomalont *et al.* 1983.

flux level, particularly in individual flares (Ilovaisky *et al.* 1980). The underlying flux level can occasionally, however, wander about apparently independently in the two bands (Ilovaisky *et al.* 1980). Substantial fluctuations in the X-ray flux level during flares occur on time scales of about one second, but the minimum fluctuation time scale in the optical band is measured at about 20 s (Petro *et al.* 1981). Assuming the optical light to be reprocessed X-ray emission, smearing on this scale implies a reprocessing site which is comparable to the size of the binary (an accretion disk?). The lack of a measured time delay between the X-ray and optical flickering also constrains the reprocessing site to be approximately symmetric about the X-ray source; it cannot be the companion for instance (Petro *et al.* 1981).

The transition to the X-ray quiescent state in Sco X-1 occurs at an optical threshold magnitude of about B=12.6 (Canizares *et al.* 1975; Ilovaisky *et al.* 1980). Below this threshold (the quiescent X-ray state) the optical brightness fluctuates much more than the X-ray flux and the correlation between the two bands is much less marked. It is possible that the flickering X-ray source becomes hidden from view, while the reprocessed emission seen in the optical is still visible. The total amplitude of variation in the 2-10 keV X-ray band is about a factor of two, but the flare amplitude is an increasing function of energy. The total range of optical variability is just over 1 magnitude in the B-band, between about 12.3 and 13.6 (Canizares *et al.* 1975).

The bi-modal behavior of the X-ray emission in Sco X-1 may also be reflected in the radio flux of the source in that radio flaring appears to be concentrated during the quiescent X-ray intervals (Canizares *et al.* 1973; Miyamoto and Matsuoka 1977). Quiescent and active states are also clearly separated in plots of X-ray spectral hardness versus flux (White *et al.* 1976; White, Peacock and Taylor 1985). White, Peacock and Taylor (1985) have modeled the X-ray spectrum of Sco X-1 with two distinct spectral components, a blackbody spectrum and a component due to comptonized bremsstrahlung emission. The X-ray flares during the active state appear to be caused by an increase in the blackbody flux, whereas in quiescence van der Klis *et al.* (1986) find an anti-correlation between blackbody and comptonized bremsstrahlung flux. Interestingly also, the frequency of quasi-periodic oscillations (QPO) observed in the X-ray emission of Sco X-1 also show a bimodal behavior (Priedhorsky *et al.* 1986), the transition between ~6 Hz and 10–20 Hz QPO corresponding to the transition between the active and quiescent X-ray states.

There is still no satisfactory explanation for the complex behavior of Sco X-1. Yet the systematic relations that have been uncovered between the various, apparently disparate, behavioral signatures of this source auger well that such a model can be developed in the future. The constraints derived from multiwavelength observations are likely to play an important role in developing our understanding of the source.

X-ray Bursts

X-ray bursts are a feature of many LMXRB. They are reviewed in detail by D. Hartmann and S. Woosley elsewhere in this volume, but some discussion is included here for completeness. X-ray bursts are believed to be caused by an explosive thermonuclear fusion reaction involving helium on the surface of a neutron star. The helium fuel is contained within material accreted from the companion star in the binary and collects on the surface of the neutron star until it reaches a critical density and temperature, when a burst occurs (Lewin and Joss 1983). The rise times of X-ray bursts are typically of order 1 s, and they decay on time scales of tens of seconds. The separation between bursts is determined by the time it takes to accrete a critical supply of helium, and a good relationship is generally found between the luminosity of a burst (i.e. the amount of fuel consumed) and the interval to the next burst. This interval is usually several hours, although bursts separated by only a few minutes are sometimes observed (probably explainable in terms of incomplete burning of the fuel reservoir).

Optical counterparts of the X-ray bursts have been detected from a number of burst sources including XB1735–444, XB1837+049, XB1636–536, Aql X-1 and XB1254–690 (see for example Lewin and Joss 1983; van Paradijs 1983). The short time scale of the burst event means that simultaneous observations at X-ray and optical wavelengths can be used to make sensitive measurements of the relative timing of the bursts in the two bands. The time delay between the X-ray and optical events can then be used to probe of the geometry of the region that is reprocessing the X-ray signal into optical light; i.e. the accretion disk or companion star.

A number of extensive X-ray/optical monitoring campaigns have been undertaken involving a number of different X-ray satellites and ground-based observatories: During world-wide coordinated burst watches in 1978, 1979 and 1980 X-ray and optical bursts were recorded simultaneous from XB1735–444, XB1837+049 and XB1636–536. A time delay of between 1 and 3 seconds is measured between the X-ray and optical bursts. Detailed modeling of the data on XB1636–536 by Pedersen *et al.* (1982) suggests that an accretion disk around the neutron star provides a plausible model for the reprocessing region, which must have a projected area of about $6 \times 10^{21} cm^2$. The mass of the companion star is also constrained to be < 2 M$_\odot$, if the neutron star is assumed to have a mass of 1.4 M$_\odot$. More extensive observations could be used to search for variations in the time delay between X-ray and optical bursts with orbital period (3.8 hours in the case of XB1636–536).

Such variations might be expected if, for instance, the disk were thicker on one side than the other as is thought to be the case in several other LMXRB (Section 4), or if reprocessed emission from the companion star were important.

7. Transient Sources

Binary X-ray sources are notoriously variable objects on virtually all time scales studied. In some systems, though, accretion onto the compact object is episodic, and they are only visible as X-ray sources during relatively brief 'transient' outbursts when the accretion rate is high. In other cases the accretion rate may become too high, so that the X-rays are absorbed or degraded in material surrounding the compact star and the X-ray source is again extinguished; or the X-ray emission may sometimes be hidden because the thickness or geometry of an accretion disk changes. Some objects exhibit periodic outbursts, often clearly due to the fact that the X-ray emitter is in an elliptical orbit around the mass-donor companion which causes the accretion rate to vary cyclically. In all these cases multifrequency observations can be of vital importance in sorting out the mechanism(s) underlying the observed changes in brightness.

Transient sources occur among both the major classes of accreting binary X-ray emitter, the low-mass and high-mass systems. In the case of the low-mass systems, optical study of the 'X-ray novae' A0620–00 and Cen X-4 has provided crucial confirmation of the basic binary model for these sources in that a K-star companion has been detected in both systems during their 'off' state (Murdin *et al.* 1980; van Paradijs *et al.* 1980). Other instances of possible detections of low-mass companion stars of X-ray transients are discussed by Murdin *et al.* (1980). McClintock and Remillard (1986) have made time resolved spectroscopic observations of the A0620–00 system and find that the companion star has a very large radial velocity shift with the 7.75 hour photometric orbital period of the system. The magnitude of this shift, which has a full amplitude of almost 1000 km s^{-1}, suggests that the X-ray source has a mass greater that the maximum allowed for a neutron star, and is probably therefore a black hole.

Hameury, King and Lasota (1986) have discussed a model for soft X-ray transients such as A0620–00 and Cen X-4 which involves an instability in the atmosphere of the mass-donor companion star due to X-ray heating. The instability sets in when the X-ray flux irradiating the companion star becomes comparable to that star's intrinsic luminosity. There follows a burst of mass transfer onto the X-ray emitting star which is stemmed only when an accretion disk forms around the compact object which is of sufficient thickness to shield the L1 point of the companion from further irradiation by X-rays. Hameury, King and Lasota (1988) compute model light curves on the basis of this model, and find that the time scales agree well with observation. The model makes specific predictions about the time history of the mass transfer pulse and the parts of the system that are illuminated by X-rays as the outburst progresses. It may be possible to test these with suitable multi-frequency observations, for example optical spectroscopy of lines produced in the mass-transfer stream. Such observations would be particularly valuable in the time leading up to an outburst.

Another LMXRB transient source is XBT0748–676, notable because it is one of only two known X-ray dip sources (section 4) which is also an eclipsing system (Parmar *et al.* 1986). Further, XBT0748–676 is also an X-ray burst source, and its transient nature has allowed both the X-ray dip morphology and the burst characteristics to be examined as a function of source intensity (Parmar *et al.* 1986; Gottwald *et al.* 1986). The optical light curve of XBT0748–676 bears a remarkable similarity to that of the eclipsing ADC source X1822–371, discussed in section 4 (Schmidtke and Cowley 1987). This is an important observation in that it adds weight to the idea that both the dip phenomenon and the modulation of X-rays from the X1822–371 ADC are caused by basically similar structure in an accretion disk.

Optical observations can also potentially be used to distinguish whether the low states in XBT0748–676 are caused by a cessation of accretion or because the accretion disk becomes thicker and hides the X-ray source. In the former case the optical emission should be much fainter during the X-ray low state, whereas in the latter instance the characteristics of the optical emission would probably not change much between the low and high X-ray states, since a large fraction of the X-ray heated disk would presumably still be visible. Such a test has already been successfully applied to the sources X2129+470 and Her X-1. The optical emission of the ADC source X2129+470 fades by about a factor of ten when the X-ray source switches off (Pietsch *et al.* 1986) suggesting that accretion stops or is markedly reduced. In contrast the optical activity of Her X-1 continues during the periodic 35-day X-ray lows (Boynton *et al.* 1983) which is conclusive evidence that these low states are caused by blocking of our line of sight to the X-ray source. Similarly, during the extended low state of Her X-1 that happened in 1983, when the normal 35-day X-ray on/off cycle ceased (Parmar *et al.* 1985), the optical activity continued unabated (Delgado, Schmidt and Thomas 1983), demonstrating that the extended low-state too occurred because we could no longer see the X-ray source, rather than because the X-ray source had ceased to emit.

Multiwavelength observations of HMXRB transients are also important, particularly for recurrent transients where in-depth studies can be made. A0535+26 /HDE 245770 is a 'typical' transient of this kind which has received substantial observational attention. Work on this system has been summarized by Giovannelli

(1985;1987; see also de Loore *et al.* 1984). The optical counterpart, HDE 245770, appears to be a typical Be star and is in orbital about an X-ray pulsar which has a spin period of 104 s (Rosenberg *et al.* 1975). Several X-ray outbursts of the system have been recorded, and analysis of ten years of observations from the *Vela 5B* satellite by Priedhorsky and Terrell (1983) indicates a modulation of the X-ray flux with a period of 111 days. This is almost certainly the orbital period of the system, and evidence for a similar periodicity has subsequently been found in optical photometry and in the equivalent width of the Hβ line (Guarnieri *et al.* 1985; Hutchings 1984; de Martino *et al.* 1985). Evidence of infrared pulsations with the 104 s X-ray spin period has also been reported (Gnedin *et al.* 1983). The basic model is that the orbit of the X-ray pulsar about its companion is very eccentric ($e > 0.3$) so that the rate of mass transfer is a function of orbital phase. The outbursts of the system are however irregular in their characteristics, and there is sometimes a delay of up to about one week between the X-ray and optical peaks. It is probable that the delay represents the time taken for material ejected from the companion to reach the orbit of the neutron star (de Loore *et al.* 1984). The irregular outburst behavior is understandable in terms of the superposition of the regular orbital 'clock' on the episodic mass ejection episodes characteristic of Be stars.

Another interesting Be star/X-ray pulsar binary is A0538–66 which is located in the LMC. This object shows regular outbursts every 16.7 days when it is in its active state, interspersed with long periods of inactivity, or quiescent states. The X-ray pulsar spin period in this system is 0.069 seconds (Skinner *et al.* 1982). A0538–66 is unique among the HMXRB transients studied in showing large changes in its optical and UV brightness commensurate with the X-ray outbursts; the amplitude of the optical outbursts is almost a factor of ten. Consequently extensive observations have been made of A0538–66 using the *IUE* satellite and various ground-based optical facilities (Charles *et al.* 1983; Densham *et al.* 1983; Howarth *et al.* 1984; Corbet *et al.* 1985). There is a consensus that the orbit of the neutron star in this system is highly eccentric, as in A0535+26, and that at periastron the critical potential of the Be star companion becomes comparable with its radius, causing a burst of mass transfer which powers the X-ray outburst. The optical/UV outburst can be understood energetically as the result of reprocessing of X-rays in an envelope around the Be star. *IUE* observations demonstrate that as the 'on-state' outburst cycle progresses the effective temperature of the Be star drops and its apparent radius increases. This suggests that the repeated close passages of the neutron star create an envelope of cloaking material around its companion, not all of which is accreted. It is not clear, however, what triggers an 'on' state in the source, or why outbursts then cease again. Particularly puzzling s the observation that the luminosity of the B star during the quiescent state is about twice as high as observed between outbursts during the 'on' state (Howarth 1984): Possibly part of the B star is shielded by an equatorial ring of ejecta during the 'on' state, and indeed the ejection of such an envelope (common in Be stars) may be the event that triggers a period of outburst activity. A number of such important questions remain concerning the behavior of A0538−66. It is clear that multiwavelength observations will play a vital role in a search for answers.

8. Conclusion

In the introduction I advanced the thesis that the need for multiwavelength studies of X-ray binaries was driven to some extent by the technical shortcomings of early X-ray instrumentation. In more recently history, ironically, and particularly in the study of LMXRB which have very faint optical counterparts, it has been the long wavelength instrumentation that has imposed the major technical limitation on multiwavelength observations. The advent of sensitive area detectors such as CCDs has greatly improved the situation in the optical and near infrared bands; indeed a significant portion of our rapidly expanding understanding of LMXRB can be directly attributed to work involving CCD detectors, particularly in the measurement of orbital periods. The major workhorse in the ultraviolet band pass, however, continues to be the *IUE* satellite and the sensitivity of this satellite is such that only a small number of the brightest LMXRB are accessible to it. This is particularly frustrating in that we are dealing with hot objects where information about the ultraviolet continuum distribution and flux variability is likely to be of great importance, and where we expect to see strong lines diagnostic of the effects of X-ray heating. Even for high-mass X-ray binaries the number of targets accessible with *IUE* to the high spectral and time resolution studies required of these objects in the ultraviolet is limited. Fortunately, we can look forward to a vast improvement in this situation when the *Hubble Space Telescope* becomes operational. It must be said, though, that even *HST* will not be ideally suited to the *simultaneous* multiwavelength coverage that is often demanded of this highly variable class of objects, because of the low Earth orbit of the observatory, and the necessity of inflexible, pre-planned observing schedules.

As the study of X-ray binaries has advanced, so multiwavelength observations have progressed from the stage of fortuitously observed coincident events to where coordinated campaigns in various wavebands are planned to investigate specific aspects of these objects. Such campaigns will continue to be expensive in time, manpower and resources for the foreseeable future, and should not be undertaken lightly if they are to retain credibility in the face of traditional demands on observatory time. Nevertheless, we are already at the stage where the various accessible wavebands can be used to target particular aspects or parts of the binary; to separately monitor, for example, the two component stars, the hot inner parts of an accretion disk and its cooler outer regions, or to separate optically thick emission regions from an optically thin envelope or wind. In this way multiwavelength observations can yield much more than the sum of their individual components, and such endeavors promise to contribute extensively to the future development of X-ray binary studies.

I am grateful to M. F. Cawley, F. Giovannelli, T. Kallman, N. Kawai, J.-P. Lasota, and S. D. Vrtilek for sending me copies of their presentations to the Taos workshop on Multiwavelength Astrophysics, and to F. A. Córdova for reading the draft of this manuscript and for encouraging me to write it. I acknowledge the support of the Royal Society.

References

Ables, J. G. 1969, *Proc. Astr. Soc. Australia*, **1**, 237.

Andrew, B., and Purton, C. 1968, *Nature*, **218**, 855.

Battistoni, G., *et al.* 1985, *Phys. Lett.*, **155B**, 465.

Becklin, E. E., Kristian, J., Neugebauer, G. and Wynn-Williams, C. G. 1972, *Nature Phys. Sci.*, **239**, 130.

Becklin, E. E., Neugebauer, G., Hawkins, F. J., Mason, K. O., Sanford, P. W., Matthews, K. and Wynn-Williams, C. G. 1973, *Nature*, **245**, 302.

Becklin, E. E. *et al.* 1974, *Ap. J.*, **192**, L119.

Bolton C. T. 1972, *Nature*, **235**, 271.

Boynton, P. E., Canterna, R., Crosa, L., Deeter, J., and Gerend, D. 1973, *Ap. J.*, **186**, 617.

Bradt, H. V., *et al.* 1975, *Ap. J.*, **194**, 445.

Braes, L. L. E. and Miley, G. K. 1972, *Nature*, **237**, 506.

Branduardi, G., Mason, K. O., and Sanford, P. W. 1978, *M.N.R.A.S.*, **185**, 137.

Canizares, C. R., Clark, G. W., Lewin, W. H. G., Schnopper, H. W., Sprott, G. F., Hjellming, R. M., and Wade, C. M. 1973, *Ap. J.*, **179**, L1.

Canizares, C. R., *et al.* 1975, *Ap. J.*, **197**, 457.

Cawley, M. F., 1987, Remarks made at the workshop on multiwavelength astrophysics, Taos, NM, August 1987.

Chadwick, P. M., Dipper, N. A., Dowthwaite, J. C., Gibson, A. I., Harrison, A. B., Kirkman, I. W., Lotts, A. P., Macrae, J. H., McComb, T. J. L., Orford, K. J., Turver, K. E., and Walmsley, M. 1985, *Nature*, **318**, 642.

Charles, P. A., Mason, K. O., White, N. E., Culhane, J. L., Sanford, P. W., and Moffatt, A. J. F. 1978, *M.N.R.A.S.*, **183**, 813.

Charles, P. A., *et al.* 1983, *M.N.R.A.S.*, **202**, 657.

Chodil, G., Mark, H., Rodrigues, R., Seward, F. D., Swift, C. D., Turiel, I., Hiltner, W. A., Wallerstein, G. and Mannery, E. J. 1968, *Ap. J.*, **154**, 645.

Clark, D. H., and Murdin, P. G. 1978, *Nature*, **276**, 44.

Corbet, R. H. D., Mason, K. O., Córdova, F. A., Branduardi-Raymont, G. and Parmar, A. N. 1985, *M.N.R.A.S.*, **212**, 565.

Córdova, F. A. 1986, *Los Alamos Science*, **No. 13**, pg 38.

Cowley, A. P., Crampton, D., and Hutchings, J. B. 1979, *Ap. J.*, **231**, 539.

Crampton, D., and Hutchings, J. B. 1981, *Ap. J.*, **251**, 604.

Danaher, S., Fegan, D. J., Porter, N. A. and Weekes, T. C. 1981, *Nature*, **289**, 568.

Davidsen, A., Malina, R. F., Smith, H., Spinrad, H., Margon, B., Mason, K. O., Hawkins, F. J. and Sanford, P. W. 1974, *Ap. J.*, **193**, L25.

Davidsen, A. and Ostriker, J. P. 1974, *Ap. J.*, **189**, 331.

Delgado, A. J., Schmidt, H. U., and Thomas, H. C. 1983, *Astr. Ap.*, **127**, L15.

Densham, R. H., Charles, P. A., Menzies, J. W., van der Klis, M., and van Paradijs, J. 1983, *M.N.R.A.S.*, **205**, 1117.

de Loore, C. *et al.* 1984, *Astr. Ap.*, **141**, 279.

de Martino, D., *et al.* 1985, *in "Multifrequency Behaviour of Galactic Accreting Sources", ed.* F. Giovannelli, SIDEREA, *pg.* 326.

Dowthwaite, J. C. *et al.* 1983, *Astr. Ap.*, **126**, 1.

Dupree, A. K. *et al.* 1980, *Ap. J.*, **236**, 969.

Eadie, G., Peacock, A., Pounds, K. A., Watson, M. G., Jackson, J. C., and Hunt, R. 1975, *M.N.R.A.S.*, **172**, 35P.

Evans, W. D., Belian, R. D., Conner, J. P., Strong, I. B., Hiltner, W. A., and Kunkel, W. E. 1970, *Ap. J.*, **162**, L115.

Fomalont, E. B., Geldzahler, B. J., Hjellming, R. M. and Wade, C. M. 1983, *Ap. J.*, **275**, 802.

Fransson, C. and Fabian, A. C. 1980, *Astr. Ap.*, **87**, 102.

Geldzahler, B. J. *et al.* 1983, *Ap. J.*, **273**, L65.

Giacconi, R., Gorenstein, P., Gursky, H., and Waters, J. R. 1967, *Ap. J.*, **148**, L119.

Giacconi, R., Gursky, H., Kellogg, E., Schreier, E., Tananbaum, H. 1971, *Ap. J.*, **167**, L67.

Gilmore, W. S., Seaquist, E. R., Stocke, J. T., and Crane, P. C. 1981, *Astron. J.*, **86**, 864.

Giovannelli, F., *et al.* 1985, *in "Multifrequency Behaviour of Galactic Accreting Sources", ed.* F. Giovannelli, SIDEREA, *pg.* 286.

Giovannelli, F., 1987, Talk given at the workshop on multiwavelength astrophysics, Taos, NM, August 1987.

Gnedin, Y. N., Khozov, G. V., and Larionov, V. M. 1983, *Astrophys. Space Science*, **93**, 207.

Gottwald, M., Haberl, F., Parmar, A. N., and White, N. E. 1986, *Ap. J.*, **308**, 213.

Gregory, P. C., Kronberg, P. P., Seaquist, E. R., Hughes, V. A., Woodsworth, A., Viner, M. R., Retallack, E. R., Hjellming, R. M. and Balick, B. 1972, *Nature (Phys. Sci.)*, **239**, 114.

Gregory, P. C. and Taylor, A. R. 1978, *Nature*, **272**, 704.

Guarnieri, A., *et al.* 1985, *in "Multifrequency Behaviour of Galactic Accreting Sources", ed.* F. Giovannelli, SIDEREA, *pg.* 310.

Hameury, J. M., King, A. R., and Lasota, J. P. 1986, *Astr. Ap.*, **162**, 71.

Hameury, J. M., King, A. R., and Lasota, J. P. 1988, *Astr. Ap.*, (in press).

Hatchett, S. and McCray, R. 1977, *Ap. J.*, **211**, 522.

Hillas, A. M. 1984, *Nature*, **312**, 50.

Hinteregger, H. F., *et al.* 1972, *Nature Phys. Sci.*, **240**, 159.

Hjellming, R. M. 1973, *Ap. J.*, **182**, L29.

Hjellming, R. M., and Johnston, K. J. 1981a, *Nature*, **290**, 100.

Hjellming, R. M., and Johnston, K. J. 1981b, *Ap. J.*, **246**, L141.

Hjellming, R. M., and Johnston, K. J. 1986, *in "Lecture Notes in Physics Vol. 266: The Physics of Accretion onto Compact Objects", eds.* K. O. Mason, M. G. Watson and N. E. White, Springer-Verlag, *pg.* 287.

Howarth, I. D., 1984, *in "Proc. 4th European IUE Conference", ESA SP-218, pg.* 449.

Howarth, I. D., Hammerschlag-Hensberge, G., and Kallman, T. 1986, *in "New Insights in Astrophysics: 8 years of UV astronomy with IUE", eds.* R. Wilson *et al.* , ESA SP-263 , *pg.* 475.

Howarth, I. D., Prinja, R. K., Roche, P. F., and Willis, A. J. 1984, *M.N.R.A.S.*, **207**, 287.

Hudson, H. S., Peterson, L. E. and Schwartz, D. A. 1970, *Ap. J.*, **159**, L51.

Hutchings, J. B.,   1984, *P.A.S.P.*, **96**, 312.

Ilovaisky, S. A., Chevalier, C., White, N. E., Mason, K. O., Sanford, P. W., Delvaille, J. P., and Schnopper, H. W. 1980, *M.N.R.A.S.*, **191**, 81.

Johnson, H. M. and Stephenson, C. B. 1966, *Ap. J.*, **146**, 602.

Johnston, K. J. *et al.* 1986, *Ap. J.*, **309**, 707.

Jones, C., Forman, W., Tananbaum, H., Schreier, E., Gursky, H., Kellogg, E., Giacconi, R. 1973, *Ap. J.*, **181**, L43.

Joss, P. C., and Rappaport, S. A. 1979, *Astr. Ap.*, **71**, 217.

Krzeminski, W. 1974, *Ap. J.*, **192**, L135.

Lamb, R. C., Godfrey, C. P., Wheaton, W. A. and Turner, T 1981, *Nature*, **296**, 543.

Levine, A., Ma, C. P., McClintock, J. E., Rappaport, S., van der Klis, M., and Verbunt, F. 1988, *Ap. J.*, (in press).

Lewin, W. H. G., and Joss, P. C.,   1983, *in "Accretion Driven Stellar X-ray Sources", ed.* W. H. G. Lewin and E. P. J. van den Heuvel, Cambridge University Press, *pg.* 67.

Li, F. K. and Clark, G. W. 1974, *Ap. J.*, **191**, L27.

Lloyd-Evans, J., Coy, R. N., Lambert, A., Lapikens, J., Patel, M., Reid, R. J. O. and Watson, A. A. 1983, *Nature*, **305**, 784.

Lucke, R., Yentis, D., Friedman, H., Fritz, G., and Shulman, S. 1976, *Ap. J.*, **206**, L25.

Lucy, L. B. and White, R. L. 1980, *Ap. J.*, **241**, 300.

Margon, B. H., Ford, H. C., Grandi, S. A., and Stone, R. P. S. 1979a, *Ap. J.*, **233**, L63.

Margon, B. H., Ford, H. C., Katz, J. I., Kwitter, K. B., Ulrich, R. K., Stone, R. P. S., and Klemola, A. 1979b, *Ap. J.*, **230**, L41.

Mark, H., Price, R. E., Rodrigues, R., Seward, F. D., Swift, C. D. and Hiltner, W. A. 1969, *Ap. J.*, **156**, 67.

Marshak, M. L. *et al.* 1985a, *Phys. Rev. Lett.*, **54**, 2079.

Marshak, M. L. *et al.* 1985b, *Phys. Rev. Lett.*, **55**, 1965.

Marshall, F. E., White, N. E., and Becker, R. H. 1983, *Ap. J.*, **266**, 814.

Mason, K. O. 1986, *in "Lecture Notes in Physics Vol. 266: The Physics of Accretion onto Compact Objects", eds.* K. O. Mason, M. G. Watson and N. E. White, Springer-Verlag, *pg.* 29.

Mason, K. O. *et al.* 1976, *Ap. J.*, **207**, 78.

Mason, K. O., Branduardi, G., and Sanford, P. W. 1976, *Ap. J.*, **203**, L29.

Mason, K. O. and Córdova, F. A. 1982, *Ap. J.*, **262**, 253.

Mason, K. O., Córdova, F. A. and White, N. E. 1986, *Ap. J.*, **309**, 700.

Mason, K. O., Hawkins, F. J., Sanford, P. W., Murdin, P. and Savage, A. 1974, *Ap. J.*, **192**, L65.

Mason, K. O., Middleditch, J., Nelson, J. E., White, N. E., Seitzer, P., Tuohy, I. R., and Hunt, L. K. 1980, *Ap. J.*, **242**, L109.

Mason, K. O., Parmar, A. N. and White, N. E. 1985, *M.N.R.A.S.*, **216**, 1033.

McClintock, J. E., and Remillard, R. A. 1986, *Ap. J.*, **308**, 110.

McClintock, J. E., Remillard, R. A., and Margon, B. 1981, *Ap. J.*, **243**, 900.

Meegan, C. A., Fishman, G. J., and Haymes, R. C. 1979, *Ap. J.*, **234**, L123.

Miyamoto, S., and Matsuoka, M. 1977, *Space Sci. Rev.*, **20**, 687.

Middleditch, J. and Nelson, J. E. 1976, *Ap. J.*, **208**, 567.

Middleditch, J., Mason, K. O., Nelson, J. E., and White, N. E. 1981, *Ap. J.*, **244**, 1001.

Milgrom, M.,   1978, *Astr. Ap.*, **208**, 191.

Molnar, L. A. 1986, *in "Lecture Notes in Physics Vol. 266: The Physics of Accretion onto Compact Objects", eds.* K. O. Mason, M. G. Watson and N. E. White, Springer-Verlag, *pg.* 313.

Molnar, L. A., and Mauche, C. W. 1986, *Ap. J.*, **310**, 343.

Molnar, L. A., Reid, M. J., and Grindlay, J. E. 1984, *Nature*, **310**, 662.

Mook, D. E. 1967, *Ap. J.*, **150**, L25.

Mook, D. E. *et al.* 1975, *Ap. J.*, **197**, 425.

Motch, C., Pedersen, H., Beuermann, K., Pakull, M. W., and Courvoisier, T. 1987, *Ap. J.*, **313**, 792.

Murdin, P. G., Allen, D. A., Morton, D. C., Whelan, J. A. J., and Thomas, R. M. 1980, *M.N.R.A.S.*, **192**, 709.

Murdin, P., Penston, M. J., Penston, M. V., Glass, I., Sanford, P. W., Hawkins, F. J., Mason, K. O., and Willmore, A. P. 1974, *M.N.R.A.S.*, **169**, 25.

Parmar, A. N., White, N. E., Giommi, P., and Gottwald, M. 1986, *Ap. J.*, **308**, 199.

Parmar, A. N., Pietsch, W., McKechnie, S., White, N. E., Trumper, J., Voges, W., and Barr, P. 1985, *Nature*, **313**, 119.

Parsignault, D. R., Gursky, H., Kellogg, E. M., Matilsky, T., Murray, S., Schreier, E., Tananbaum, H., Giacconi, R. and Brinkman, A. C. 1972, *Nature (Phys. Sci.)*, **239**, 123.

Pedersen, H., *et al.* 1982, *Ap. J.*, **263**, 325.

Pelling, R. M., 1973, *Ap. J.*, **185**, 327.

Petro, L. D., Bradt, H. V., Kelley, R. L., Horne, K., and Gomer, R. 1981, *Ap. J.*, **251**, L7.

Pietsch, W., Steinle, H., Gottwald, M., and Graser, U. 1986, *Astr. Ap.*, **157**, 23.

Porter, N. A. 1984, *Nature*, **312**, 347.

Pravdo, S. H., White, N. E., Kondo, Y., Becker, R. H., Boldt, E. A., Holt, S. S., Serlemitsos, P. J., and McCluskey, G. E. 1980, *Ap. J.*, **237**, L71.

Prendergast, K. H. and Taam, R. E. 1974, *Ap. J.*, **189**, 125.

Priedhorsky, W. B., Hasinger, G., Lewin, W. H. G., Middleditch, J., Parmar, A. N., Stella, L., and White, N. E. 1986, *Ap. J.*, **306**, L91.

Priedhorsky, W. B., and Terrell, J. 1983, *Nature*, **303**, 681.

Pringle, J. E. 1974, *Nature*, **247**, 21.

Rappaport, S. and Joss, P. C. 1983, *in "Accretion Driven Stellar X-ray Sources"*, ed. W. H. G. Lewin & E. P. J. van den Heuvel, Cambridge University Press, *pg.* 1.

Reppin, C., Pietsch, W., Trumper, J., Voges, W., Kendziorra, E., and Staubert, R. 1979, *Ap. J.*, **234**, 329.

Rosenberg, F., Eyles, C. J., Skinner, G. K., and Willmore, A. P. 1975, *Nature*, **256**, 628.

Samorski, M. and Stamm, W. 1983, *Ap. J.*, **268**, L17.

Sanford, P. W. and Hawkins, F. J. 1972, *Nature (Phys. Sci.)*, **239**, 135.

Sandage, A. R., Osmer, P., Giacconi, R., Gorenstein, P., Gursky, H., Waters, J., Bradt, H., Garmire, G., Sreehantan, B. V., Oda, M., Osawa, K., and Jugaku, J. 1966, *Ap. J.*, **146**, 315.

Sato, N., Hayakawa, S., Nagase, F., Masai, K., Dotani, T., Inoue, H., Makino, F., Makishima, K. and Ohashi, T. 1986, *Publ. Astron. Soc. Japan*, **38**, 731.

Schreier, E., Levinson, R., Gursky, H., Kellogg, E., Tananbaum, H., and Giacconi, R. 1972, *Ap. J.*, **172**, L79.

Schmidtke, P. C., and Cowley, A. C. 1987, *Astron. J.*, **92**, 375.

Seaquist, E. R., Garrison, R. F., Gregory, P. C., Taylor, A. R., and Crane, P. C. 1979, *Astron. J.*, **84**, 1037.

Seaquist, E. R., Gilmore, W., Johnston, K. J., and Grindlay, J. E. 1982, *Ap. J.*, **260**, 220.

Seward, F. D., Page, C. G., Turner, M. J. L., and Pounds, K. A. 1976, *M.N.R.A.S.*, **175**, 39P.

Seward, F. D., Grindlay, J. E., Seaquist, E. R., and Gilmore, J. E. 1980, *Nature*, **287**, 806.

Skinner, G. K., Bedford, D. K., Elsner, R. F., Leahy, D., Weisskopf, M. C., and Grindlay, J. 1982, *Nature*, **297**, 568.

Spencer, R. E.,   1979, *Nature*, **282**, 483.

Spencer, R. E., Swinney, R. W., Johnston, K. J., and Hjellming, R. M. 1986, *Ap. J.*, **309**, 694.

Stella, L., Priedhorsky, W., and White, N. E. 1987, *Ap. J.*, **312**, L17.

Stephenson, C. B., and Sanduleak, N. 1977, *Ap. J. Suppl.*, **33**, 459.

Stewart, G. C., *et al.* 1987, *M.N.R.A.S.*, **228**, 293.

Tananbaum, H., Gursky, H., Kellogg, E., Levinson, R., Schreier, E. and Giacconi, R. 1972, *Ap. J.*, **174**, L143.

Tananbaum, H., Gursky, H., Kellogg, E., Giacconi, R. and Jones, C. 1972, *Ap. J.*, **177**, L5.

van der Klis, M., Stella, L., White, N. E. Jansen, F., and Parmar, A. N. 1987, *Ap. J.*, **316**, 411.

van der Laan, H. 1966, *Nature*, **211**, 1131.

van Paradijs, J. 1983, *in* "*Accretion Driven Stellar X-ray Sources*", *ed.* W. H. G. Lewin and E. P. J. van den Heuvel, Cambridge University Press, *pg.* 189.

van Paradijs, J., Verbunt, F., van der Linden, T., Pedersen, H., and Wamsteker, W. 1980, *Ap. J.*, **241**, L161.

Watson, M. G., Willingale, R., Grindlay, J. E., and Seward, F. 1983, *Ap. J.*, **273**, 688.

Webster, B. L. and Murdin, P. G. 1972, *Nature*, **235**, 37.

Whelan, J. A. J. *et al.* 1977, *M.N.R.A.S.*, **181**, 259.

Willingale, R., King, A. R., and Pounds, K. A. 1985, *M.N.R.A.S.*, **215**, 295.

White, N. E., Becker, R. H., Boldt, E. A., Holt, S. S., Serlemitsos, P. J., and Swank, J. H. 1981, *Ap. J.*, **247**, 994.

White, N. E. and Holt, S. S. 1982, *Ap. J.*, **257**, 318.

White, N. E., Kallman, T. R. and Swank, J. H. 1983, *Ap. J.*, **269**, 264.

White, N. E., Mason, K. O., and Sanford, P. W. 1976, *M.N.R.A.S.*, **176**, 201.

White, N. E., Peacock, A., and Taylor, B. G. 1985, *Ap. J.*, **296**, 475.

Willmore, A. P., Mason, K. O., Sanford, P. W., Hawkins, F. J., Murdin, P., Penston, M. V. and Penston, M. J. 1974, *M.N.R.A.S.*, **169**, 7.

The Classical Nova Outburst

Sumner Starrfield

Department of Physics, Arizona State University and
Theoretical Division, Los Alamos Scientific Laboratory

1. Introduction

The classical nova outburst occurs on the white dwarf component in a close binary system. Nova systems are members of the general class of cataclysmic variables and other members of the class are the Dwarf Novae, AM Her variables, Intermediate Polars, Recurrent Novae, and some of the Symbiotic variables. Although multiwavelength observations have already provided important information about all of these systems, in this review I will concentrate on the outbursts of the classical and recurrent novae and refer to other members of the class only when necessary.

A Cataclysmic Variable (hereafter: CV) is commonly assumed to contain a Roche Lobe filling secondary, on or near the main sequence, losing hydrogen-rich material through the inner Lagrangian point onto an accretion disk that surrounds a white dwarf primary. The additional defining characteristic of the AM Her and Intermediate Polars is the presence of a magnetic field on the white dwarf which can be either strong enough to affect the flow of gas from the secondary (AM Her) or somewhat weaker and only capable of disrupting the accretion disk close to the white dwarf's surface (intermediate polar). The presence of a *strong* magnetic field will also affect the progress of the classical nova outburst and this problem is just now beginning to be addressed since Nova V1500 Cygni 1975 has recently been discovered to be an AM Her variable. One must also be aware that there are systems which violate the above defining characteristics and are still considered to be members of the class. For example, U Sco, a recurrent novae, appears to be transferring material that is helium rich. Not only does this violate the mass transfer condition, it also makes it rather unlikely that the secondary is close, in an evolutionary sense, to the main sequence. The physical process that drives the secondary into filling its Roche Lobe is still not known nor is it understood what physical process is acting, in the accretion disk, to remove angular momentum from the gas and allow it to spiral onto the white dwarf.

Given the above caveats about the basic system, our understanding of the cause of the nova outburst rests on strong theoretical foundations. The theoretical studies have shown that a gradually accumulating shell of hydrogen-rich material on a massive white dwarf will be unstable to a thermonuclear runaway and the simulations of this process reproduce most of the observed features of the nova outburst. The calculations further imply that the energetics of the outburst, and thus the type of nova, is sensitive not only to the abundances of the CNONe nuclei, but also depends on the mass of the accreted shell, white dwarf mass, and accretion rate.

The organization of this review is as follows: in Section 2, I present a brief summary of a nova outburst in order to emphasize the gross details of its evolution with time. Section 3 follows with a discussion of the thermonuclear runaway theory as the cause of the outburst. In Section 4, I discuss the observations of novae that have been done in various wavelength intervals and emphasize the basic multiwavelength nature of the studies. Included in that section is a brief description of the novae in outburst at the time this review was written (January 1988). I end with a summary and discussion.

2. A Summary of the Nova Outburst

A nova outburst is classified, according to the rate of decline from maximum, either as 'fast' or 'slow.' The initial eruption of a fast nova is very rapid, with the major part of the rise to visual maximum taking place in a day or less. During the rising branch of the light curve, the nova exhibits spectral features corresponding to an optically thick, expanding shell. As the expansion is very rapid and the bolometric luminosity is nearly constant, the effective temperature smoothly declines and reaches a minimum of 4000K-7000K at visual maximum (Gallagher and Ney 1976; Gallagher and Starrfield 1978). The maximum in optical light is determined by the temperature and density at which hydrogen recombines and the opacity drops. After this time, the photosphere moves inward in mass and the surface layers continue to expand.

It is only during the brief period around maximum that pure absorption line or P-Cygni profiles are seen and are potentially accessible to analysis (Williams *et al.* 1981). The duration of the photospheric phase depends upon the speed class of the nova. In fast, luminous, novae the primary shell may become optically thin in a few days; thereafter, the optical spectral region is dominated by bremsstrahlung and hydrogen bound-free emission (Gallagher and Ney 1976; Ennis *et al.* 1977; Gehrz 1988; Martin 1988). In a slower, lower optical luminosity nova, continuous mass loss can maintain a low temperature photosphere for several months or years (Bath 1978; Ney and Hatfield 1978). The speed class of a nova can be determined rapidly since the nova decline rate and expansion velocity are correlated. An expansion velocity can easily be measured from line widths on a low dispersion spectrogram.

The early post-maximum decline in the optical is the result of the expanding shell relaxing to a stable luminosity and the redistribution of the luminosity into the ultraviolet as the shrinking photosphere moves inward in mass. The stable luminosity is set by the Eddington limit (Gallagher and Starrfield 1976, 1978; Sparks, Starrfield, and Truran 1978; Wu and Kester 1977; Truran 1982; Starrfield 1986, 1987, 1988), which is $M_{Bol} = -7$ for a typical nova and is determined by the white dwarf mass. While the bolometric luminosities of the fastest, brightest novae are initially brighter than this, their luminosity drops to this value soon after maximum. In contrast, novae with maximum absolute visual magnitudes fainter than -7 show little change in bolometric luminosity after maximum; the initial decline in the visual can be completely attributed to flux redistribution into the ultraviolet and EUV.

During the declining branch of the light curve, novae undergo important changes in physical conditions. Because the hardening of the radiation field from the central source produces an ionization front moving out through the expanding ejecta, because the density of matter is high compared to most nebulae, and because the process is time dependent; the spectrum can become exceedingly complex, especially for novae with significant post-maximum stellar winds (Gallagher 1978). In many novae a thermal infrared excess now appears as a direct result of grain formation in the ejecta (Geisel *et al.* 1970; Gallagher 1977; Ney and Hatfield 1978; Gehrz *et al.* 1980, 1984, 1986; Gehrz, Grasdalen, and Hackwell 1985; Mitchell and Evans 1984; see Gehrz 1988 for a review). As the nova begins its final decline, the ionization level in the ejecta increases to moderately high levels before approaching conditions similar to those found in planetary nebulae. It is not known when or how a nova ends its outburst, but on a time scale of (at most) a few years, mass loss decreases and the nova returns to its quiescent luminosity (Gallagher and Holm 1974; Ney and Hatfield 1978; Starrfield 1979, MacDonald, Fujimoto, and Truran 1985; Starrfield 1986; Ögelman, Krautter, and Beuermann 1987).

3. A Nova as a Thermonuclear Event

In this section we assume that the nova system is a close binary with one member a white dwarf and the other member a larger, cooler star that fills its Roche lobe. Because it fills its lobe, any tendency for the secondary to grow in size as a result of evolutionary processes or for the lobe to shrink from angular momentum losses, by some (unknown) mechanism, will cause a flow of gas through the inner

Lagrangian point into the lobe of the white dwarf. The high angular momentum of the transferred material causes it to spiral into an accretion disk surrounding the white dwarf. Some viscous process, also unknown, acts to transfer mass inward and angular momentum outward through the disk so that a fraction of the material lost by the secondary will ultimately end up on the white dwarf.

The accreted layer grows in thickness until it reaches a temperature that is high enough for thermonuclear burning of hydrogen to begin at the bottom. The further evolution of nuclear burning on the white dwarf now depends upon the mass and luminosity of the white dwarf, the rate of mass accretion, and the chemical composition of the reacting layer. If the bottom of the accreted layer is degenerate, then the simulations of this phenomenon resemble the observations of the nova outburst (Starrfield, Truran, and Sparks 1978; Sparks, Starrfield, and Truran 1978; Gallagher and Starrfield 1978; Starrfield 1986, 1987, 1988; Starrfield, Sparks, and Truran 1985, 1986; Starrfield and Sparks 1987; Shaviv and Starrfield 1987; Starrfield, Sparks, and Shaviv 1988). Observations of novae ejecta also imply that there is mixing of core material into the accreted layer so that the final chemical composition will reflect a combination of core plus accreted material (Sparks, Starrfield, and Truran 1988).

If the material is degenerate enough, a thermonuclear runaway (hereafter: TNR) will occur, and the temperatures in the accreted envelope will grow to values exceeding 10^8K (see Starrfield 1988 and references therein). During the early part of the evolution, the lifetimes of the CNO nuclei against proton captures are very much longer than the decay times for the β^+-unstable nuclei [$\tau(^{13}N) = 863s$, $\tau(^{14}O) = 102s$, $\tau(^{15}O) = 176s$, $\tau(^{17}F) = 92$ s] so that these nuclei can decay and their daughters capture another proton in order to keep these reactions cycling. As the temperature increases in the shell source, the CNO nuclei lifetimes against proton captures continuously decreases until, at temperatures of $\sim 10^8$K, they become shorter than the β^+-decay lifetimes. Therefore, once the temperature in the shell source exceeds 10^8K, the β^+-unstable nuclei become abundant and *important*. I emphasize that it is the operation of the CNO reactions at high temperatures and densities that produces large amounts of the β^+-unstable nuclei in the envelope. Their presence now imposes severe constraints on the energetics of the outburst. First, because the wait for them to decay effectively halts any further rise in energy generation and, second, because it is their decay at late times which provides the kinetic energy for ejecting the shell and the luminous energy radiated during the outburst.

In more detail, one of the most important results from the hydrodynamic simulations has been the identification of the role played by the four β^+-unstable nuclei in the outburst. These four nuclei: ^{13}N, ^{14}O, ^{15}O, and ^{17}F not only affect the energy generation but influence the entire outburst. As soon as the temperatures in the shell source exceed 10^8K, the abundances of these nuclei increase to where they severely impact the nuclear energy generation in the envelope since every proton capture must now be followed by a waiting period before the β^+ decay occurs and another proton capture can occur. I also note that all of the computer simulations show that during the evolution to peak temperature a convective region has formed just above the shell source and grown to include virtually the entire accreted envelope. Therefore, at the peak of the outburst the most abundant of the CNO nuclei in the envelope will be the β^+-unstable nuclei.

This will have several effects on the subsequent evolution. First, since the energy production in the CNO cycle comes from proton captures followed by a β^+-decay, at maximum temperature the rate at which energy is produced depends only on the half-lives of the β^+-unstable nuclei and numbers of the CNO nuclei initially present in the envelope. This is because the CNO reactions do not create new nuclei, but only redistribute them among the various CNO isotopes (Starrfield, Truran, Sparks, and Kutter 1972). Second, since the convective turn-over time scale is usually about 10^2 sec near the peak of the TNR, a significant fraction of the β^+-unstable nuclei can reach the surface without decaying and the rate of energy generation at the surface will exceed 10^{12} to 10^{13} erg gm^{-1} s^{-1} (Starrfield 1988). Third, convection throughout the envelope will bring unburned CNO nuclei into the shell source, when the temperature is rising very rapidly, and this will keep the CNO nuclear reactions operating far from equilibrium.

Once peak temperature is reached and the envelope begins to expand, the simulations of the outburst, which include a detailed calculation of the abundance changes with time of the nuclei, show that the rate of energy generation declines only as the abundances of the β^+-unstable nuclei decline since their decay is neither temperature nor density dependent (see Starrfield 1988 for a detailed review). The numerical calculations done with the CNO nuclei enhanced (Starrfield, Truran, and Sparks 1978; Starrfield and Sparks 1987) show that more than 10^{47}erg are released into the envelope after its expansion has begun. The envelope reaches radii of more than 10^{10}cm before all of the ^{13}N has disappeared. Therefore, the decays of the β^+-unstable nuclei provide the delayed source of energy which is ultimately responsible for both ejecting the shell and producing the luminous output of the outburst. Finally, since these nuclei decay when the temperatures in the envelope have declined to values that are too low for any further proton captures to occur, the final isotopic ratios in the ejected material will not agree with those ratios predicted from studies of equilibrium CNO burning.

Up to this point, the discussion has not depended upon enhanced CNO nuclei in the envelope but has been based on the hypothesis that in order for an outburst to occur the shell source will be degenerate enough so that the peak temperature exceeds 10^8 K. If this occurs, the effects of the β^+-unstable nuclei become inevitable. However, the observational fact that the CNO nuclei are enhanced in the ejecta also requires them to be enhanced in the nuclear burning region. All arguments about the effects of the β^+-unstable nuclei are only strengthened if the CNO nuclei are enhanced. If this occurs, peak energy generation is increased and more energy will be stored for release at late times in the outburst. Our simulations have shown that the presence of enhanced CNO nuclei in the envelope is required in order to produce a fast nova outburst. *No calculation done with only a solar mixture has been successful in reproducing a realistic fast nova.*

The theoretical calculations show that this evolution will release enough energy to eject material with expansion velocities that agree with observed values and that the predicted light curves produced by the expanding material can agree quite closely with the observations (Sparks, Starrfield, and Truran 1978; Starrfield, Truran, and Sparks 1978; Starrfield, Sparks, and Truran 1974a, 1974b, 1985, 1986; Prialnik, Shara, and Shaviv 1978, 1979; MacDonald 1980; Prialnik; *et al.* 1982; Starrfield, Sparks, and Truran 1985, 1986; Starrfield, Sparks, and Shaviv 1988).

The nucleosynthesis that occurs during the outburst will enhance ^{13}C, ^{14}N, ^{15}N, ^{17}O, and ^{7}Li and these nuclei will be ejected by the nova explosion (Starrfield, Truran and Sparks 1978; Lazareff *et al.* 1979; Starrfield *et al.* 1978; Audouze *et al.* 1979; Hillebrandt and Thielemann 1982; Wiescher *et al.* 1986). It has also been predicted that the ^{26}Al anomaly could be a result of nuclear burning during the nova outburst (Arnould *et al.* 1980). These predictions, along with the observational confirmation of non-solar CNO abundances in nova ejecta, demand that novae be included in studies of galactic nucleosynthesis.

Observational studies of recent novae have reported very large enhancements of neon (V1500 Cyg: Ferland and Shields 1978; V693 CrA: Williams *et al.* 1985; V1370 Aql: Snijders *et al.* 1987). Gehrz, Grasdalen, and Hackwell, (1985) have reported the discovery of [Ne II] emission at 12.8μm in Nova Vul #2 1984 and Gehrz *et al.* (1986) report the condensation of SiO_2 grains in the same nova. These results, in combination with the *IUE* spectra that also show strong neon lines (Starrfield *et al.* 1988, in preparation), imply that at least four of the recent nova outbursts have ejecta rich in neon. The most likely explanation is that some novae are ejecting material which has been processed to neon and beyond during the prior evolution of the white dwarf (Starrfield, Sparks, and Truran 1986).

The hydrodynamic simulations also predict that there will be a phase of constant luminosity following the initial outburst. The cause of this phase is that only a fraction of the accreted envelope is ejected during the initial explosion. The remaining material (anywhere from 10% to 90%) quickly returns to quasistatic equilibrium with a radius extending to $\sim 10^{11}$cm. Because the shell source is still burning at the bottom of the envelope, the luminosity is $\sim L_{Ed}$ and, therefore, the effective temperature exceeds 10^5K. This material will slowly be ejected by radiation pressure driven mass loss (Starrfield, Sparks, and Shaviv 1988), but as long as some of it is still present on the white dwarf, the remnant will radiate at a constant luminosity.

X-ray, ultraviolet, and infrared studies confirm that novae do exhibit this constant luminosity phase following the initial rise to maximum (Wu and Kester 1977; Ney and Hatfield 1978; Stickland *et al.* 1981, Sparks *et al.* 1982; Snijders *et al.* 1984; Williams *et al.* 1985; Ögelman, Krautter, and Beuermann 1987). The radiated energy, the effective temperature, and the time scale of this phase of the outburst provide fundamental data about the white dwarf (Gallagher and Starrfield 1976, 1978; Starrfield 1979, 1980, 1988; Truran 1982; Sion and Starrfield 1986; MacDonald, Fujimoto, and Truran 1985). These data have been used to show that the masses of white dwarfs in binary systems range from $\sim 0.6 M_\odot$ (DQ Her 1934) to $\sim 1.2 M_\odot$ or even higher (Nova V1500 Cyg 1975 and U Sco 1979). The inferred mass for DQ Her appears to be in substantial agreement with the values determined from radial velocity studies (Smak 1980; Young and Schneider 1980).

Recent calculations have shown that a recurrent nova outburst can occur as a result of a TNR. By examining the consequences of accretion of hydrogen-rich material with a solar abundance of the CNO nuclei onto a massive white dwarf at high rates of mass accretion, Starrfield, Sparks and Truran (1985) and Starrfield, Sparks, and Shaviv (1988) found that a TNR resulted after anywhere from three to thirty-three years of evolution and that the evolution resembled U Sco. Sion and Starrfield (1986) have also modeled the observed behavior of Z And. In other calculations Shaviv and Starrfield (1987, 1988) have investigated the effects of the accretion energy on the progress of the runaway. Their initial results show that profound changes to the simulations occur when this energy is included. Further studies are in progress.

Finally, a calculation of accretion onto oxygen, neon, and magnesium white dwarfs has produced extremely violent outbursts in which the entire accreted envelope was ejected at high velocities (Starrfield, Sparks, and Truran 1986). This study was an attempt to simulate the novae that are ejecting enhanced neon.

4. Observations of Novae i. Outburst

Up until recently, the most comprehensive studies of novae were done only in the optical and there exists a rich literature of such studies. Most of the early data has already been summarized in the classical reviews of Payne-Gaposchkin (1957), McLaughlin (1960), and Gallagher and Starrfield (1978). A more recent review can be found in a book that will soon appear which is completely devoted to both observational and theoretical studies of the nova outburst (Bode and Evans 1988). In addition, some material on novae can be found in the proceedings of the 1987 CV workshop (Drechsel, Kondo, and Rahe 1987).

In this section, I first present a summary of the phases of the outburst, follow with a summary of observations of recent novae, and then conclude with a discussion of the results from each wavelength interval. The ordering of the last part of this section was determined by the dates when the first data on novae were obtained in that wavelength interval. Some of the main features of novae which have proved particularly interesting are given below:

The Phases of the Outburst

1. Premaximum: The spectrum in the optical is usually dominated by broad absorption lines and emission lines are either weak or absent. Spectral types are B to A, although some novae have been observed to have a later spectral class; for example, RR Pic was listed as having a class of F and Nova Vul 1987 was probably cooler. In addition, the optical spectrum of V 1500 Cyg 1975, one day before maximum, was that of a B2Ia star with unusual absorption line strengths for the C, N, and O elements (Boyarchuk *et al.* 1977). Spectra obtained in the ultraviolet also show a cool continuum but there are usually strong emission lines of FeII and other abundant elements with similar ionization potentials superimposed on this continuum. The unusual strengths, as compared to normal stars, of the C, N, and O lines seen in Nova V1500 Cygni 1975, are rather common (McLaughlin 1960). The theoretical studies imply that the anomalous strengths are actually caused by abundance enhancements.

It can, therefore, be very useful and exciting to catch a nova on the rise. Unfortunately, there was a three day delay in observing V1500 Cyg with *Copernicus* and no ultraviolet detection resulted since by that time the photosphere had cooled to about 5000K (Jenkins *et al.* 1977). The same problem also happened during our recent attempts to study Nova Vul 1987 with the *IUE* Satellite. Its ultraviolet continuum was dropping so rapidly that I was unable to obtain decently exposed short wavelength spectra.

An important advance in the understanding of the early spectra of novae has occurred as a direct result of the outburst of SN 1987a in the LMC. It was quickly realized that the first ultraviolet spectra of SN 1987a strongly resembled ultraviolet spectra of novae taken near maximum except that the lines were much broader in the supernova (Kirshner *et al.* 1987). As a direct result, we are now developing techniques that will allow us to apply modern theoretical developments in spherical, expanding, stellar atmospheres to analyses of novae spectra. We will be able to obtain abundances by a completely new method from the nebular studies.

2. Maximum: At maximum the Principal Spectrum appears. The features that are seen evolve continuously in velocity as the receding photosphere moves inward in mass and many of the absorption lines seen at maximum can eventually be identified with the emission lines in the nebular shell (McLaughlin 1960; Gallagher and Starrfield 1978). Optical observations show very strong P-Cygni profiles in the lines of hydrogen, Fe II, and possibly other singly ionized metals. As the nova begins its decline, the continuum fades more rapidly then the emission lines and the P-Cygni profiles disappear. After a short time, the emission lines dominate the spectrum. They have rather complex profiles and suggest that material has been ejected in clouds, blobs, or rings.

3. Early Decline: New absorption line systems now appear (in the optical) at high velocities compared to the principal absorption line system. These systems are called the diffuse enhanced and consist of broad features that may separate (as time passes) into many sharp subcomponents. At still later times in many novae, very highly ionized features appear in absorption at even larger velocities than the diffuse enhanced systems. These systems are collectively called the Orion spectrum (based on their similarity to young OB stars) and are optical analogues of the 'sharp, narrow, absorption' components seen in the ultraviolet spectra of the most luminous O and B stars that exhibit strong stellar winds. However, the Orion features always seem to remain broad and do not separate into sharp components. The excitation of the emission lines increases steadily with time and it is not uncommon to find lines of triply or quadruply ionized elements in either the optical or ultraviolet spectra. Many novae show coronal features at late stages in the outburst; one recent example is Nova Vul 1984 #2 (Greenhouse *et al.* 1988).

4. Transition: During this stage the ionization increases to levels of 50-60 ev and the electron densities decrease to values of 10^8 to 10^{10} cm^{-3}. We can now use the techniques developed for the analysis of planetary nebulae and quasars to determine the variation of electron density and temperature with time and, also, the elemental abundances (Ferland and Shields 1978; Lance, McCall, and Uomoto 1988, hereafter LMU). Since novae are time dependent, this phase represents an interesting exploration of nebular physics and allows us to use the time variation as an additional constraint in determining the abundances.

5. Final Decline: This phase is marked by the return of the nova to minimum and, depending upon the speed class, can last for years to decades. Few optical observations and no ultraviolet observations have been obtained during this stage since novae are faint and have been thought to be uninteresting. Recent *EXOSAT* observations of Nova Muscae 1983, Nova Vul 1984 #1 and #2, and RS Oph (Mason *et al.* 1986; Ögelman, Krautter, and Beuermann 1987) imply that the turn off

Table 1 Galactic Novae in Outburst in 1988

Nova constellation	Year of outburst	α			δ (1950.0)		
Muscae	1983	11h	49m	35s	-66°	55'	43"
Cen	1986	14h	32m	13s	-57°	24'	31"
U Sco	1987	16h	19m	37s	-17°	45'	43"
Vul	1987	19h	02m	32s	21°	41'	39"
V1370 Aql	1982	19h	20m	50s	02°	24'	00"
PW Vul	1984	19h	24m	03s	27°	15'	54"
RR Tel	1947	20h	00m	18s	-55°	52'	00"
Vul #2	1984	20h	24m	41s	27°	40'	40"
And	1986	23h	09m	48s	47°	12'	01"

time scale for a classical nova is about 2-3 years and during that time the central source is radiating at $L \sim L_{Ed}$ with an effective temperature of $\sim 3 \times 10^5$K. On the other hand, optical spectra of Nova Muscae obtained in January 1987 suggest that the central source is still hot and luminous (Krautter and Williams 1988, in preparation).

The theoretical calculations predict that the hydrostatic remnant is now undergoing an intense period of mass loss in combination with a gradually increasing effective temperature (Starrfield, Truran, and Sparks 1978; Sparks, Starrfield and Truran 1978; Starrfield 1979; Truran 1979, 1982; Starrfield 1986, 1988; MacDonald, Fujimoto, and Truran 1985). The remaining accreted envelope has reached a stage where it is undergoing equilibrium nuclear burning and its bolometric magnitude can be used to estimate the mass of the white dwarf (Truran 1982). Finally, it is the return to minimum that sets the stage for accretion to resume (if it has not begun already) and, therefore, marks the beginning of the evolution to another outburst.

Observations of Recent Novae

Table 1 gives a list of all novae currently in outburst. Many of them are still bright and can be studied with small telescopes; more data is welcomed. The first classical nova studied with the *IUE* satellite was V1668 Cygni. It was a moderately fast nova which reached a V_{max} of ~ 6.2 on September 12, 1978. The data were analyzed by Stickland *et al.* (1979, 1981) who restricted their analysis to the nebular phase and found that the total CNO abundance in the ejecta was about 30 times solar and that nitrogen was about 200 times solar. They reported an expansion velocity ~ 800 km s^{-1}, an ejected mass of $\sim 6 \times 10^{-5} M_\odot$, and a kinetic energy of 6×10^{44} ergs. These results are in excellent agreement with calculations of Starrfield, Truran, and Sparks (1978) for a CNO outburst on a $1.0 M_\odot$ white dwarf.

In 1981, V693 CrA was discovered to be in outburst at ~7 mag. Optical spectra were published both by Brosch (1982) and by Williams *et al.* (1985). Williams *et al.* (1985) also analyzed the *IUE* data and presented both a ultraviolet line list and line fluxes over the duration of the outburst. Brosch (1982) determined an ejection velocity of ~2200 km s^{-1} from the FWHM of the hydrogen lines. This high a value was also obtained from low dispersion *IUE* data. In addition, a recent analysis of a high dispersion image shows CIV 1550Åwith the blue edge of the P-Cygni profile extending to a velocity of almost 8000 km s^{-1} (Sion *et al.* 1986). The most important finding of the *IUE* analysis (Williams *et al.* 1985) is that the abundances of all the intermediate mass elements from nitrogen to aluminum are enhanced over a solar mixture by a factor of about 100 (by number).

Early the next year, Nova V1370 Aql was discovered in outburst and the group led by Seaton and Snijders obtained *IUE*, optical, infrared, and radio data for this nova (Snijders *et al.* 1987). This nova was also observed in the optical by Andrillat (1983) and Rosino, Iijima, and Ortolani (1983) and in the infrared by Gehrz *et al.* (1984), Williams and Longmore (1984), and Bode *et al.* (1984). Snijders *et al.* (1987) report that this nova ejected ~$5 \times 10^{-6} M_\odot$with expansion velocities of ~10^4km s^{-1}. The abundances determined for the ejecta were as unusual as reported for V693 CrA. Neon was the most abundant element in the ejecta and elements up to sulfur were enhanced. The infrared data showed an excess at ~10μm but no excess at ~20μm (Gehrz *et al.* 1984). Snijders *et al.* (1984) reported a grain mass of the same order of magnitude as the gas mass so that this material is an important part of the ejecta. Starrfield, Sparks, and Truran (1986) explain this as an outburst on an ONeMg white dwarf (but see Wiescher *et al.* 1986).

Krautter *et al.* (1984) reported on optical, ultraviolet, and infrared data of Nova Muscae 1983. They found a distance of ~5kpc and noted that it was radiating at ~L_{Ed} for a $1.0M_\odot$ white dwarf in agreement with the TNR predictions. Analysis of high dispersion ultraviolet and optical spectra showed that this nova ejected material in a large number of discrete clouds (Krautter *et al.* 1984). An abundance analysis showed that He/H was enhanced over solar, N/C was ~20, and N/O was ~2.4. These values are characteristic of hot hydrogen burning. Pacheco and Codina (1985) determined an expansion velocity of ~1000km s^{-1} and confirmed, from optical spectra, the overabundance of He, and CNO. They also found an overabundance of Fe which is difficult to understand in terms of the TNR theory although iron enhancements have been suggested for other novae (V1500 Cygni: Ferland and Shields 1978; V1370 Aql: Snijders *et al.* 1984). This nova was also detected by *EXOSAT* (Ögelman, Beuermann, and Krautter 1984; see also Ögelman, Krautter, and Beuermann 1987). This satellite observed three novae and found they were all radiating with $T \sim 3 \times 10^5 K$ and $L \sim L_{Ed}$ at late times in their outbursts (see below).

The most recent slow nova was PW Vul (1984 #1). Optical and infrared data for this nova are presented in Kenyon and Wade (1986) who find a distance of ~ 1.2 kpc and $M_v \sim 5.5$. They also report He/H of ~ 0.13 and that oxygen is enhanced in the ejecta. They did not find a neon enhancement. The *IUE* data are now being reduced and analyzed (see Starrfield and Snijders 1987 for some representative *IUE* spectra of this nova).

Another nova was discovered in outburst in Vul late in 1984 (Nova Vul 1984 #2). By January 1988, it has declined to ~ 13.8 mag and is still being observed in the ultraviolet, optical, infrared, and radio. The ejected material is rich in oxygen, neon, and magnesium so that this must be a fourth member of the oxygen, neon, and magnesium class of outbursts (Starrfield, Sparks, and Truran 1986). The *IUE* data showed that [Ne V] 3346Åappeared during the fall of 1985 and that in most of the ultraviolet spectra [Ne IV] 1602Åis the strongest line in the short wavelength spectral region. In addition, [Ne IV] 2422Åis also present which, in combination with the infrared results (Gehrz, Grasdalen, and Hackwell 1985; Gehrz *et al.* 1986) is strong evidence for enhanced neon in this nova.

The 1979 outburst of U Sco, a recurrent nova, was like that of a classical nova with some very important differences. Optical data for this outburst were analyzed by Barlow *et al.* (1981) who found that helium was twice hydrogen by number. The analysis of the *IUE* spectra was given in Williams *et al.* (1981) who reported that nitrogen was overabundant while carbon and oxygen probably were not. The CNO abundances can be explained by solar CNO material being processed through a hot hydrogen burning region. U Sco suffered another outburst in 1987 and the data obtained during this outburst indicated that material was ejected at very high velocities and that nitrogen was probably enhanced in the ejecta (Sekiguchi *et al.* 1988). Unfortunately, no *IUE* data were obtained because it was too close to the sun.

Simulations of both outbursts of U Sco, involving accretion of material containing hydrogen, have been calculated by Starrfield, Sparks, and Truran (1985), Sparks, Starrfield, and Truran (1986), Truran *et al.* (1988), and Starrfield, Sparks, and Shaviv (1988). Studies of the accretion disk, both before and after the outburst (Barlow *et al.* 1981, Williams *et al.* 1981, Hanes, 1985), show only lines due to helium so this recurrent nova appears to be transferring helium-rich material from the secondary and ejecting both hydrogen and helium (Starrfield, Sparks, and Shaviv 1988).

RS Oph was discovered to be in outburst in January 1985. There has already been a workshop devoted to the analysis of the 1985 outburst (Bode 1986) and we shall not review it any further. Discussions of the relationship of this outburst to the outbursts of classical novae can be found in Sparks, Starrfield, and Truran (1986), Starrfield, Sparks, and Truran (1985), and Starrfield, Sparks, and Shaviv (1988). An alternative view can be found in Livio, Truran, and Webbink (1986). Finally, V394 CrA suffered an outburst in August 1987. Both optical and *IUE* data were obtained during the outburst and are now being analyzed. Initial reductions suggest that this object may resemble U Sco. Observations are still in progress for Nova Her 1987, Nova And 1986, Nova Cyg 1986, and Nova V394 CrA 1987.

Although this review is primarily concerned with observations of classical novae in outburst, I briefly note some recent studies of novae whose outbursts occurred some time in the past. A description of the outburst can usually be found in Payne-Gaposchkin (1957). GK Per was studied by Bianchini and Sabbadin (1983) who found a ultraviolet continuum distribution characteristic of an accretion disk with a mass accretion rate of $\sim 2 \times 10^{16}$ gm s^{-1}. However, the continuum slopes that they obtained do not agree with predictions of standard accretion disk theory. Duerbeck *et al.* (1980) and Rosino, Bianchini, and Rafanelli (1982) have studied RR Pic and report that it shows very high ionization emission lines and that the continuum slope fits a temperature of $\sim 3 \times 10^4$K.

The ultraviolet spectra of T CrB have been analyzed by Duerbeck *et al.* (1980) and Cassatella *et al.* (1982) who find large changes in the ultraviolet flux from one spectrum to another. The ultraviolet luminosity varies from $\sim 5 L_\odot$ to $40 L_\odot$. Kenyon and Webbink (1984) attempted to fit the continuum by simulations of accretion onto main sequence stars at various rates. This binary system consists of an M giant and a massive ($1.6 M_\odot$: Kenyon and Garcia 1986) compact component. The spectra obtained for DQ Her have been analyzed by Ferland *et al.* (1984) who report an electron temperature of less than 500K, which agrees with the optical studies of the expanding nebula (Williams *et al.* 1978).

V603 Aql is the brightest old nova in the sky and a prime target for continuing optical, infrared, and ultraviolet studies. Ferland *et al.* (1982) showed that the abundances in the accretion disk are quite close to solar implying that the material being transferred from the secondary is normal. This is very strong evidence in support of the requirements that the enhanced abundances seen in novae ejecta must come from the core of the white dwarf (Sparks, Starrfield, and Truran 1988; Starrfield 1988).

Optical Studies

Much of our present understanding of the nova outburst comes from optical studies and if one includes recent studies of the novae which were observed to outburst in 1670 (CK Vul: Shara and Moffatt 1982) and 1783 (WY Sge: Shara and Moffatt

1983; Kenyon and Berriman 1987), one then realizes that the study of novae predates the birth of modern astrophysics. The earliest data on novae consisted mainly of qualitative descriptions of photographic spectra since, for the most part, these spectra were not calibrated. Besides these spectra, there was considerable effort placed on obtaining light curves. A large number of light curves for novae can be found in Payne-Gaposchkin (1957) and it was the analysis of these light curves that suggested that there was a relationship between speed class (rate of decline from maximum) and absolute magnitude. Recent studies have capitalized on this information and have tried to put it on a firm observational foundation (Cohen and Rosenthal 1983; Cohen 1985).

It is the published work on Nova V1500 Cygni 1975 (Ferland and Shields 1978; Ferland, Lambert, and Woodman 1986a,b; LMU), Nova V1370 Aql 1982 (Snijders *et al.* 1987), and Nova U Sco 1979 (Barlow *et al.* 1981; Williams, *et al.* 1981) that has provided the most detailed analyses of a nova outburst. In addition, Martin (1988) has reanalyzed published spectra taken during the 1934 outburst of DQ Her and has provided a modern interpretation of its behavior.

Although it is better to combine optical data with data taken in other wavelength intervals, one can still obtain the extinction to a nova using only the optical data. For example, for a bright nova, one can use the Balmer decrement, a Paschen to Balmer line ratio, the helium triplet line ratio, interstellar absorption line equivalent widths, or interstellar polarization (Ferland 1977; Ferland, Lambert, and Woodman 1986a,b). Once the extinction is determined, comparison of the nova to nearby stars can give an immediate, if not too accurate, distance estimate. Better estimates can be obtained from the expansion parallax method using infrared data (see the discussion in the infrared section and Gehrz 1988).

The most important information that one can obtain from the optical data are the abundances of the light elements determined from emission line fluxes. Most of the modern spectrophotometric and photometric studies provide calibrated fluxes for both emission lines and continuum points that, in addition to abundances, can also be used to determine the distance, amount of ejected mass, and kinetic energies of the ejecta. If data are taken over a long enough time during the outburst, one can then obtain information about the underlying object, rate of decline to quiescence, and the time when the nova returns to quiescence.

The emission lines also contain a wealth of information about the conditions in the ejected envelope and the nature of the central star. The basic trend, as the nova declines, is for lines requiring lower electron densities and higher electron temperatures to gradually dominate the spectrum (Gallagher and Starrfield 1978; LMU). The cause is obvious: the shell is expanding and the effective temperature of the central source is either rising or remaining constant at values exceeding $3 \times 10^5 \mathrm{K}$

(Gallagher and Starrfield 1976; LMU; Ögelman, Krautter, and Beuermann 1987).

In addition to the optical forbidden lines commonly seen in novae such as [OII] 3727Å, [OIII] 4363Å, [NIII] 4640Å, [OIII] 4959Å, and [OIII] 5007Å; some novae are observed to show coronal line emission. Coronal lines normally seen in the optical include, but are not limited to, [Fe VII] 6087Å, [FeX] 6374Å, and [FeXI] 7892Å. Nova Muscae 1983 still shows very strong coronal line emission over four years after the initial explosion (Krautter and Williams 1988, in preparation). V1500 Cygni showed infrared coronal lines as does Nova Vul 1984 #2. The cause of coronal line emission is not yet known but X-ray data may indicate that we are seeing emission from shocks in the ejected material. On the other hand, photoionization from a very hot central source, $T > 3 \times 10^5 K$, may be sufficient to explain these lines very late in the outburst (Krautter and Williams 1988, in preparation).

Forbidden line emission from a source as complex as a nova in outburst provides great difficulty in both analysis and interpretation (see for example Ferland and Shields 1978; LMU). The problems involve large scale departures from spherical symmetry and inhomogeneities in the expanding material. The problems become more tractable at late stages in the outburst (generally when most observers have lost interest) when the ejecta are completely ionized and recombination rather than collisional processes dominate. The most accurate optical abundance determinations for novae are the studies of Williams *et al.* (1978,1979), Gallagher *et al.* (1980), and their collaborators (see Starrfield 1988) in which they analyzed the resolved shells of old novae and applied standard techniques previously developed to study planetary nebulae.

As in planetary nebulae, one can solve for the electron density and temperature by simultaneously solving a system of equations involving various line ratios as a function of time. The most useful in novae are the ratios: [OIII] (4959Å+ 5007Å) / [OIII] (4363Å); [OIII] (4959Å+ 5007Å) / [NeIII] (3869Å); and [OIII] (4959Å+ 5007Å) / HeI (5876Å). These lines are generally strong and can be measured with high signal to noise. When applied as a function of time, they are constrained by the fact that the elemental abundances have to be constant. A detailed discussion of V1500 Cyg and the references to the necessary constants can be found in LMU.

Nova V1370 Aql 1982 has also been observed and analyzed in great detail (Snijders *et al.* 1987). I will discuss the ultraviolet data in a later section, here I concentrate only on the optical. They used a variety of line ratios to obtain gas phase conditions and abundances and were helped by the great variety of ions that were present in the gas. For example, they found CII, CIII, and CIV, NII to NIV, OIII, NeIII, NeIV, and NeV plus other elements. The large number of lines present in the spectra allowed them to determine electron temperature and density as a function of time. For day 156 after the outburst, they found that the electron temperature was about 10^4K, the electron density was $\sim 3 \times 10^8$ cm^{-3}, and the light elements from carbon to iron were very enhanced over solar. Starrfield, Sparks, and Truran (1986) have interpreted this event as occurring on an oxygen, neon, magnesium white dwarf. However, it is also important to note that the large range in ionization potential suggests that the gas in this nova was clumpy and the physical conditions varied from one clump to another. This is a very common feature in novae.

When this review appears in print, all three recent novae in Vul and Nova Muscae 1983 will still be in outburst and may have only recently reached the nebular stage. The optical behavior of each one of these novae differ from each other, therefore, data are still being obtained and analyzed.

Ultraviolet Studies

Two recent reviews of the ultraviolet studies of novae (done mainly with the *IUE* Satellite) have recently appeared (Starrfield and Snijders 1987; Starrfield 1986). A number of novae have now been observed in the ultraviolet and the data have proved to be very important in our understanding of the outburst. Over the past few years, ultraviolet photometry and/or spectra have been obtained for FH Ser (Gallagher and Code 1974), V1500 Cygni (Wu and Kester 1976, Jenkins *et al.* 1977), Nova Cygni 1978 (Cassatella *et al.* 1979; Sparks *et al.* 1980; Stickland *et al.* 1981), U Sco (Williams *et al.* 1981), Nova CrA 1981 (Sparks *et al.* 1982; Williams *et al.* 1985), Nova Aql 1982 (Snijders *et al.* 1984, 1987), RS Oph (Snijders 1986), Nova Vul 1984 #1 (Starrfield *et al.* 1988, in preparation), and #2 (Starrfield *et al.* 1988, in preparation).

The characteristics of the outburst of each one of these novae have been unique and qualitatively different in the ultraviolet. In addition, the two recurrent novae studied with the *IUE* Satellite, U Sco and WZ Sge, exhibited completely different outburst behavior in the visual (Barlow *et al.* 1981) and in the ultraviolet (Williams *et al.* 1981). Therefore, it seems clear that ultraviolet observations must be obtained for all bright novae in order to improve our understanding of these exciting objects.

Elemental abundances, mostly from ultraviolet data, have already been determined for both the remnant and the ejecta (Gallagher and Starrfield 1978, Stickland *et al.* 1981; Williams *et al.* 1981, 1985; Snijders *et al.* 1984, 1987; Starrfield and Snijders 1987). Two separate methods exist for determining abundances: (a) spectrum synthesis can be done for the photospheric absorption and emission lines that appear near maximum, and (b) nebular theory can be applied to the emission lines which appear a few weeks after the beginning of the outburst. It is a fact of atomic physics that most resonance lines of ions of the cosmically most abundant elements fall in the ultraviolet, whereas lines from excited states occur in the visible. Accurate abundances can be obtained from the P-Cygni lines (and their time dependent behavior) from CNO in the ultraviolet (Williams *et al.* 1981).

In addition, there are lines available in the ultraviolet that originate from many ions that do not radiate in the visible. For example, reliable carbon abundances for novae did not exist before the *IUE* because carbon has no prominent forbidden lines in the optical region. In the ultraviolet, however, we have found very strong lines from transitions such as: CII 1335Å, CIII 1909Å, and CIV 1549Å. Other lines that are seen include Lyα, HeII 1640Å, NV 1240Å, OI 1302Å, and OIV 1400Å, besides lines of other ions that have shown abundance enhancements (novae line lists can be found in Williams *et al.* 1985 and Snijders *et al.* 1987). None of the ultraviolet resonance lines suffer collisional de-excitation at the densities characteristic of novae ejecta following the outburst. Therefore, abundances can be deduced from line intensities (Williams *et al.* 1981, 1985; Stickland *et al.* 1981, Snijders *et al.* 1987).

It is the existence of the *IUE* that has made it possible to determine abundances in novae for atoms such as neon, magnesium, silicon, and aluminum. The fact that three of the recent novae observed with the *IUE* have shown massive enhancements of these atoms in their ejecta is very important and, consequently, we have proposed that a fraction of novae arise from outbursts on oxygen-neon-magnesium white dwarfs rather than carbon-oxygen white dwarfs (Starrfield, Sparks, Truran 1986). This result has important implications concerning the evolution of massive stars in binary systems.

Infrared Studies

An important review of infrared studies of novae in outburst will soon appear (Gehrz 1988; hereafter G88). Since Gehrz and his collaborators have made the largest series of observations of novae, I will only briefly summarize the infrared work on novae. The key point about the infrared studies is that they are now the only way to detect the formation and study the evolution of dust in the ejecta. Dust formation was first proposed by McLaughlin (1936) to explain the deep depression in the light curve of DQ Her (called the 'transition'). However, he and others later provided other suggestions for this behavior (McLaughlin 1949, 1960) and this idea fell into disfavor.

It was not until 1970 that Geisel, Kleinmann, and Low (1970) and Hyland and Neugebauer (1970) found, from infrared observations, that FH Ser had formed an optically thick dust shell. Since that time, detailed infrared studies have been published for FH Ser (Geisel *et al.* 1970), V1229 Aql (Geisel *et al.* 1970), V1500 Cygni (Gallagher and Ney 1976; Ennis *et al.* 1977; see G88 for a complete discussion of this remarkable nova), NQ Vul (Ney and Hatfield 1978; Gehrz *et al.* 1977), LW Ser (Gehrz *et al.* 1978), V1668 Cyg (Gehrz *et al.* 1980), V1370 Aql (see G88 for a discussion), PW Vul (Gehrz *et al.* 1987), and Nova Vul 1984 #2 (Gehrz *et al.* 1985a,b, 1986; Greenhouse *et al.* 1988). Fragmentary data exist for some other novae (G88).

The discovery and interpretation of the infrared emission in novae has had major effects on our understanding of the progress of the nova outburst. First, the detection of grain formation in the ejecta has led to an improvement in our understanding of how the grain growth process occurs in an astrophysical environment. In addition, the infrared studies show that most novae form carbon grains with the two exceptions (V1370 Aql and Nova Vul 1984 #2) being the most unusual novae ever studied. The total energy emitted by the grains, which are reradiating the ultraviolet energy emitted by the remaining accreted layers on the hot white dwarf, quickly reaches a value that is roughly equivalent to the optical luminosity at maximum. This last is the strongest demonstration of the constant luminosity phase predicted by the theoretical studies of the outburst.

Finally, one can use the infrared emission at two different phases, the early black-body expansion phase and the early part of the dust condensation phase, to determine an expansion parallax to the nova. This is the most accurate early distance method (G88). At late times the direct expansion of the shell can usually be measured (Cohen 1985).

The infrared development of a nova generally exhibits the following features. The initial observations, if obtained early enough in the outburst, show that it is radiating like a black body expanding with time. Because it is optically thick, Ney and Hatfield (1978) refer to this as the pseudophotospheric expansion phase. G88 uses the term 'fireball' since he regards this phase as analogous to the early development of terrestrial thermonuclear weapon explosions. This analogy is not correct, however, since a fireball is produced by the explosive energy from the weapon heating the surrounding atmosphere (Glasstone and Dolan 1977, see Chapter 2).

Gallagher and Ney (1976) used the existence of this phase to determine the distance to V1500 Cygni. In order to determine the distance to the nova, one measures the rate of the angular expansion of the photosphere and combines it with the doppler velocity of the lines observed at that time. Given both that the measured velocities correspond to the material that is producing the black body emission and that this material was ejected with spherical symmetry, one can determine a very accurate distance. This expansion parallax method has now been applied to all novae that were observed early enough in the outburst to detect the expanding photosphere.

At some time after the expanding material begins to go optically thin at infrared wavelengths, the nova enters the free-free expansion phase (Gallagher and Starrfield 1978). Gehrz *et al.* 1974, Ennis *et al.* (1977), Gehrz *et al.* (1980a,b), and G88 show that the density of the expanding shell can be determined from the wavelength where the optically thin free-free emission turns over into the Rayleigh-Jeans tail. This result is also well known from radio studies.

Following this phase, some two to three months into the outburst, most novae begin to develop an infrared excess. The assumption commonly made is that this excess is caused by the formation and growth of grains in the expanding shell which then reradiate the ultraviolet energy from the hot, luminous white dwarf. Note, however, that Bode and Evans (1980, 1981) tried to explain the infrared excess by light echoes from pre-existing material. This hypothesis was unsuccessful in explaining many of the features of the infrared excess and has fallen into disfavor.

One nova that did not show an excess until very late in its outburst was V1500 Cygni. It was proposed that the excess was caused not by grains but by strong emission from [NeII] at $12.8\mu m$ (Ferland and Shields 1978b). The presence of this line in novae was finally confirmed in Nova Vul 1984 #2 and was instrumental in showing that neon was overabundant in this nova (Gehrz *et al.* 1986; Starrfield, Sparks, and Truran 1986).

When first detected, the infrared excess attributed to grains exhibits a continuum black body temperature slightly exceeding 1000K. If this material has just formed, then this temperature is considerably below a value of $\sim$2000K at which the grains are normally expected to form (G88). As time passes, the grain emission slowly decreases in temperature to values around 800K. One explains the variation in temperature by the formation of small particles which are inefficient radiators. They then slowly grow to a size that exceeds that of normal interstellar particles $(0.01\mu m$-$0.03\mu m)$. The composition of the grains is usually thought to be some form of amorphous carbon since the infrared excess fails to show any features due to silicates (G88).

However, as in any discussion of novae, there are two exceptions to the above discussion. The first was V1370 Aql 1982. This nova was already known from the early *IUE* studies to be unusual because it ejected material with a very strange abundance distribution (Snijders *et al.* 1984, 1987). Infrared observations showed that it also exhibited an infrared excess with a broad continuum peaked at about $8\mu m$ but superimposed on this broad excess was a feature at $10\mu m$ that has yet to be identified. More recently, Nova Vul 1984 #2 was observed to form silicate grains since its infrared emission showed both the $10\mu m$ and the $20\mu m$ features seen in other astronomical objects. It showed no continuum excess and its infrared emission never became optically thick (Gehrz *et al.* 1986).

This introduces another interesting point about the infrared excesses of novae:

some novae show optically thick infrared emission from grains while some novae exhibit only optically thin emission. In both cases they develop infrared emission, characteristic of grain formation, but in the optically thin case the amount of energy radiated by the grains never approaches the values emitted by the nova at maximum optical light. In contrast, in the optically thick case, the infrared emission in the novae increases until it is radiating nearly the same amount of energy as was seen early in the outburst. In addition, the optically thin emitting novae do not show a deep depression in their light curves and sometimes the optical light curve shows no indication, whatsoever, that dust has formed in the ejecta. A possible explanation of this phenomenon is that the region where the dust is forming has been ejected asymmetrically and does not completely block the ultraviolet light from the central source (G88).

Ney and Hatfield (1978), observing NQ Vul, used the gradual growth in size of the dust forming region to apply the black body expansion parallax method to this nova. Since that time, it has been applied to all the novae that showed an infrared excess characteristic of grain formation. Although one is still faced with the problem of determining the appropriate doppler velocity to use in the solution, it is a very accurate technique for determining nova distances. It will be interesting to check the values determined from this method with the nebular expansion studies done in the optical (Cohen and Rosenthal 1983; Cohen 1985) when the expanding shells of these novae are finally resolved.

At about the same time as the infrared excess from grains becomes significant, some novae are seen to show emission from coronal lines (G88). It has been known for decades that the optical spectral studies of novae show coronal line emission in the infrared but these lines were not detected in the infrared until the outburst of V1500 Cygni (Gallagher and Starrfield 1978). Since that time other novae have been found to exhibit coronal line emission with the most notable case being Nova Vul 1984 #2 (Greenhouse *et al.* 1988). Greenhouse (1987, private communication) has also found coronal line emission from [Ca VIII], [Si VI], and [Si VII] in Nova Cyg 1986 and Nova Her 1987. G88 speculates that coronal line emission is probably ubiquitous in novae but is generally hidden by the grain emission. The lines have an excitation temperature of $\sim 10^6$ K and their presence will constrain the modeling of the physical conditions in the gas at the time that they are seen. Some novae were also detected by *IRAS* (Dinerstein 1986; Callus *et al.* 1987). The observations of Nova Muscae were consistent with free-free emission from a hot gas. Finally, in the spirit of this review, I note that G88 concludes his infrared review with a plea for multiwavelength observations of novae.

Radio Studies

The first radio studies of novae date from around the time of the first infrared studies of novae. A review of work up to the early 1980's can be found in Bode and Evans (1988). Until very recent deep VLA surveys of novae (Hjellming 1987; private communication), only 5 novae had been detected and studied in the radio: HR Del 1967 (Hjellming and Wade 1970), FH Ser (Hjellming and Wade 1970), V1500 Cyg (Hjellming *et al.* 1979), V1370 Aql (Snijders *et al.* 1987), and Nova Vul 1984 #2 (Taylor *et al.* 1987).

Except for Nova Vul 1984 #2 (as usual), the observations of novae during their optical decline are consistent with free-free emission from an expanding cloud of ionized gas. The observations of Nova Vul 1984 #2 are somewhat different in that there appears to have been emission from two ejection phases. The first from emission from a shock propagating through the ejected shell, and the second from emission caused by the remaining material on the white dwarf being ejected by a wind at a rate of $\sim 10^{-5}$ $M_\odot yr^{-1}$ (Taylor *et al.* 1987).

The importance of the radio studies are that they delineate the kinematics of the expanding shell at large distances from the central source. The radio data indicate that the material ejected during the nova outburst has the form of a thick shell with strong velocity and density gradients. These characteristics can also be seen in some of the optical images of old novae such as Nova DQ Her 1934 (Williams *et al.* 1978). The radio light curves of novae (Hjellming *et al.* 1979) usually evolve from optically thick phases where the radio emission is coming from an optically thick 'photosphere', through a transition phase where successively deeper layers of the shell begin to dominate the emission. In the late stages of the outburst, the radio emission is dominated by optically thin radiation from the entire nova shell. The shell radio emission is dominantly free-free unless there is significant interaction with external gas. Reynolds and Chevalier (1984), based upon a prediction by Chevalier (1977) that nova shells might produce non-thermal emission due to interaction of their ejecta with the interstellar medium, found a partial shell of synchrotron emission associated with Nova GK Persei 1901. The recurrent nova RS Oph produced its spectacular radio source (Padin *et al.* 1985, Hjellming 1986) as a result of the interaction of the nova ejecta with the circumbinary gas produced by the wind from the red giant binary companion. This same interaction was used to explain the X-ray data obtained with *EXOSAT*

The observations of Nova Vul #2 (Taylor *et al.* 1987) indicate that the ejected material interacted with external gas in a non-spherical geometry. They were able to resolve the radio emitting ejecta at two different epochs with the high resolution VLA, obtaining an expansion rate which, when combined with measured expansion velocities of $\sim 10^3 km\ s^{-1}$, gave a distance of ~ 3.6 kpc. They also interpreted their observations as showing that material was ejected in a 'ring.' For reasonable assumptions about the physical conditions in the ejecta they report that the nova ejected $\sim 8 \times 10^{-4} M_\odot$ expanding with a kinetic energy of $\sim 8 \times 10^{45}$ ergs. These values are high for an outburst that is predicted to have occurred on a massive oxygen, neon, magnesium white dwarf (Starrfield, Sparks, and Truran 1986) and Taylor *et al.* discuss this fact.

The amount of mass involved in the explosion is what would be expected for a low mass white dwarf and it is possible that this system has already suffered many nova explosions. If such were the case, then the amount of heavy elements in the ejected material suggests that the white dwarf has lost a lot of core material. If its mass has been whittled down to about $0.9 M_\odot$, then it could accrete and eject a larger amount of mass then could a $1.3 M_\odot$ white dwarf. In addition, other factors such as white dwarf luminosity and mass accretion rate will also effect the characteristics of the outburst. Starrfield, Sparks, and Truran (1986) pointed out that they could increase or decrease the accreted mass by decreasing or increasing the mass accretion rate. It will be useful to do some additional calculations of outbursts on oxygen, neon, magnesium white dwarfs.

X-ray Studies

In spite of attempts by *Einstein* (Becker and Marshall 1981), novae were not detected in outburst at X-ray wavelengths until *EXOSAT* observed emission from Nova Muscae 1983 (Ögelman, Beuermann, and Krautter 1984). This was followed shortly thereafter by *EXOSAT* observations of X-rays from Nova Vul 1984 #1 and #2 (Ögelman, Krautter, and Beuermann 1987). In addition, RS Oph, a recurrent nova, was also detected by *EXOSAT* (Mason *et al.* 1986). A review of the X-ray emitting properties of CV's can be found in Córdova and Mason (1983). The lack of a detection by *Einstein* was probably an unfortunate coincidence. Clearly, an object radiating with luminosities $\sim L_{Ed}$ and which has a radius of $\sim 10^9$ cm, will emit copious amounts of soft X-rays. In addition, material ejected during the nova outburst is expected to be moving at velocities exceeding 10^3 km s^{-1}. This high a velocity is enough to produce X-rays from shocks produced by the expanding material running into circumbinary material (Brecher, Ingham, and Morrison 1977; Bode and Kahn 1985).

The observations reported for the three classical novae are best understood if we assume that the X-ray flux is coming from the white dwarf. The observations then suggest that it has a temperature of $\sim 3 \times 10^5$ K and a luminosity which corresponds to a $\sim 1.0 M_\odot$ white dwarf radiating at the Eddington limit (Ögelman, Krautter, and Beuermann 1987). Nevertheless, as emphasized by Ögelman, Beuermann, and Krautter (1984) and Ögelman, Krautter, and Beuermann (1987), the data are not consistent with the predictions of the thermonuclear runaway model. The calculations of Starrfield, Sparks, and Truran (1985) show that the effective temperature should reach much higher values than found in the X-ray observations which were done at late stages in the outburst. The temperatures found above are more consistent with low mass (large radius) white dwarfs than with the high mass white dwarfs predicted by the thermonuclear runaway theory (Starrfield, Truran, and Sparks 1978; Starrfield 1979). Note, however, that recent optical spectra suggest that the central source has continued to increase in temperature but the luminosity is uncertain (Krautter and Williams 1988, in preparation).

The temperatures predicted by the simulations of the nova outburst are a very strong function of the radius of the remnant which is a poorly predicted quantity. The calculated evolutionary sequences predict that the hot white dwarf will remain luminous until all of the remaining accreted layers have been lost. Currently, there are no detailed calculations that show either how the remnant turns off or on what timescale. Obviously, the larger the remnant, at a constant luminosity, the lower the temperature. In addition, the existence of a radiation pressure driven wind will reduce the measured effective temperature. Therefore, the temperature discrepancy does not seem like a serious problem with the thermonuclear runaway theory; just one that needs more study. It also shows that using polytropes to describe the structure of the remnant envelope is a very poor approximation (MacDonald, Fujimoto, and Truran 1985). They should have used the X-ray data to scale their solution.

Ögelman, Krautter, and Beuermann (1987) may have also measured the turn off time scale for Nova Muscae 1983 since their last observation, 900 days after maximum light, suggests that its X-ray flux had decreased by more than a factor of two from maximum (However, note the above remarks that the central source may still be hot and luminous). This is an important measurement because, previously, we could only guess how long the nova would take to turn off. It will be interesting to examine optical data taken around the same time and see if there are any features that are correlated with the X-ray turnoff. As emphasized in an earlier section, one of the most important theoretical problems, yet to be solved, is the turnoff timescale and the cause of the turnoff. Their results show, however, that if the dominant cause of mass ejection of the envelope is by a stellar wind, then the presence of a wind will seriously affect the white dwarf mass estimates.

The X-ray studies of RS Oph were just as important (Mason *et al.* 1986) as the radio and the same model was used to interpret both sets of data. RS Oph is a recurrent nova that has outbursts about every 18 years. The system contains both a white dwarf and a red giant that is losing mass both into the Roche lobe of the compact star and into the region surrounding the binary. Unlike the classical nova systems, in which there appears to be little gas surrounding the system, the gas ejected from the compact object in RS Oph must penetrate the circumbinary material which comes from the red giant wind. As the ejected gas collides with the circumbinary material, a shock forms and it will produce X-rays if the velocity of the expanding material is large enough (Brecher, Ingham, and Morrison 1977; Bode and Kahn 1985; O'Brien, Kahn, and Bode 1986). Mason *et al.* reported that their observations, made early in the outburst, were consistent with the predictions of X-ray emission from a blast wave penetrating circumbinary material. This explanation was supported by the observations of the broad widths of the optical emission lines that indicated expansion velocities for the ejected gas of $\sim$3000km s^{-1} (Bruch 1986). In addition, the optical spectra showed coronal line emission from [FeX] 6374Åand [Fe XIV] 5303Åwhich indicated temperatures of nearly a keV. Nevertheless, at late times the observed X-ray fluxes fell more rapidly than predicted by Bode and Kahn (1985) and various causes for this deviation are discussed by Mason *et al.* (1986). Their last X-ray data point for RS Oph (Mason *et al.* 1986) was obtained more than 200 days after the optical outburst. It showed that, at that time, a source still existed in RS Oph, and had a temperature $\sim$3$\times$10^{5}K. This is strong evidence for the existence of a white dwarf in this system and this temperature is remarkably close to the values that *EXOSAT* found for the three *classical novae* during their outbursts. However, the luminosity of RS Oph was less than expected for a massive white dwarf radiating at the Eddington limit and suggests that the white dwarf had finally ejected all of the accreted layers and was now cooling back to quiescence. This determination of a cooling time scale for the white dwarf provides a strong constraint for further calculations of thermonuclear runaways on massive white dwarfs.

One wavelength interval yet to provide anything but upper limits is the γ-ray. Although we have determined upper limits to the predicted emission in this interval (Truran, Starrfield, and Sparks 1978), no positive detection has resulted (Leising and Clayton 1987). It is appropriate to note, however, that more sensitive detectors and emission from the oxygen, neon, magnesium outbursts may make this detection more probable.

5.   Summary and Discussion

The nova outburst is an exciting event that requires, if not demands, that observations be made in all wavelength intervals. An important point is that studies of the nova outburst have broad implications for studies of stellar evolution, stellar atmospheres, the interstellar medium, galactic nucleosynthesis, and the extragalactic distance scale. It is also clear that every wavelength interval has contributed important information to our understanding of the nova outburst. Because of the rapid variations in the central source and the expanding material, it is even better when observations can be made either simultaneously or close together in time. For example, simultaneous measurements of HeII 4686Åand 1640Åwill provide an estimate of the reddening while measurements of the X-ray flux and the optical emission lines can tell us the characteristics of the expanding nebula and provide information about the turn off time scale.

It is important to begin observing a nova as early in the outburst as possible. A combination of observations at all wavelengths can provide a rapid estimate of the distance to the nova, the extinction, and the characteristics of the expanding photosphere. When recent studies of theoretical expanding, spherical, stellar atmospheres are applied to early spectra of a nova, it will become possible to determine elemental abundances by a completely independent method from nebular analyses.

During the early decline phase of the outburst it is important to monitor the ultraviolet, optical, and infrared fluxes and determine both the energetics and the rate of mass loss which, together, control the radiation characteristics (Gallagher and Starrfield 1978; Gehrz 1988; LMU). The existence of continuing mass loss from the hot white dwarf remnant is accessible through observations of the ultraviolet resonance lines associated with ions normally seen in the Orion spectrum of novae (e.g., NIII, CIII, CIV, NV, Si IV etc.). The presence of these lines has allowed direct estimates of mass loss rates during the early stages of the outburst.

At late times in the outburst, the X-ray results have shown that the hot white dwarf is still radiating at $\sim L_{Ed}$ and its temperature is $\sim 3 \times 10^5$K. They also imply that the white dwarf in Nova Muscae 1983 was beginning to cool back to quiescence about three years after the initial explosion. However, recent optical spectra imply that the nebula is still being photoionized by a source that is hotter than 3×10^5K and quite luminous (Krautter and Williams 1988, in preparation). If true, then the nova has not yet turned off and we must find some other way to explain the X-ray data. Nevertheless, the X-ray temperature and luminosity should be useful in the construction of time dependant nebular models that can predict the optical behavior of a nova as it is turning off.

The infrared studies have proved to be very important in understanding how dust forms and evolves with time in an astrophysical environment. It is already known that grains form in novae as very small particles with temperatures of ~ 1000K. They then grow to large sizes and cool to temperatures of ~ 800K. The data show that all but two of the well studied novae have formed grains of amorphous carbon. The ejecta of the two anomalous novae were very oxygen rich and they probably formed grains that were either silicate rich (V1370 Aql) or consisted of the normal silicate material found in other astronomical sites (Nova Vul 1984

#2). It also appears that distance estimates, obtained from the infrared excess during the phase when the dust luminosity is increasing, are very accurate. Recent studies of ultraviolet and optical emission lines have shown that there are two classes of novae; one which occurs on carbon, oxygen white dwarfs and one that occurs on oxygen, neon, magnesium white dwarfs. Because the stellar evolution processes that produce the two types of white dwarfs are quite different, the implications concerning the evolution of binary stars are quite profound and discussed in Starrfield, Sparks, and Truran (1986).

The hydrodynamic calculations of thermonuclear runaways in the accreted hydrogen-rich envelopes of white dwarfs, have been very successful in reproducing the gross features of the nova outburst: ejected masses, kinetic energies, and light curves. More important, these calculations predicted: (1) that enhanced CNO nuclei would be found in the ejecta of fast novae, (2) that the isotopic ratios of the CNO nuclei would be far from solar, (3) that there should be a post maximum phase of constant luminosity lasting for months, or longer, and (4) that the properties of the outburst should be strong functions of the mass of the white dwarf. As discussed in Starrfield (1987, 1988; and in this review), observational confirmation of each of these points has now appeared in the literature. The theoretical studies of the nova outburst have also identified the factors that strongly influence the characteristics of the outburst. These are: (1) the white dwarf mass, (2) the envelope mass, (3) the white dwarf luminosity, (4) the rate of mass accretion, and (5) the chemical composition of the envelope.

The prediction that the ejecta of novae would be enhanced in CNO nuclei has been confirmed by many observational studies at all wavelengths (Sneden and Lambert 1975; Williams *et al.* 1978, 1981, 1985; Williams and Gallagher 1979; Ferland and Shields 1978; Tylenda 1978; Gallagher *et al.* 1980; Stickland *et al.* 1981; Snijders *et al.* 1984, 1987). In addition, Sneden and Lambert (1975) determined that the $^{12}C/^{13}C$ isotopic ratio (in combination with the extreme overabundance of carbon) in DQ Her strongly supported a thermonuclear runaway as the cause of the outburst.

I am grateful to the many people who have contributed to this review: F. Córdova, G. Ferland, J. Gallagher, R. Gehrz, R. Hjellming, S. Kenyon, J. Krautter, J. Liebert, E. Ney, G. Shaviv, E. Sion, G. Sonneborn, W. Sparks, L. Stryker, J. Truran, R. Wade, R. M. Wagner, R. Wehrse, R. Williams, and C.-C. Wu. I would like to thank R. Gehrz for sending me his Annual Review chapter ahead of publication and F. Córdova, R. Hjellming, W. Sparks, and S(usan) Starrfield for their comments on and additions to an earlier version of this manuscript. I am also grateful to Dr.'s George Bell, Arthur Cox, Stirling Colgate, Michael Henderson, and Jay Norman for the hospitality of the Los Alamos National Laboratory and a generous allotment of computer time. Some of the material presented in this review was obtained through the facilities of the Boulder RDAF which is supported by NASA grant NAS5-28731 to the University of Colorado. Terry Armitage's help is gratefully acknowledged. I would also like to acknowledge continuing support from the National

Science Foundation through grant AST85-16173 to Arizona State University, from NASA through grant NAG5-41 to Arizona State University, and from the DOE.

References

Andrillat, Y. 1983, *M.N.R.A.S.*, **203**, 5p.

Arnould, M., Norgaard, H., Thielemann, F.-K. and Hillebrandt, W. 1980, *Ap. J.*, **237**, 1016.

Audouze, J., Lazareff, B., Sparks, W.M. and Starrfield, S. 1979, in *The Elements and Their Isotopes in the Universe*, ed. M. Gabriel (Liege: University of Liege), p. 53.

Barlow, M.J., Brodie, J.P, Brunt, C.C., Hanes, D.A., Hill, P.W., Mayo, S.K., Pringle, J.E., Ward, M.J., Watson, M.G., Whelan, J.A.J., and Willis, A. J. 1981, *M.N.R.A.S.*, **195**, 61.

Bath, G.T. 1978, *M.N.R.A.S.*, **182**, 35.

Boyarchuk, A.A., Galkina, T.S., Krasnobabtsev, V.I., Rachkovskaya, T.M., and Shakhovskaya, N.I. 1977, *Sov. A. J.*, **21**, 257.

Becker, R.H., and Marshall, F.E. 1981, *Ap. J. (Letters)*, **244**, L93.

Bianchini, A. and Sabbadin, F. 1983, *Astr. Ap.*, **125**, 112.

Bode, M. F. 1986, *RS Oph(1985) and the Recurrent Nova Phenomenon* (Utrecht: VNU Science Press).

Bode, M. F., and Evans, A. N. 1980, *M.N.R.A.S.*, **193**, 21.

Bode, M. F., and Evans, A. N. 1981, *M.N.R.A.S.*, **197**, 1055.

Bode, M. F., and Evans, A. N. 1988, *The Classical Nova* (New York: Wiley), in press.

Bode, M. F., Evans, A.N., Whittet, D.C.B., Aitken, D.K., Roche, P.F., and Whitmore, B. 1984, *M.N.R.A.S.*, **207**, 897.

Bode, M. F., and Kahn, F. D. 1985, *M.N.R.A.S.*, **217**, 205.

Brecher, K., Ingham, W. H., and Morrison, P. 1977, *Ap. J.*, **213**, 492.

Brosch, N. 1982, *Astr. Ap.*, **107**, 300.

Bruch, A. 1986, in *RS Oph(1985) and the Recurrent Nova Phenomenon* (Utrecht: VNU Science Press), p 13.

Callus, C. M., Evans, A., Albinson, J. S., Mitchell, R.M., Bode, M.F., Jameson, R.F., King, A.R., and Sherrington, M. 1987, *M.N.R.A.S.*, **229**, 539.

Cassatella, A., Benvenuti, P., Clavel, J., Heck, A., Penston, M., Selvelli, P.L., and Macchetto, F. 1979, *Astr. Ap.*, **74**, L18.

Cassatella, A., Patriarchi, P., Selvelli, P. L., Bianchi, L., Cacciari, C., Heck, A., Perryman, M.,and Wamsteker, W. 1982, in *Third European IUE Conference*, ed. E. Rolfe, A. Heck, and B. Battrick, ESA SP-176 (E. S. A., Noordwijk, The Netherlands), p. 229.

Chevalier, R.A. 1977, *Astr. Ap.*, **59**, 289.

Cohen, J. G. 1985, *Ap. J.*, **292**, 90.

Cohen, J. G., and Rosenthal, A. J. 1983, *Ap. J.*, **268**, 689.

Córdova, F.A. and Mason, K.O. 1983, in *Accretion Driven Stellar X-ray Sources*, ed. W.H.G. Lewin and E.P.J. van den Heuvel (Cambridge: Cambridge University Press).

Dinerstein, H. L. 1986, *A. J.*, **92**, 1381.

Drechsel, H., Kondo, Y., and Rahe, J. 1987, *Cataclysmic Variables. Recent Multifrequency Observations and Theoretical Developments*, (Dordrecht: Reidel).

Duerbeck, H. W., Klare, G., Krautter, J., Wolf, B., Seitter, W. C., and Wargau, W. 1980, in *Proceedings of the Second European IUE Conference*, ESA SP 157, p. 91.

Ennis, D., Becklin, E.E., Beckwith, S., Elias, J., Gately, I., Matthews, K., Neugebauer, G., and Willner, S.P. 1977, *Ap. J.*, **214**, 33.

Ferland, G. J. 1977, *Ap. J.*, **215**, 873.

Ferland, G.J., Lambert, D.L., McCall, M.L., Shields, G.A., and Slovak, M.H. 1982, *Ap. J.*, **260**, 794.

Ferland, G.J., Lambert, D.L., and Woodman, J. H. 1986a, *Ap. J.* Supp., **60**, 375.

Ferland, G.J., Lambert, D.L., and Woodman, J. H. 1986b, *Ap. J.* Supp., **62**, 939.

Ferland, G.J. and Shields, G.A. 1978a, *Ap. J.*, **226**, 172.

Ferland. G.J., and Shields, G.A. 1978b, *Ap. J. (Letters)*, **224**, L15.

Ferland, G.J., Williams, R.E., Lambert, D.L., Shields, G.H., Slovak, M., Gondhalaker, P. M., and Truran, J. W. 1984, *Ap. J.*, **281**, 194.

Gallagher, J. S. 1977, *A. J.*, **82**, 609.

Gallagher, J. S. 1978, *Ap. J.*, **221**, 211.

Gallagher, J. S., and Code, A. F. 1974, *Ap. J.*, **189**, 303.

Gallagher, J.S., and Holm, A.V. 1974, *Ap. J. (Letters)*, **189**, L23.

Gallagher, J.S., and Ney, E.P. 1976, *Ap. J. (Letters)*, **204**, L35.

Gallagher, J.S., Hege, E.K., Kopriva, D.A., Williams, R.E., and Butcher, H.A. 1980, *Ap. J.*, **237**, 55.

Gallagher, J.S. and Starrfield, S.G. 1976, *M.N.R.A.S.*, **176**, 53.

Gallagher, J.S. and Starrfield, S.G. 1978, *Ann. Rev. Astr. and Ap.*, **16**, 171.

Gehrz, R. D. 1988, *Ann. Rev. Astr. and Ap.*, in press(**G88**).

Gehrz, R.D., Grasdalen, G.L., Greenhouse, M., Hackwell, J.A., Hayward,T., and Bentley, A.F. 1986, *Ap. J. (Letters)*, **308**, L63.

Gehrz, R.D., Hackwell, J.A., Grasdalen, G.L., Ney, E.P., Neugebauer, G., and Sellgran, D. 1980b, *Ap. J.*, **239**, 570.

Gehrz, R.D., Grasdalen, G.L., and Hackwell, J.A. 1985, *Ap. J. (Letters)*, **298**, L163.

Gehrz, R.D., Grasdalen, G.L., and Hackwell, J.A., and Ney, E. P. 1980a, *Ap. J.*, **237**, 855.

Gehrz, R.D., Hackwell, J. A., and Jones, T. W. 1974, *Ap. J.*, **191**, 675.

Gehrz, R.D., Ney, E.P., Grasdalen, G.L., Hackwell, J.A., and Thronson, H.A. 1984, *Ap. J.*, **281**, 303.

Gehrz, R. D., Ney, E. P., Grasdalen, G. L., Hackwell, J. A., and Thronson, H. A. 1984, *Ap. J.*, **281**, 303.

Gehrz, R. D., Grasdalen, G. L., and Hackwell, J. A. 1985, *Ap. J. (Letters)*, **298**, L47.

Geisel, S.L., Kleinmann, D.E., and Low, F.J. 1970, *Ap. J. (Letters)*, **161**, L101.

Glasstone, S., and Dolan, P.J. 1977, *The Effects of Nuclear Weapons, Third Edition*, (Washington: U.S. Department of Energy).

Greenhouse, M. A., Grasdalen, G. L., Hayward, T. L., Gehrz, R.D., Jones, T.J. 1988, *A. J.*, in press.

Hanes, D.A. 1985, *M.N.R.A.S.*, **213**, 443.

Hillebrandt, W. and Thielemann, F.-K. 1982, *Ap. J.*, **255**, 617.

Hjellming, R. M., van Gorkom, J. H., Taylor, A. R., Seaquist, E. R., Padin, S., Davis, R. J., and Bode, M. F. 1986, *Ap. J. (Letters)*, **307**, L71.

Hjellming, R. M., and Wade, C.M. 1970, *Ap. J. (Letters)*, **162**, L1.

Hjellming, R. M., Wade, C.M., Vandenberg, N.R., and Newell, R. T. 1979, *A. J.*, **84**, 1619.

Hyland, A. R., and Neugebauer, G. 1970, *Ap. J. (Letters)*, **160**, L177.

Jenkins, E.B., Snow, T.P., Upson, W.L., Starrfield, S.G., Gallagher, J.S., Friedjung, M., Linskey, J.L., Anderson, R.L., Henry, R.C., and Moos, H.W. 1977, *Ap. J.*, **212**, 198.

Kenyon, S. J., and Berriman, G. 1987, preprint.

Kenyon, S. J., and Garcia, M. R. 1986, *A. J.*, **91**, 125.

Kenyon, S. J., and Wade, R. 1986, *Pub. A.S.P*, **98**, 935.

Kenyon, S. J., and Webbink, R. F. 1984, *Ap. J.*, **279**, 252.

Kirshner, R.P., Sonneborn G., Crenshaw, D.M., and Nassiopoulos, G.E. 1987, *Ap. J.*, **320**, 602.

Krautter, J., Beuermann, K., Leitherer, C., Oliva, E. Moorwood, A.F.M., Deul, E., Wargau, W., Klare, G., Kohoutek, L, van Paradijs, J., and Wolf, B. 1984, *Astr. Ap.*, **137**, 307.

Lance, C.M., McCall, M.M., and Uomoto A. K. 1988, *Ap. J. Supp.*, in press (**LMU**).

Lazareff, B., Audouze, J., Starrfield, S., and Truran, J. W. 1979, *Ap. J.*, **288**, 875.

Leising, M. D., and Clayton, D. D. 1987, *Ap. J.*, **323**, 159.

Livio, M., Truran, J. W., and Webbink, R. 1986, *Ap. J.*, **308**, 736.

MacDonald, J. 1980, *M.N.R.A.S.*, **191**, 933.

MacDonald, J., Fujimoto, M. Y., and Truran J. W. 1985, *Ap. J.*, **294**, 263.

Martin, P.G. 1988, in *The Classical Nova*, ed. M. Bode and N. Evans, (Wiley, New York), in press.

Mason, K.O., Córdova, F. A., Bode, M. F., and Barr, P. 1986, in *RS Oph and the Recurrent Nova Phenomenon*, ed. M. F. Bode (VNU Press; Utrecht), p 167.

McLaughlin, D. B. 1936, *Ap. J.*, **84**, 104.

McLaughlin, D. B. 1949, *Pub. A.S.P*, **61**, 74.

McLaughlin, D. B. 1960, in *Stellar Atmospheres: Stars and Stellar Systems VI*, ed. J. S. Greenstein (Chicago :University of Chicago Press), p. 585.

Mitchell, R.M., and Evans, A. 1984, *M.N.R.A.S.*, **209**, 9.45.
Ney, E. and Hatfield, B. F. 1978, *Ap. J. (Letters)*, **219**, L111.
O'Brien, Kahn, F. D., and Bode, M. F. 1986, in *RS Oph and the Recurrent Nova Phenomenon*, ed. M. F. Bode (VNU Press; Utrecht), p 177.
Ögelman, H., Beuermann, K., and Krautter, J. 1984, *Ap. J. (Letters)*, **287**, L31.
Ögelman, H., Krautter, J., and Beuermann, K. 1987, *Astr. Ap.*, **177**, *110*.
Pacheco, J. A. de Freitas and Codina, S.J. 1985, *M.N.R.A.S.*, **214**, 481.
Padin, S., Davis, R.J., and Bode, M.F. 1985, *Nature*, **315**, 306.
Payne-Gaposchkin, C. 1957, *The Galactic Novae*, (New York: Dover).
Prialnik, D., Livio, M., Shaviv, G. and Kovetz, A. 1982, *Ap. J.*, **257**, 312.
Prialnik, D., Shara, M., and Shaviv, G. 1978, *Astr. Ap.*, **62**, 339.
Prialnik, D., Shara, M., and Shaviv, G. 1979, *Astr. Ap.*, **72**, 192.
Reynolds, S. P. and Chevalier, R. A. 1984, *Ap. J. (Letters)*, **281**, L33.
Rosino, L., Iijima, T., and Ortolani, S. 1983, *M.N.R.A.S.*, **205**, 1069.
Rosino, L., Bianchini, A., and Rafanelli, P. 1982, *Astr. Ap.*, **108**, 243.
Seaquist, E.R. and Palimaka, J. 1977, *Ap. J.*, **217**, 781.
Sekiguchi, K., Feast, M. W., Whitelock, P. A., Overbeek, M. D., Wargau, W., and Spencer-Jones, J. 1988, *M.N.R.A.S.*, in press.
Shara, M. M., and Moffatt, A. F. J. 1982, *Ap. J. (Letters)*, **258**, L41.
Shara, M. M., and Moffatt, A. F. J. 1983, *Ap. J.*, **264**, 560.
Shaviv, G., and Starrfield, S. 1987, *Ap. J. Lett*, **321**, L51.
Shaviv, G., and Starrfield, S. 1988, *Ap. J.*, submitted.
Sion, E. M., Starrfield, S., Van Steenberg, M. E., Sparks, W.M., Truran, J. W., and Williams, R. E. 1986, *A. J.*, **92**, 1145.
Sion, E. M., and Starrfield, S. 1986, *Ap. J.*, **303**, 130.
Smak, J. 1980, *Acta Astron.*, **30**, 267.
Sneden, C. and Lambert, D. L. 1975, *M.N.R.A.S.*, **170**, 533.
Snijders, M. A. J. 1986, in *RS Oph(1985) and the Recurrent Nova Phenomenon*, ed. M. Bode (Utrecht: VNU press), p. 51.
Snijders, M. A. J., Batt, T.J., Seaton, M.J., Blades, J.C., and Morton, D. C. 1984, *M.N.R.A.S.*, **211**, 7p.
Snijders, M.A.J., Batt, T.J., Roche, P. F., Seaton, M. J., Morton, D. C., Spoelstra, T.A.T., and Blades, J. C. 1987, *M.N.R.A.S.*, **228**, 329.
Sparks, W. M., Starrfield, S., and Truran, J. W. 1978, *Ap. J.*, **220**, 1063.
Sparks, W. M., Starrfield, S., and Truran, J. W. 1986, in *RS Oph(1985) and the Recurrent Nova Phenomenon*, ed. M. F. Bode (Utrecht: VNU Press), p. 39.
Sparks, W.M., Starrfield, S., and Truran, J. W. 1988, in *Atmospheric Phenomena*, ed. K. Nomoto, in press.
Sparks, W.M., Starrfield, S., Williams, R.E., Truran, J.W., and Ney, E.P. 1982, in *Advances in Ultraviolet Astronomy: Four Years of IUE Research*, ed. Y. Kondo, J.M. Mead, R.D. Chapman (NASA Conference Publication 2238), p. 478.
Sparks, W. M., Wu. C.-C., Holm, A., and Schiffer, F. H. 1980, *Highlights of Astronomy*, **5**, 105.
Starrfield, S. 1979, in *White Dwarfs and Variable Degenerate Stars*, ed. H.M. Van Horn and V. Weidemann (Rochester: University of Rochester), p. 274.
Starrfield, S. 1980, in *Stellar Hydrodynamics*, ed. A.N. Cox, and D.S. King, *Space Science Reviews*, **27**, 635.
Starrfield, S. 1986, in *Radiation Hydrodynamics*, ed. D. Mihalas, and K.-H. Winkler, (Dordrecht: Reidel), p. 225.
Starrfield, S. 1987, in *New Insights in Astrophysics: Eight Years of Ultraviolet Astronomy with IUE*, ed. E. Rolfe, ESA press, (Nordwijk) SP-263, p. 239.
Starrfield, S. 1988, in *The Classical Nova*, ed. N. Evans, and M. Bode, (New York: Wiley) in press.
Starrfield, S. and Snijders, M. A. J. 1987, in *Scientific Accomplishments of the IUE Satellite*, ed. Y. Kondo (Dordrecht: Reidel), p. 377.
Starrfield, S., and Sparks, W. M. 1987, in *Cataclysmic Variables. Recent Multifrequency Observations and Theoretical Developments*, ed. H. Drechsel, Y. Kondo, and J. Rahe, (Dordrecht: Reidel), p. 379.

Starrfield, S., Sparks, W.M., and Shaviv, G. 1988, *Ap. J. (Letters)*, in press.
Starrfield, S., Sparks, W. M. and Truran, J. W. 1974a, *Ap. J. Suppl.*, **28**, 247.
Starrfield, S., Sparks, W. M. and Truran, J. W. 1974b, *Ap. J.*, **192**, 647.
Starrfield, S., Sparks, W. M., and Truran, J. W. 1985, *Ap. J.*, **291**, 136.
Starrfield, S., Sparks, W. M., and Truran, J. W. 1986, *Ap. J. (Letters)*, **303**, L5.
Starrfield, S., Truran, J. W. and Sparks, W. M. 1978, *Ap. J.*, **226**, 186.
Starrfield, S., Truran, J.W., Sparks, W.M. and Arnould, M. 1978, *Ap. J.*, **222**, 600.
Starrfield, S., Truran, J.W., Sparks, W.M. and Kutter, G. S. 1972, *Ap. J.*, **176**, 169.
Stickland, D.J., Penn, C.J., Seaton, M.J., Snijders, M.A.J., and Storey, P. J. 1981, *M.N.R.A.S.*, **197**, 107.
Taylor, A. R., Seaquist, E. R., Hollis, J. M., and Pottasch, S. R. 1987, *Astr. Ap.*, **183**, 38.
Truran, J. W. 1982, in *Essays in Nuclear Astrophysics*, ed. C.A. Barnes, D.D. Clayton, and D. Schramm (Cambridge: Cambridge U. Press), p. 467.
Truran, J. W., Livio, M., Hayes, J., Starrfield, S., and Sparks, W. M. 1988, *Ap. J.*, in press.
Truran, J. W., Starrfield, S., and Sparks, W. M. 1978, in *Gamma Ray Spectroscopy in Astrophysics*, ed. T. L. Cline and R. Ramaty, (NASA Technical Memorandum 79619), p. 315.
Tylenda, R. 1978, *Acta Astr.*, **28**, 333.
Wiescher, M., Gorres, J., and Thielemann, F.-K., and Ritter, H. 1986, *Astr. Ap.*, **160**, 56.
Williams, P. M., and Longmore, A. J. 1984, *M.N.R.A.S.*, **207**, 139.
Williams, R.E. and Gallagher, J. S. 1979, *Ap. J.*, **228**, 482.
Williams, R.E., Ney, E.P., Sparks, W.M., Starrfield, S., and Truran, J. W. 1985, *M.N.R.A.S.*, **212**, 753.
Williams, R.E., Sparks, W.M., Gallagher, J.S., Ney, E.P., Starrfield, S., and Truran, J. W. 1981, *Ap. J.*, **251**, 221.
Williams, R. E., Woolf, N. J., Hege, E. K., Moore, R. L., and Kopriva, D. A. 1978, *Ap. J.*, **224**, 171.
Wu, C.-C. and Kester, D. 1977, *Astr. Ap.*, **58**, 331.
Young, P. and Schneider, D. P. 1980, *Ap. J.*, **238**, 955.

X-Ray and γ-Ray Bursts

D. Hartmann and S. E. Woosley

Board of Studies in Astronomy and Astrophysics
University of California at Santa Cruz

1. Introduction

The long wavelength emission known to accompany X-ray bursts and believed to accompany γ-ray bursts is thought to be due to reprocessing of hard photons originating from a neutron star. Therefore, simultaneous observations extending from radio to γ-ray wavelengths allow us to probe the geometry and the physical conditions of plasma in the neutron star vicinity, be it the magnetosphere, a companion star, or an accretion disk.

Only a few years after the discovery of X-ray bursts (Grindlay *et al.* 1976; Belian *et al.* 1976) simultaneous emission in the optical and X-ray wavebands had been detected from these sources (Grindlay *et al.* 1978; McClintock *et al.* 1979; Hackwell *et al.* 1979). In the case of γ-ray burst sources (Klebesadel, Strong, and Olsen 1973) we have yet to be so fortunate, though indirect evidence exists from historic plate collections that these sources may also emit simultaneously at optical wavelengths (Schaefer 1981; Schaefer *et al.* 1984a; Moskalenko *et al.* 1988). Indeed, the case may be made that the reason that γ-ray burst sources remain so mysterious is to a large extent a consequence of the unknown multi-wavelength ($\lambda\lambda\lambda$ hereafter) properties of these transients. Imagine the state in which the study of advanced stages of stellar evolution and supernovae would find itself were our entire data base limited to the neutrinos from the explosion ($\sim 99\%$ of the energy).

We begin this chapter with a discussion of results obtained from previous $\lambda\lambda\lambda$ studies of X-ray bursts (§2) and γ-ray bursts (§3). Simultaneous (during outburst) and non-simultaneous observations are briefly summarized for both types of transient sources. Theoretical models are reviewed in §4 and in §5 we discuss some of the $\lambda\lambda\lambda$ observing programs and facilities presently active or in the planning stage. Throughout we stress, as have many others, the importance of coordinated simultaneous $\lambda\lambda\lambda$ observations for our understanding of the nature and underlying physics of X-ray and γ-ray bursters. When studying any kind of astronomical object, it is advantageous to avoid "wavelength chauvinism" but for transients it is absolutely essential to study the event at all wavelengths.

2. X-ray Bursts

X-ray burst sources are recognized as a subgroup of the high luminosity Population II X-ray sources (comprised of galactic bulge and globular cluster sources) which are believed to be compact binary systems containing weakly magnetic neutron stars accreting from low mass companion stars that are filling their Roche lobe. X-ray burster sources do not pulse and in contrast to X-ray pulsars, which are known to be strongly magnetized neutron stars in a binary system with companion stars more massive than ~ 1 $M_\odot$(with the exception of 4U 1626-67; Levine *et al.* 1987), the faintness of the optical counterparts of most low mass X-ray binaries rules out giants, supergiants, and early-type main sequence stellar companions. Typically one has companion masses less than 1 $M_\odot$. The bulk of the light detected from these systems results from reprocessing of X-rays in an accretion disk. Therefore, the $\lambda\lambda\lambda$ analysis of these systems, both in the quiescent state and during outburst, can give valuable insights into the geometry and the physical properties of accretion flows around compact objects. In the following sections we present the basic properties of X-ray bursts, summarize the status of non-simultaneous $\lambda\lambda\lambda$ analysis of optically identified systems, and discuss simultaneous $\lambda\lambda\lambda$ observations.

Salient features

The observed properties of X-ray bursts have been extensively reviewed (Lewin and Joss 1981, 1983; van Paradijs 1983; Joss and Rappaport 1984; Lewin 1985; Matsuoka 1985; Hayakawa 1985; Melia and Joss 1986; Melia 1987). Here we briefly summarize the observational and theoretical status from the $\lambda\lambda\lambda$ point of view. Figure 1 shows the distribution of burst sources on the celestial sphere. The strong concentration towards the galactic center and the fact that many of them (about 30%) are found in globular clusters indicates that they are associated with Population II. In fact, X-ray bursters in globular clusters are overabundant compared to the galactic bulge. Located at distances of order of 10 kpc, most X-ray burst sources have a detectable persistent X-ray luminosity. During outburst the X-ray flux typically rises by about one order of magnitude to peak values $F_x \sim 5 \times 10^{-8}$ erg cm^{-2} s^{-1} on time scales between 0.1 and 10 s. However, it is not clear if these X-ray fluxes imply that X-ray bursts emit at super-Eddington rates. Fitted blackbody spectral temperatures at burst maximum in excess of the theoretical Eddington limit, $T_{Edd} \sim 2$ keV, and reasonably accurate distance determinations to some globular cluster sources leading to luminosities up to an order of magnitude larger than the Eddington limit provide evidence for super-Eddington fluxes. However, calculations of burst spectra that take into account the hardening due to electron scattering should (Woosley 1982) and in fact do indicate (van Paradijs 1982; Czerny and Sztajno 1983; Ebisuzaki, Hanawa, and Sugimoto

1983, 1984; Hoshi 1984; London, Taam, and Howard 1984, 1986; Ebisuzaki and Nomoto 1985, 1986; Ebisuzaki 1987; Babul and Paczynski 1987) that spectral (color) temperatures can exceed effective temperatures by as much as 50%. Insofar as any inferred "super-Eddington" fluxes are based upon spectra alone, there is no problem. There does exist a problem at a more fundamental level, however, when the bolometric flux plus a reasonable estimate of the source distance implies super-Eddington luminosity. For example, bursts from the source XB 1905+00 (e.g., Chevalier and Ilovaisky 1987) exceed L_{Edd} by a factor of about 6 if the distance estimate of D $\sim$ 25 kpc, based on an absolute magnitude $M_v \sim +2$ of its optical counterpart, is correct. If in fact the luminosity exceeds L_{Edd}, no hydrostatic solution for the atmosphere exists and mass loss is driven. Models of radiation-driven winds that include general relativistic effects (e.g., Paczyński and Prószyński 1986) indicate that all the super-Eddington energy flux is used to gently blow off matter with a terminal velocity of 0.01 c, and that the radiative luminosity never exceeds the Eddington limit by more than one percent. However, even a magnetic field as weak as 10^8 gauss may strongly affect the wind through magnetic confinement ($a T_{Edd}^4 \sim B^2/8\pi$) of the emitting plasma, allowing it to radiate at super-Eddington rates. If the luminosity exceeds L_{Edd} by less than a few percent the above mentioned observations of bursts from XB 1905+00 yield a distance of $\sim$ 10 kpc implying $M_v \sim +4$, which is much fainter than the value 1.2 ± 1.1 suggested by van Paradijs (1981, 1983) as an average value for low-mass X-ray binaries (see the chapter by K. Mason).

The decay back to the persistent flux following an X-ray burst occurs on time scales of order of 1 to 100 s (varies significantly with energy in the sense that the decay proceeds more rapidly at higher energies). Bursts of unusually long duration ($\gg$100 s) have been seen from GX 17+2 (Tawara *et al.* 1984b; Sztajno *et al.* 1986). Bursts with X-ray tails lasting much longer than 5 minutes have been reported from several sources (Swank *et al.* 1977; Lewin *et al.* 1984; Gottwald *et al.* 1987b; Czerny *et al.* 1987). Burst repetition may be either regular or erratic with average recurrence times between hours and days. However, several bursts with separations of only a few minutes have been observed (Lewin *et al.* 1976; Murakami *et al.* 1980b; Pedersen *et al.* 1982b; Inoue 1984; Matsuoka 1985; Gottwald *et al.* 1987a; Langmeier *et al.* 1987) and burst-inactive phases frequently extend up to several months. Many sources that show a strong variability in their recurrence behavior do *not* show a corresponding variability in the accretion rate as determined by the persistent X-ray flux, which is hard to understand within the context of the current interpretation of X-ray bursts as thermonuclear explosions of a critical amount of matter accreted from a low-mass companion star filling its Roche lobe (see § 4).

NGP

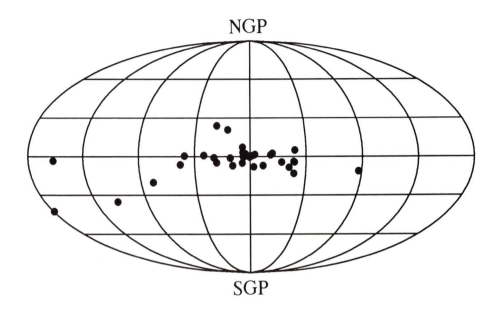

SGP

Figure 1 Sky map, in galactic coordinates, of X-ray burst sources for which reasonably accurate positions are known (Table 1). About 30% of the burst sources are located inside globular clusters.

It is debated whether or not bursts in and out of globular clusters have the same global properties (luminosity function, etc.) or if they constitute a distinct class (e. g., Grindlay 1985; Verbunt, van Paradijs, and Elson 1984). This question bears on the evolutionary origin of X-ray burst sources in and out of clusters. The relative overabundance of bursters inside clusters may be the result of effective tidal capture of neutron stars in the condensed cores of globular clusters (Fabian, Pringle, and Rees 1975). The field sources may then be survivors of cluster evaporation (see Goodman and Hut 1985 for a review of the dynamics of star clusters) or tidal disruption in the galactic disk (Grindlay 1984, 1985, 1986, 1987; but see van Paradijs and Lewin 1985).

Averaged over long periods of time one finds that the ratio of average persistent X-ray luminosity to the time-averaged burst luminosity is, with a few important exceptions, of order 10^2 (Lewin *et al.* 1977). Simultaneous optical/X-ray bursts with an energy ratio $E_{opt}/E_x = R_{ox} \sim 10^{-4}$ were observed from a few sources (discussed below in more detail). The X-ray spectra in outburst are approximately that of a blackbody having maximum temperature of order 2-3 keV and softening in the declining phase. Photospheric radii are estimated to be of order 10 km, consistent with theoretical estimates of neutron star radii (Swank *et al.* 1977; Hoffman, Lewin, and Doty 1977; van Paradijs and Lewin 1987). The detection of absorption lines at E ∼ 4 keV from MXB 1636-53 (Waki *et al.* 1984; Turner and Breedon 1984), if interpreted as gravitationally redshifted absorption lines from iron peak elements, lends further support to the neutron star association and has been used to constrain the neutron star mass-radius relationship (Waki *et al.* 1984; Fujimoto and Taam 1986). However, the observed line equivalent widths exceeding 100 eV and a possible confusion with detector intrinsic Xenon features argue against the reality of the 4 keV lines (Foster, Ross, and Fabian 1987). Emission line features at E ∼ 6.4 - 6.7 keV, observed in several burst sources by EXOSAT, are plausibly explained by Fe XXVI recombination in a hot $(10^{7-8}$ K) accretion disk corona created by X-ray illumination (White *et al.* 1986). The observed spectral softening in the decay phase of the burst is consistent with a cooling blackbody at approximately constant radius. However, when the burst is sufficiently strong photospheric radius expansion has been observed in form of the "classical" double peaked bursts (Lewin *et al.* 1976; Hoffman, Cominsky, and Lewin 1980; Basinska, Lewin, and Sztajno 1984) and the possibly related "bursts with precursor" (Hoffman *et al.* 1978; Tawara *et al.* 1984ab; Lewin, Vacca, and Basinska 1984). These bursts exhibit the double peak structure predominately at high energies but not in the bolometric flux. Recently observed bursts from MXB 1636-53 that do show double peaks (or even multiple peaks) in the bolometric flux (Sztajno *et al.* 1985; Pennix, van Paradijs, and Lewin 1987) do not exhibit substantial radius expansion. A subclassification of Type I bursts into *bright* bursts (with radius expansion) and *faint* bursts (without radius expansion) was suggested by Sugimoto, Ebisuzaki, and Hanawa (1984). Recent calculations by Ebisuzaki and Nakamura (1988) confirm the hypothesis that a hydrogen-rich envelope is ejected during a bright class burst and retained during a faint one. Using the fact that during a bright class burst the neutron star magnetic field apparently is not strong enough to confine the envelope, Ebisuzaki and Nakamura derive upper limits of about 3×10^{10} gauss for the magnetic field strength.

Because the highly eccentric orbit of EXOSAT allowed uninterrupted monitoring of X-ray sources over long periods of time, precise data on the relation between X-ray burst recurrence behavior and burst energetics were obtained for MXB 1636-53 (e.g., Lewin *et al.* 1987a). Aside from large deviations for individual bursts, a strong linear correlation was observed between time integrated burst flux (fluence) and time separation to the previous burst, confirming and extending earlier results of Hoffman, Lewin, and Doty (1977) and Basinska *et al.* (1984). A similar correlation was also observed in the optical (Pedersen *et al.* 1982b).

Most of the characteristic properties summarized above strictly apply only to Type I bursts, which are caused by unstable thermonuclear burning of accreted plasma on the neutron star surface (i.e., §4; Woosley and Taam 1976; Joss 1977;

Maraschi and Cavaliere 1977). A distinct class of X-ray bursts (Type II) is observed from MXB 1730-335 (rapid burster). The unique properties of the rapid burster, which exhibits both types of bursts (Hoffman, Marshall, and Lewin 1978), are discussed in detail by Joss and Rappaport (1984), Lewin and Joss (1983), and Lewin (1985).

$\lambda\lambda\lambda$ studies of X-ray burst sources in quiescence

Optical identifications of X-ray burst sources do not come easy. Because of their concentration towards the galactic center, severe extinction and reddening makes identification difficult. Furthermore, typical magnitudes are of order $m_V \sim 18$ (Table 1) so that crowding of faint sources in the field becomes a problem. None of the burst sources near the centers of globular clusters or the galactic center has yet been identified optically. The optical appearence is dominated by emission from X-ray heated gas in the accretion disk, outshining low-mass companion stars by many orders of magnitude. Generally no stellar absorption lines are observed. Direct spectral classification of companion stars was obtained for two burst sources that also exhibit soft X-ray transients indicating mid-K type main sequence companions (see below). VLA studies of GX 17+2 revealed a variable radio counterpart coincident with a G-type star that may be the triple companion of this compact X-ray binary system (Grindlay and Seaquist 1986). The binary nature of X-ray burst sources is further revealed by the presence of periodic dips in the persistent X-ray flux of some sources (Walter *et al.* 1982; White and Swank 1982) (see Figure 2 for time histories of several "X-ray dippers"), which are interpreted as increased absorption due to blobs of matter in the rim of the accretion disk (e.g., White and Mason 1985; White 1986; Courvosier *et al.* 1986; Parmar *et al.* 1986). Complete X-ray eclipses in X-ray burst sources are rare (Cominsky and Wood 1984; Parmar *et al.* 1986) and optical variability, if detected, is relatively weak (e.g., 0.15 magnitudes for V926, the optical counterpart of XB 1735-444: Corbet *et al.* 1986; van Amerongen, Pedersen, and van Paradijs 1987). This can be understood as a consequence of shielding of the companion star by a geometrically thick accretion disk (Milgrom 1978). Although Milgrom's model is probably correct, the presence of periodic optical/X-ray modulation and eclipses in a few sources (Table 1) tells us that the structure and geometry of X-ray burst systems is more complex. Scattering of X-rays in an extended accretion disk corona and large scale density inhomogeneiities can cause correlated optical/X-ray flux variations that recur periodically. Binary periods of X-ray bursters obtained from X-ray absorption dips and optical photometry appear to be of order of hours (van Paradijs 1983; White and Mason 1985; White 1986), but can be as short as 11 minutes (Stella, Priedhorsky, and White 1987). Several X-ray burst sources exhibit long-

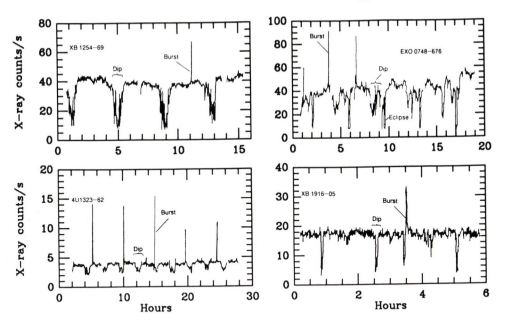

Figure 2 Light curves for X-ray burst sources that display periodic dips in the persistent X-ray flux ("X-ray dippers"). Inferred orbital periods of X-ray burst sources are of order of hours. (reproduced with the permission of White 1986 and Parmar *et al.* 1986).

term quasi-periodic cycles with periods between 0.5 and 2 years (Priedhorsky and Holt 1987).

A clear signature of many counterparts is their UV excess relative to normal stars. The most distant X-ray burst source identified this way is perhaps XB 1905+000 at D ∼ 20 kpc (Chevalier, Ilovaisky, and Charles 1985). The spectra are characterized by a blue continuum with few emission lines (predominantly N III λ4640 and He II λ4686) superimposed on it, and Balmer lines are generally absent (Canizares, McClintock, and Grindlay 1979; Bradt and McClintock 1983; van Paradijs 1983). From the theoretical interpretation of these spectra in terms of X-ray heated gas one concludes that the He II feature and the Balmer lines, if present, are produced by radiative recombination and that the N III λ4640 line blend is caused by the Bowen fluorescence mechanism (McClintock, Canizares, and Tarter 1975; Hatchet, Buff, and McCray 1976; see also Kallman and McCray 1982 for the theory of X-ray nebular models). This interpretation was independently confirmed by the detection of O III lines predicted by the Bowen model (Kallman and McCray 1980).

Some of the X-ray bursts originate from sources that are also known to be soft X-ray transients (SXT's). These X-ray transients often repeat on time scales of order of a few years, brighten by a factor of 10^2 to 10^6 within a few days, maintain the high state for weeks or even months and then decay within a few weeks. Persistent luminosities range from 10^{36} to 10^{38} erg s^{-1}. Spectral temperatures are of order of several keV. Their X-ray and optical properties have been reviewed by White, Kaluzienski, and Swank (1984) and Bradt and McClintock (1983) respectively. Although only few sources of this kind are known, there is good evidence that they are members of the low mass X-ray binary class. Theoretical explanations of soft X-ray transients invoke either accretion disk instabilities, such as the ones discussed in conjunction with dwarf novae (Canizzo, Ghosh, and Wheeler 1983; Lin and Taam 1984), or mass transfer instabilities in the secondary star due to X-ray illumination from the neutron star (Osaki 1985; Hameury, King, and Lasota 1986,1987,1988). During outburst the optical emission is dominated by the accretion disk, but during the decay phase spectral features revealing late type main sequence companion stars have been detected in several sources (Cen X-4: van Paradijs *et al.* 1980; Aql X-1: Charles *et al.* 1980; Thorstensen *et al.* 1986). Recent optical observations of the soft X-ray transient source EXO 0748-676 revealed eclipses in phase with the X-ray eclipses (Crampton *et al.* 1986). The observed accretion disk spectrum of EXO 0748-676 is similar in its general properties to those described above, but shows weak emission lines of H and He II which vary in velocity, intensity, and profile throughout the orbital cycle. Unfortunately, the companion star is too dim (V$\sim$ 23) for a detailed spectral classification.

Simultaneous $\lambda\lambda\lambda$ observations of X-ray bursts

The property of X-ray bursts that allowed simultaneous $\lambda\lambda\lambda$ observations soon after their discovery is their "reliability." With a typical recurrence time of less than a day, coordinated observing programs have a good chance of observing a burst. Most coordinated $\lambda\lambda\lambda$ monitoring programs have been carried out in the optical band and the X-rays. The first successful observations of this kind utilized the 1.5m CTIO telescope and the SAS 3 satellite and resulted in the detection of one simultaneous optical/X-ray burst from the source MXB 1735-44 (Grindlay *et al.* 1978; McClintock *et al.* 1979). Optical monitoring with the 2.3m WIRO photometer and X-ray observations of SAS 3 shortly thereafter resulted in a second detection of simultaneous $\lambda\lambda\lambda$ emission from MXB 1837+05 (Ser X-1) (Hackwell *et al.* 1979). With the advent of the Japanese satellite *Hakucho* a world wide coordinated burst watch was called for (Lewin, Cominsky, and Oda 1979). Optical coverage with the Danish 1.5m and the ESO 3.6m telescopes at LaSilla during 1979 and 1980 resulted in the detection of 10 additional optical/X-ray bursts from MXB 1636-53 (Pedersen *et al.* 1982a,b; Lawrence *et al.* 1983; Matsuoka *et al.* 1984). For one of these bursts multi-color (UBVR) data were obtained (Lawrence *et al.* 1983). The track of this burst in a color-color diagram was used to determine the temporal evolution of the temperature of the plasma responsible for the optical emission ($T_{max} \sim 7 \times 10^4$K). The distance to the source was determined to be of order of 5 kpc. More recently two additional optical/X-ray events from MXB 1636-53 were reported from simultaneous observations with EXOSAT, *Tenma*, and the ESO 3.6m telescope (Trümper *et al.* 1985; Turner *et al.* 1985). Figure 3 shows the optical/X-ray time histories of one of these events.

From the observations carried out to date the following characteristics of simultaneous optical/X-ray bursts are obtained: (i) Temporal burst profiles are similar in the optical and the X-ray band, but the optical emission profile is significantly wider ("smeared out"); (ii) The optical flux is delayed with respect to the X-rays by typically 1-4 seconds, although one burst does not appear to be delayed at all (Turner *et al.* 1985); (iii) The optical emission can't be explained by simple extrapolation of the X-ray spectrum to longer wavelengths; (iv) The ratio of optical to X-ray energy emitted during outburst is $R_{ox} \sim 10^{-4}$; (v) The optical emission is to first order characterized by a blackbody of maximum temperature $T \sim 5 \times 10^4$K; (vi) X-ray features lasting $\gtrsim$ 2s reappear in the optical band (Lawrence *et al.* 1983; Cominsky, London, and Klein 1987).

These characteristics are consistent with the picture of X-ray reprocessing in a geometrically thick accretion disk rather than the surface of the companion star, but the available data are not yet sufficient to determine the exact geometry and physical conditions of the reprocessing region. However, the interpretation of the data depends strongly on the assumptions made about the reprocessing physics. Because extreme variability in the illuminating X-ray flux occurs on short time scales the assumption of steady state reprocessing is in general not justified. Time dependent reprocessing has been considered by Chester (1978) and Alme and Wilson (1974) for the case of Her X-1, which has small amplitude 1.24 s pulsations. The general problem of large amplitude fluctuations together with short time scales typical for X-ray and γ-ray bursts has been addressed only recently (London and Cominsky 1983; London 1984; Cominsky, London, and Klein 1987). The crucial physical questions are the characteristic time scales and the spectral distribution of the reprocessed light. Geometrical effects and the microphysics of radiative reprocessing cause a delay and a smearing of the optical signal with respect to the X-ray burst. The important effects can be sorted out by considering several characteristic time scales (London 1984): (i) the burst duration ($\sim$1-100 s), (ii) the light travel time from the X-ray source to the reprocessing site of order 1-10 s for typical low-mass X-ray binaries, (iii) the geometrical smearing time due to the finite size of the reprocessing region, (iv) the hydrodynamic time scale measuring the importance of the effect of dynamic changes of the reprocessor due to the

Table 1: Locations and $\lambda\lambda\lambda$ data of X-ray burst sources

Source	l	b	Association	P_{orb} Hours	Magn.	Remarks[a]
0323+02	180.5	-42.4			V=16.4	UV
0512-40	244.4	-35.1	NGC 1851			gc
0614+09	200.8	-3.4	V1055 Ori		V=18.5	NIII
0748-68	280.3	-20.0		3.8	V=16-23	SXRT;D;E;HII;Hem;NIII
1254-69	303.5	-6.4		3.9	V=19.1	D;HeII;Hem;NIII
1323-62	306.9	+0.3		2.9		D
1455-31	332.2	+23.9	Cen X-4	8.2	V=13-19	SXRT;HII;Hem;K3-K7 V
1516-56	322.6	+0.9	Cir X-1			id. uncertain
1608-52	330.8	-0.8	QX Nor		V=18-20	SXRT
1636-53	332.8	-4.7	V801 Ara	3.8	V=17.5	OX;UV;M0 V
1659-29	353.9	+7.3		7.1	V=18.3	E;D;HeII;NIII
1702-42	343.8	-1.2				
1705-44	343.3	-2.3		1.3		
1708-23	0.6	+9.7				
1715-32	354.0	+3.1				IR
1724-30	356.2	+2.3	Terzan 2			gc;QPO
1728-33	354.2	-0.06	Grindlay 1			gc;IR
1730-33	354.8	-0.1	Liller 1			gc;QPO
1732-30	357.4	+1.1	Terzan 1			gc
1735-44	345.9	-7.0	V926Sco	4.65	V=17.5	OX;R;HeII;Hem;NIII
1742-29	359.9	-0.04	GCX-1			
1743-29	359.6	-0.4	GCX-2			
1743-28	0.8	+0.4	GCX-3			
1744-26	2.2	+1.0	GX3+1			QPO
1745-24	3.8	+1.7	Terzan 5			gc
1747-37	353.6	-5.0	NGC 6441			gc
1812-12	18.0	+2.4	Ser X-2		K=7.0	
1813-14	16.5	+1.3	GX17+2	1.4	V=17.5	QPO;G8III;R
1820-30	2.8	-7.8	NGC 6624	0.19		gc;QPO;R
1832-23	10.5	-6.9				
1837+04	36.1	+4.9	Ser X-1		V=19.2	OX;HeII
1850-08	25.3	-4.3	NGC 6712			gc
1906+00	35.0	-3.8			V=20.5	UV
1909+00	35.7	-4.2	Aql X-1	1.3d	V=15-20	OX;SXRT;HeII;NIII;G7-K3 V
1916-05	31.3	-8.4		0.83	R=22	D
1940-04	35.3	-13.2				
2143+38	87.4	-11.3		9.8d		

[a] remarks: Companion star spectral type is indicated if known; QPO = quasi periodic oscillation; OX = simultaneous optical/X-ray bursts; D = dips in the X-ray flux; E = eclipsing; UV = ultraviolet excess; R = associated radio source; SXRT = soft X-ray transient; gc = globular cluster source; GCX = galactic center X-ray source; Hem = H emission lines; He II = He II line at $\lambda4686$; N III = N III emission at $\lambda4640$.

X-ray illumination, and (v) the reprocessing time scale, t_{rep}, measuring the time for X-ray burst photons to be absorbed plus the time it takes for the thermalized

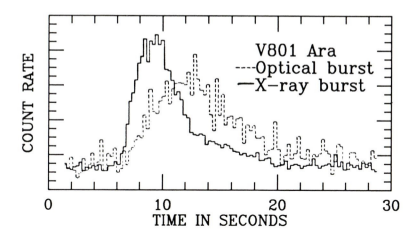

Figure 3 Counting rates of a simultaneous optical/X-ray burst from MXB 1636-53 (with permission from Trümper *et al.* 1985).

energy to re-emerge. Calculations of the reprocessed flux by Melia, Rappaport, and Joss (1986) that included hydrodynamical effects and by Cominsky, London, and Klein (1987), who neglected the dynamic response of the illuminated atmosphere, appear to be in agreement to within a factor of two. An accurate estimate of the reprocessing time t_{rep} is essential in order to use the observed delay between optical light and X-rays to determine the geometry of the system.

At photon energies less than about 30 keV the energy deposition is dominated by K-shell photoionization of metals and can be specified by a unique function of the average deposition column density, m_γ (London 1984), in which the penetration depth of soft X-rays is approximately proportional to the 3rd power of the photon energy (Figure 4). In contrast, illumination by γ-rays with energies above 30 keV leads to a roughly constant deposition column density of ~ 10 g cm^{-2}.

The absorption time for X-rays is generally much shorter than the burst duration and the deposited energy is thermalized by Coulomb collisions on an even shorter time scale. Thus, the reprocessing time t_{rep} is mainly determined by the radiation transfer of the deposited energy back to the surface. Pedersen *et al.* (1982a) estimated for the bursts from MXB 1636-53 that the reprocessing time was much less than the burst duration, which was recently replicated by fully implicit time-dependent calculations (Cominsky, London, and Klein 1987). Thus, the observed delays are mainly due to geometric effects. This is consistent with a disk size of order 7×10^{10}cm obtained from disk models with temperature $T \sim 28000$ K (from the continuum spectrum) combined with observed average optical brightness M$_V \sim 1.2 \pm 1.1$ for low-mass X-ray binaries (van Paradijs 1981, 1983).

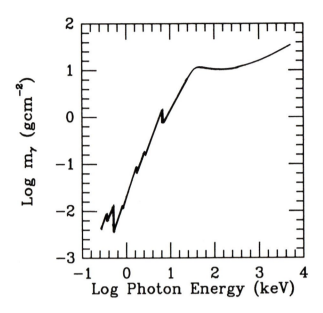

Figure 4 Deposition column density as a function of the incident photon energy (reproduced from London 1984).

As summarized by T. Kallman during the Taos workshop, contemporaneous data allow a test of the reprocessing model and a crude probing of reprocessor geometry and properties (not only for X-ray burst sources but for X-ray binaries in general; see the discussion by K. Mason elsewhere in this volume).

Recently, optical flares with durations of order of 10 s or 10 min apparently not caused by X-ray reprocessing in an accretion disk have been detected from MXB 1735-444 and GX17+2 (Imamura *et al.* 1987).

3. Multi-Waveband Analysis of γ-Ray Bursts

Gamma-ray bursts are high energy transient events (duration $\sim$ 0.1-10 s) that occur randomly on the celestial sphere. With three exceptions there is no clear evidence for more than one burst from a given source. Since they were first reported in the literature (Klebesadel, Strong, and Olsen 1973) little progress has been made towards a theoretical understanding of these enigmatic sources. Extreme variability and absorption and emission features in a number of γ-ray burst spectra as well as several properties of the unusual event that occurred on March 5, 1979 have led to the generally but not universally accepted view that γ-ray bursts originate from the vicinity of strongly magnetic neutron stars (Woosley 1982, 1984a; Lamb 1984), but many basic questions remain unanswered. For example, the distance to a typical source is not known. Though most researchers place them in the general vicinity of our own galaxy, some consider the possibility of their extragalactic origin (Atteia and Hurley 1986; Paczynsky 1986; Babul, Paczynski, and Spergel 1987; Goodman 1986).

Another question that has no satisfactory answer is whether or not the neutron star is a member of a binary system. Since their discovery it was realized that identification of these sources with objects emitting at other wavelengths would result in significant progress towards answering this question. Unfortunately to this date no object detected at any other wavelength has been identified unambiguously with a γ-ray burster. Gamma-ray burst counterparts at long wavelengths might be identified during their quiescent phases or during outburst. In the following sections we present some basic properties of γ-ray bursts, summarize the status of the non-simultaneous $\lambda\lambda\lambda$ analysis of those γ-ray burst source regions that have small enough error boxes to allow deep searches for quiescent counterparts, and discuss recent simultaneous $\lambda\lambda\lambda$ observations.

Salient γ-ray burst features

The γ-ray burst phenomenon has been extensively discussed in a series of conference proceedings (Lingenfelter, Hudson, and Worrall 1981; Vedrenne and Hurley 1983, Hurley and Vedrenne 1986; Woosley 1984a; Liang and Petrosian 1986) and will be reviewed in a forthcoming issue of the *Annual Reviews of Astronomy and Astrophysics* (Higdon and Lingenfelter 1988). Here we briefly summarize the observational status of γ-ray bursts from the $\lambda\lambda\lambda$ point of view. About 400 γ-ray bursts have been detected to date. The repetition time scale for the vast majority of burst sources appears to be longer than about 1-10 years (Atteia *et al.* 1987a; Schaefer and Cline 1985). Occurring, but not neccessarily being detected, at an approximate rate of $\lesssim 10^2$ bursts per year with fluences above 10^{-6} erg cm^{-2} (or perhaps several thousand with fluences exceeding 10^{-8} erg cm^{-2}; Bewick *et al.* 1975; Beurle *et al.* 1981; Meegan, Fishman, and Wilson 1985), γ-ray bursts appear to be the most common form of emission of observable galactic neutron stars (Hurley 1987b). For comparison consider the known ~ 100 bright galactic X-ray sources (Rappaport and Joss 1983), $\sim$35 X-ray bursters (Matsuoka 1985; see also Table 1), $\sim$20 X-ray pulsars (White 1987), and ~ 450 radio pulsars (Manchester 1987). Three exceptional sources with multiple recurrences may constitute a novel class of γ-ray bursts (see below). Figure 5 shows the isotropic galactic distribution of γ-ray bursts.

The bulk of the γ-ray burst energy is carried by photons of several 100 keV (Figure 6). Observations by the SMM satellite show that substantial emission above 10 MeV is a common feature of γ-ray bursts (Matz *et al.* 1985; Share

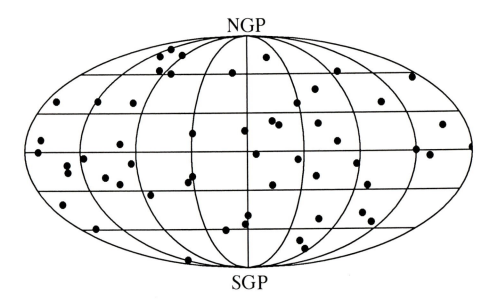

Figure 5 Distribution of γ-ray burst sources in galactic coordinates. Coordinates are taken from the catalog of Atteia *et al.* (1987a). The points plotted correspond to burst sources with well determined locations, sources with two alternative positions or that have only a known annulus are not shown.

et al. 1986). There is observational evidence for low energy ("optical") emission accompanying γ-ray bursts (see below). Such emission is expected on theoretical grounds if either the neutron star reprocesses the γ-rays in its magnetosphere (Hartmann, Woosley, and Arons 1987, 1988), the surface layers of a companion star (London and Cominsky 1983; Melia, Rappaport, and Joss 1986; Cominsky, London, and Klein 1987), or in an accretion disk (Epstein 1985; Michel 1985; Melia 1987b). Observed γ-ray fluences vary between 10^{-7} and 10^{-3} erg cm^{-2}, implying typical luminosities of order of 10^{39} D$^2_{kpc}$ erg s^{-1}. However, the distance scale for γ-ray bursters is not known to within a factor of 100 or more, which corresponds to an uncertainty of at least 10^4 in the energy requirements for the burst mechanism! A solution to this problem may await the yet to be determined identification of γ-ray bursts with another class of known objects.

Although the reality of spectral features at low energies E~ 50 keV (cyclotron resonance) and high energies E~ 450 keV (redshifted e$^\pm$ annihilation), observed in a fraction of bursts, is controversial, they lend support to the belief that γ-ray bursts originate on or near neutron stars. Whether or not the neutron stars that produce γ-ray bursts are members of binary systems, like the X-ray bursters, is not clear. Lacking substantial persistent X-ray flux and optical counterparts

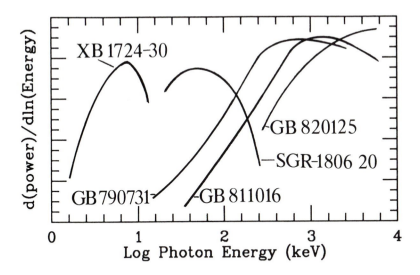

Figure 6 Emitted power per logarithmic bandwidth of photon energy of X-ray bursts and γ-ray bursts (adopted from Epstein 1986). Only about 2% of the total energy of γ-ray bursts falls in the X-ray band. The recently discovered repeating source SGR 1806-20 is intermediate between the two classes.

no compelling evidence for binary membership is readily available. Here again simultaneous λλλ observations may hold the key.

What the rapid burster (MXB 1730-335) is for the X-ray burst sources, the famous "March 5, 1979" event (GB790305b) is for γ-ray bursts (Mazets *et al.* 1979, 1980). Though this event is frequently exploited to support the paradigm of neutron star origin of γ-ray bursts, it is important to realize that GB790305b is unique in many respects (Cline 1980, 1982; Evans *et al.* 1980). The soft emission following the initial super-outburst of this source clearly revealed a periodicity of ∼ 8.1 s (Mazets *et al.* 1979; Cline *et al.* 1980) and a spectrum like that of X-ray pulsars. Despite many attempts to determine periodicities in the light curves of other γ-ray bursts only few sources were found to have statistically significant periodicities (GB771029: 4.2 s Wood *et al.* 1981; GB840805: 2.2 s (smooth) and 5 s (spiky) Kouveliotou *et al.* 1988). The fast rise time and large fluence of GB790305b allowed the determination of a precise position (box area ∼ 0.1 arcmin^2; Cline *et al.* 1982) which is apparently associated with the distant (∼ 55 kpc) supernova remnant N49 in the LMC.

The recently discovered repeating source SGR 1806-20 (Atteia *et al.* 1987b;

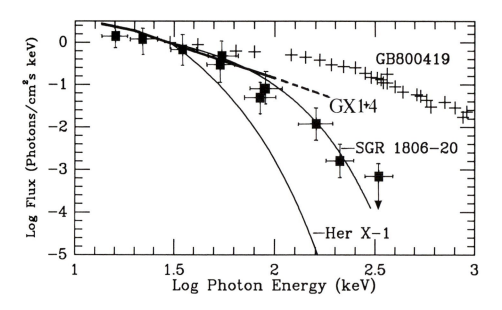

Figure 7 Average spectrum of the soft repeater SGR 1806-20 (filled squares) fitted by a simple exponential with kT = 30 keV (Hartmann, Woosley, and Arons 1987), compared to typical spectra from X-ray pulsars such as Her X-1 (Voges 1985) and GX 1+4 (White, Swank, and Holt 1983) and the spectrum of the "classical" γ-ray burst GB800419 (crosses) (Dennis *et al.* 1982; Nolan *et al.* 1984).

Laros *et al.* 1987; Kouveliotou *et al.* 1987) may be the third member of a novel class of γ-ray bursts, the "Soft Gamma Repeaters" (SGR). Members of this class (GB790305b = SGR 0526-66, GB790107 = SGR 1806-20, GB790324 = SGR 1900+14) share the following characteristics: (i) repetition, (ii) soft spectra (kT ∼ 30 keV), (iii) association with high-density regions (galactic bulge or LMC), (iv) short durations, and (v) simple time histories. The continuum spectrum of SGR 1806-20 shows a remarkable similarity with the spectra of X-ray pulsars (Figure 7). On other hand, the high energy component of "classical" γ-ray bursts is not present in SGR 1806-20. As discussed during the Taos workshop by Fenimore, reasonable fits to the spectrum can be obtained with a simple exponential F ∝ exp(-E/kT) with spectral temperature, T, of order of a few tens of keV, and a variety of other spectral forms, but not for a single temperature blackbody. Optically thin thermal bremsstrahlung fits including an energy dependent Gaunt factor give T∼ 40 keV (Atteia *et al.* 1987b).

Non-simultaneous λλλ studies of γ-ray bursts

To perform a meaningful λλλ search for quiescent γ-ray burst counterparts precise localizations are essential ("error boxes" of less than 1 arcmin^2). Burst arrival time analysis utilizing the detectors on the widely separated members of the interplanetary network (Cline *et al.* 1980) has provided a few precise localizations. We summarize in the following subsections the $\lambda\lambda\lambda$ searches performed in most major wavebands. Unfortunately, the result of all the efforts presented below is that no γ-ray burst source has yet been identified unambiguously with a source of quiescent radiation in any of these bands. The exception may be the much debated association of GB790305b with the LMC supernova remnant N49 (Cline 1980, 1982; Evans *et al.* 1980). The $\lambda\lambda\lambda$ content of γ-ray burst error boxes is summarized in Table 2.

A. Radio

Hjellming and Ewald (1981) used the VLA to study the quiescent flux from the GB781119 error box. They detected two sources in this field but no association was claimed. More recent VLA studies of several additional error boxes by Barat *et al.* (1984a) and Schaefer (1986) resulted in a few source detections, but again did not yield a convincing radio counterpart.

B. Infrared

No candidate for a quiescent γ-ray burst counterpart at wavelengths longer than a micron was detected in the search of a total of 32 error boxes (Apparao and Allen 1982; Schaefer and Ricker 1983; Schaefer 1986; Schaefer *et al.* 1987c). The observations involved several ground based facilities (KPNO, IRTF, CTIO) and the IRAS satellite. The most constraining results occurred in two cases where K-band magnitudes of possible candidates must be larger than $m_K \sim 19$ (Schaefer *et al.* 1987c). The authors conclude that these observations are in contradiction with expected IR fluxes from accretion disk models (Rappaport and Joss 1985), the neutron star white dwarf binary scenario of Tremaine and Zytkow (1986), and the thermonuclear model (Woosley and Wallace 1982; Hameury 1985) if the accreted matter stems from a companion star. Models that invoke a cold disk (Epstein 1985; Melia 1987b) predict IR fluxes much smaller than the obtained limits.

C. Optical

To date, only a few γ-ray burst sources have been localized to better than 20 arcmin^2 by γ-ray instruments alone. The total number of accurate localizations will not increase by much in the near future, because the Interplanetary Network of γ-ray detectors (Cline *et al.* 1980) no longer exists. The determination of new error boxes that allow deep CCD work to be performed is postponed until new $\lambda\lambda\lambda$ observatories, such as the HETE system (see the discussion by G Kriss), the Soviet experiment *Tournesol (Sunflower)* on GRANAT (Smith 1987 Mazets *et al.* 1987), or a rebuild Network (*e.g.*, consisting of γ–ray detectors or SIGMA/GRANAT, BATSE/GRO, and SMM near earth, and ULYSSES (Jupiter) PHOBOS (Mars), and PVO (Venus) in interplanetary space) are operative. Several deep optical studies have been carried out for some of the smaller γ-ray burst error boxes (see references after Table 2). The result is often phrased in terms of "emptiness of the box down to some limiting magnitude m_V^*" (~ 24 in some extreme cases). A word of caution is in order here. First, error boxes are rarely completely empty, but galaxies and other "normal" objects are often rejected as possible counterparts. For example, the brightest stellar-like object in the small error box of GB790406 is an inconspicuous M dwarf of magnitude R~ 22. However, there is no *a priori* reason to assume that the optical counterparts of γ-ray bursts should be unusual in any way. As long as we do not know what a γ-ray burst looks like in quiescence, one should always consider the total content of the error box and not introduce a bias based upon theoretical preconceptions. Secondly, in a number of cases the detection of an optical transient inside the error box (detection well separated in time from the occurrance of the γ-ray burst) has led to restricted searches of much smaller regions around the optical positions. This strategy, though justified as a reasonable first order approach, leads to an incompleteness in the search. To illustrate this point, consider the error box of GB791105b in which Schaefer *et al.* (1984a) discovered the optical transient OT1901. A subsequent search for high proper motion objects in this box (Ricker Vanderspek, and Ajher 1986) concentrated on an area surrounding the position of OT1901, but neglected a large fraction of the γ-ray error box, because it was too large for a reasonable search. Only the detection of truly simultaneous $\lambda\lambda\lambda$ flashes justifies the search to be carried out in the smallest common error box. Systematic studies including or even exceeding the complete 3σ error regions are needed and first steps in this direction have been taken at the Naval Observatory in Flagstaff (Jennings and Vrba 1987, private communication).

D. X-rays

Searches for associations between γ-ray bursts and catalogued x,γ-ray emitters (steady and variable sources) have not been successful (with the exception of the apparent association of GB790305b with N49 in the LMC). Deep exposures of several γ-ray burst error boxes with instruments on the *Einstein* satellite (0.5-5 keV) resulted in a marginal (3.5σ) detection of one X-ray point source in the field of GB781119 (Grindlay *et al.* 1982; Pizzichini *et al.* 1981, 1986) but subsequent observations with the EXOSAT satellite (0.02-2.5 keV) did not confirm this detec-

Table 2 Multi-wavelength content of small γ-ray burst error boxes

Burst	Area arcmin^2	Radio	Infrared	Optical	X-ray	Refs.
781119	10	3D	1D	2D(m~17);OT1928	1D?	1;2;4;5;7;9;11
	0.1		RU(K=19)	R 2D(R~24) PM		13;16;21-28
790113	78			R (V~ 22)		18
	0.05		RU(K=18)	OT1944		4;8;20
790305b	2		1D(K~17)	N49	N49	1;16;26;29
	0.1		R N49	R N49	R N49	3-4;30
790325b	2		1D G star	104 Her	U	3-4;14;24;27;31
790331	20			Mira Fy Aql		15;31
				R refl. neb.		
790406	0.3		U(J=18)	9D R~23	U	1;10;12;17
				7 gal. 2 stars		23;32
790418	2.9					33
790613	0.8	U		PM; 9D(R~19-22)		6;19;34
791105b				OT1901		35
	0.1		RU(K=19)	RPM(R~16-21)		4;8;19
791116	0.7					35

Notes: OT19xx indicates the detection of an optical transient occuring in the year 19xx inside the (3σ) γ-ray error box. D = detection without identification or association with detections in other bands. U = upper limits only. R indicates a restricted search covering only a fraction of the error box. PM indicates that a proper motion study was carried out. UV = UV excess.

References: (1) Apparao and Allen 1982 (2) Schaefer and Ricker 1983 (3) Schaefer 1986 (4) Schaefer *et al.* 1987c (5) Hjellming and Ewald 1981 (6) Hjellming 1983; priv comm in (34) (7) Schaefer 1981 (8) Schaefer *et al.* 1984 (9) Pedersen *et al.* 1983 (10) Motch *et al.* 1984 (11) Schaefer *et al.* 1983 (12) Hurley 1983 (13) Schaefer and Ricker 1983 (14) Hartmann *et al.* 1988 (15) Hartmann and Pogge 1987 (16) Fishman, Duthie, and Dufor 1981 (17) Chevalier *et al.* 1981 (18) Barat *et al.* 1984b (19) Ricker, Vanderspek, and Ajhar 1986 (20) Schaefer 1986 (21) Cline *et al.* 1981 (22) Pizzichini *et al.* 1981 (23) Pizzichini *et al.* 1986 (24) Boer *et al.* 1986 (25) Grindlay *et al.* 1982 (26) Helfand and Long 1979 (27) Boer *et al.* 1987 (28) Cline *et al.* 1981 (29) Evans *et al.* 1980 (30) Cline *et al.* 1982 (31) Laros *et al.* 1985 (32) Laros *et al.* 1981 (33) Atteia *et al.* 1987 (34) Barat *et al.* 1984a (35) Cline *et al.* 1984

tion (Boer *et al.* 1986, 1987). The non-detection of a substantial X-ray flux places severe constraints on the neutron star surface temperature, T_0, or alternatively the mass accretion rate from

$$T_0 \sim 1.4 \times 10^9 \left(\frac{\dot{M}}{p} \right)^{1/4} \quad K,$$

where $\dot{M}$ is the accretion rate in $M_\odot y^{-1}$ and p is the emitting fraction of the neutron star surface. However, deviations of the X-ray spectrum from that of a blackbody significantly change the obtained limits (Romani 1987). Figure 8 shows some of the most constraining temperature limits obtained to date. In the

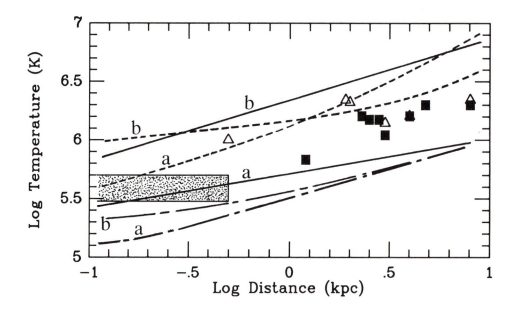

Figure 8 Typical result of X-ray observations of γ-ray burst sources in the quiescent state (compiled from Pizzichini *et al.* 1986; Boer *et al.* 1986; Romani 1987). Dependent upon assumptions about accretion geometry and spectral shape in the X-ray band (usually blackbody) the upper limits on point source fluxes obtained from *Einstein* and EXOSAT observations constrain the mass accretion rate (or neutron star surface temperature) and distance scale of γ-ray burst sources. Shown are results for the error boxes of (i) GB781104b (———) for p=1 (a) and p=10^{-3} (b) (ii) GB790406 (— · —) for p=1 assuming a blackbody spectrum (b) and a non-blackbody spectrum (Romani 1987) (a); and (iii) GB781119 (– – –) for p=10^{-3} observed with *Einstein* (b) and EXOSAT (a). For all curves an average hydrogen column density of $N_H \sim 10^{20}$ cm^{-2} has been assumed. For comparison, *Einstein* upper limits for pulsars within ~ 500 pc (Helfand 1983) (dotted box) and *Einstein* upper limits (filled squares) and detections of X-ray point sources (triangles) for young supernova remnants (Tsuruta 1985) are indicated.

context of the thermonuclear model (*e.g.*, Woosley 1982; Woosley and Wallace 1982; Hameury *et al.* 1985) accretion rates of order 10^{-15} M$_\odot$y^{-1} imply source distances in excess of ~ 500 pc for GB781119. However, possible effects due to beaming of the γ-rays, time dependent accretion (the γ-ray burst perhaps inhabits accretion for some time after the burst), and emission outside the X-ray windows (in particular for accretion rates as low as 10^{-17} M$_\odot$y^{-1} that result in temperatures of order 10^5 K) make these constraints less stringent. The AXAF mission (e.g., Weisskopf 1987), expected to be able to detect even some nearby old neutron stars (Wilson 1987) is likely to improve the X-ray limits significantly.

Simultaneous λλλ studies of γ-ray bursts

A. γ-rays

High energy γ-rays (E $\gtrsim$ 1 GeV) incident on the top of the earths atmosphere generate electromagnetic cascades that could be detected on the ground. The Cerenkov detector at Ootacamund, India, monitored the sky for about 1.2 yrs (Bhat *et al.* 1981). During this time six potentially detectable γ-ray bursts occured. No corresponding detection was made, which implies that either there are no γ-rays emitted simultaneously at GeV energies, or high energy γ-ray emission exists but the differential energy spectrum is steeper than E$^{-2.5}$ in the range 0.1 to 5 GeV.

B. Radio

The first search for radio emission in coincidence with γ-ray bursts was carried out between 1970 and 1973 (Baird *et al.* 1975, 1976). The experiment, operating mainly at 151 MHz, was an almost continuous all-sky survey with a sensitivity limit of $\sim 10^{-12}$ erg cm^{-2} event^{-1} in a 1 MHz bandwidth. It was originally designed to catch radio bursts from supernova explosions and included 5 widely spaced stations. The archived data were used to perform a search for radio spikes in coincidence with 19 γ-ray bursts then discovered by the Vela satellite. Two searches, one in an interval between -1 and + 10 hours and another within ± 10 min, failed to yield any "simultaneous" radio/γ bursts. This places an upper limit on the ratio of energy emitted in the radio band to the total γ-ray burst energy of R$_{r\gamma} \sim 10^{-8}$ (Cavallo and Jelly 1975).

Between July 1976 and May 1979 an automated all sky monitoring campaign at 151 and 468 MHz was carried out continuously at IFC/CNR-Milano and TESRE/CNR-Bologna (Mandolesi *et al.* 1977; Inzani *et al.* 1978, 1982). During that period 65 γ-ray burst events were used to search for coinciding (within ± 10 minutes) radio bursts. No compelling evidence for correlated radio/γ-ray transients was obtained at a sensitivity level of $\sim 10^{-13}$ erg cm^{-2} s^{-1} MHz^{-1} level.

C. The soft X-ray – γ-ray connection

Typical observations of γ-ray bursts cover the photon energy range between a few tens of keV and a few tens of MeV. From the few simultaneous X-ray observations that have been obtained previous to the *Ginga* mission (Wheaton *et al.* 1973; Metzger *et al.* 1974; Cline *et al.* 1979; Gilman *et al.* 1980; Terrell *et al.* 1982; Hueter 1984; Katoh *et al.* 1984; Laros *et al.* 1984) we know that no more than about 2% of the burst energy is radiated below $\sim$30 keV. Also Helfand and Vrtilek (1983) have conducted a search for simultaneous X-ray bursts in the Columbia Astrophysics Laboratory *Einstein* IPC (0.1-4.0 keV) data base. No detection was found which limits the ratio $R_{x\gamma} = E_x/E_\gamma$ to less than 3%. Recent searches for fast X-ray transients in the data bases of Ariel 5 (Pye and McHardy 1983), HEAO 1 A-1 (Ambruster *et al.* 1983; Ambruster and Wood 1986) and HEAO 1 A-2 (Connors, Serlemitsos, and Swank 1986) resulted in the detection of seven soft X-ray events that were either spatially *or* temporally correlated with γ-ray bursts or had characteristics suggesting an association with (unobserved) faint γ-ray bursts. However, for none of these events is the association with γ-ray bursts beyond doubt.

Preliminary results from recent simultaneous soft X-ray to γ-ray (1.5-400 keV) observations of γ-ray bursts with *Ginga* (Makino *et al.* 1987) were reported at the Taos workshop by Murakami. Since the satellite launch on February 5, 1987 the γ-ray burst detector on board *Ginga* (formerly ASTRO-C) registered several bursts that showed a soft X-ray "tail" lasting for 50-100 s after the turn-off of the hard γ-rays. There was essentially no delay between the onset of soft X-ray and the γ-ray emission ($\delta t_{x\gamma} \leq 0.5$ s). The data from the *Ginga* satellite provide strong support for the thermonuclear model of γ-ray bursts (see § 4).

The spectra gradually softened and a preliminary analysis of GB870303 showed that a blackbody fit is acceptable for the spectrum of the soft "tail". The ratio of energy emitted at X-ray energies (1.5-10 keV) to the total energy (1.5-500 keV) was found to be less than about 10%, consistent with previous estimates of this ratio, *i.e.*, $R_{\gamma o}$ = E(below 30 keV)/E(above 30 keV) $\sim$ 2%. This X-ray deficit places severe constraints on the burst emission mechanism and source geometry because of the reprocessing of γ-rays to X-rays in the neutron star surface (Epstein 1986; Imamura and Epstein 1987). The emission mechanism must predominantly produce γ-rays and only a small fraction of the energy emitted from the burst site can be thermalized on the neutron star surface. If the γ-rays are emitted isotropically the limit on thermalization implies that the source region cannot be very close to the stellar surface (Imamura and Epstein 1987). The relative paucity of X-rays in γ-ray burst spectra argues against theoretical models in which either the emission mechanism is too efficient in producing X-rays, such as the synchrotron process, or the burst site is too close to the neutron star surface, as in the thermonuclear model, so that reprocessing in the stellar surface generates a large X-ray flux.

On the other hand, the soft X-ray tail appears to be well fit by a blackbody spectrum of temperature $T \sim 1.5$ keV. Combining this value with the observed flux $F = 9 \times 10^{-9}$ erg cm^{-2} s^{-1} Murakami estimates the radius of the emission region rom

$$\mathcal{F} = \sigma\, T^4 \left(\frac{R}{D}\right)^2 \sim 1.1 \times 10^{-9} \left(\frac{R}{1km}\right)^2 \left(\frac{D}{1kpc}\right)^{-2} \left(\frac{T}{1keV}\right)^4 \quad erg\,cm^{-2}s^{-1} \quad .$$

The measurements indicate that R(km) = 1.3 D(kpc) which implies, if the emission omes from the surface of a neutron star with R = 13 km, that the distance o GB 870303 cannot exceed 10 kpc! In the more likely case that the X-ray ·mitting region covers a small fraction, p, of the neutron star surface the distance ·stimate is decreased by $p^{1/2}$. Combining this result with the observed cooling ime scale (Woosley 1982), Murakami estimates the mass of the cooling plasma as $M \sim p \times 10^{20-22}$g, which together with the observed fluence S = 2×10^{-5} erg cm^{-2} mplies an energy generation efficiency of $\epsilon \sim 10^{-7}$ erg/nucleon. These values are ll consistent with the predictions of the thermonuclear model for γ-ray bursts.

Although these results are preliminary they already indicate that quite restric- ive conclusions can be drawn from simultaneous X-ray and γ-ray observations. The *Ginga* experiment will certainly provide many more simultaneous λλλ spectra of γ-ray bursts in an energy range where previous data were virtually non-existing.

D. Optical flashes from γ-ray bursts

Early attempts to detect optical flashes from γ-ray burst locations (Grindlay *et al.* 1974), utilizing meteor patrol pictures of the prairie network, gave lower limits for he ratio of energy emitted in the γ-range to the energy emitted in the optical band $R_{\gamma o} = E_\gamma/E_o \sim 100$). More recent studies of similar nature (Halliday *et al.* 1978; Hudec *et al.* 1984, 1986, 1987; Hudec 1985) gave similar limits. The intriguing liscovery of three historical optical flashes within γ-ray burst error boxes on a subset of the archival photograph collection at Harvard (Schaefer 1981; Schaefer *et al.* 1984a) was the first indication that optical emission may indeed accompany γ-ray bursts. Under the assumption that the optical bursts lasted one second and :hat the assumed γ-ray luminosity was comparable to that of the more recently observed γ-ray burst, $R_{\gamma o}$ was found to be of order 900. Additional candidates were found for GBS1703+01 (Scholz 1984), GBS1901+14 (Greiner and Flohrer 1985; see also Greiner *et al.* 1987 and Flohrer *et al.* 1986), and GBS 1938+38 ᐧMoskalenko *et al.* 1988). Three optical flashes (occurring in 1946 March 28, 1946 August 31, and 1954 April 27) were observed from an identical position near, but not inside, the error box of GB790325b (Hudec *et al.* 1988). However, it is not likely :hat the source of the recurrent optical flashes, designated OTS 1809+314, and :hat of the γ-ray burst are causally connected (Hartmann *et al.* 1988; Laros 1988). No optical counterpart brighter than $m_v \sim 22$ was detected on deep CCD images of :he position of OTS 1809+314 (Hartmann *et al.* 1988). Deep CCD surveys of the

1928-OT/1978-GRB field (Schaefer, Seitzer, and Bradt 1983; Pedersen *et al.* 1983) revealed no quiescent counterpart brighter than about 23rd magnitude. A search of about 1500 hours of archival plates covering 10 γ-ray burst error boxes (Atteia *et al.* 1985) did not reveal optical flashes of similar type as the ones reported by Schaefer *et al.*. Two optical flash candidates, Hertzsprung 1900 and Popovic 1911 have been noted by Klemola (1983), but Schaefer (1983) demonstrated that at least the Hertzsprung 1900 event was caused by a plate defect. Another "archival" optical flash of ~ 0.25 seconds duration that occured during television observations of meteors from NASA's airborne Auroral Expedition in 1969 was reported by Wdowiak and Clifton (1985). In this case the counterpart was determined to be the double star β Camelopardalis. Brown and Fried (1987) monitored this source photometrically for about 50 hours with too independent telescopes at sites separated by about 5 miles. No bursts were detected during the period of monitoring.

Photometric monitoring of the site of the repeating 1979 March 5 event led to the detection of 3 optical flashes (Pedersen *et al.* 1984) out of which 2 are probably caused by a satellite glint and a meteor. However, the flash of February 8, 1984 is a strong candidate for a true optical flash from this γ-ray burst source (Schaefer *et al.* 1987a). This conclusion is also supported by a recent study indicating no coincidence between the optical flashes and known satellite positions (Maley 1987). The absence of simultaneous γ-ray bursts on these occasions places an upper limit of about 10^4 on $R_{\gamma o}$. The worldwide "multi-frequency March 5, 1979 Burst Watch" around New Year 1984/85, suggested by Holger Pedersen, did observe two radio and two optical transients, but none could be confirmed by simultaneous observations with two or more instruments (see Hurley 1987a for a summary of the watch campaign). If one considers the four spacecrafts operating at that time with γ-ray burst detector sensitivities of about 10^{-6} erg cm^{-2} a limit of $R_{\gamma o} \leq 6 \times 10^3$ is implied (Bisnovatyi-Kogan and Illiaronov 1986). The March 5, 1979 source, as well as a few other bursts with small error boxes are monitored for optical flashes by the so-called "γ-ray burst monitoring system" (GMS) (Pedersen 1987). Another ground-based monitoring program utilizes a wide-FOV video camera at Pic du Midi and Haute Provence (Atteia 1987). The non-detection of an optical flash simultaneous with the γ-ray burst GB830313 gives a lower limit of $R_{\gamma o} \sim 70$ for this particular event (Atteia 1987).

Furthermore, there is the recently discovered "Perseus flasher" which was tentatively associated with γ-ray burst sources though the observed flash rate of order of hours argued against this interpretation (MacRobert 1985a,b; Katz *et al.* 1986). Searches for optical flashes from this source on the plates from the Ondrejov collection (Hudec *et al.* 1986) and the Sonneberg collection (Greiner *et al.* 1987), archival meteor patrol photographs taken by the MORP network (Halliday, Feldman, and Blackwell 1987), photographic monitoring of the source for 38.0 hours with the 0.76m telescope at the Manastash Ridge Observatory (Garnavich and Temple 1987a,b), 30 hours of photographic monitoring with the 61-inch reflector at the U.S. Naval Observatory Flagstaff station (Brown 1987), 70 hours of search with the Explosive Transient Camera (ETC) system (Vanderspek, Zachary, and Ricker 1987), as well as 76 hours of monitoring with several different telescopes and instruments of the NLARC (Corso, Ringwald, and Harris 1987) all gave negative results. Also EXOSAT X-ray observations of the source region found no detectable X-ray point source (Lewin *et al.* 1987a). A possible non-astronomical origin of these flashes due to reflections of sunlight off tumbling satellites was suggested by a number of investigators (Maley 1987; Schaefer *et al.* 1987b; Vanderspek, Zachary, and Ricker 1987). Finally, to add one more to a rapidly growing list, Warner (1986) drew attention to still another historical observation of a transient optical event that had been carried out by Eduard Heise in 1850. Warner also points out that there is a large background contaminating the data sets of optical burst observations, historical ones as well as wide-field CCD searches presently pursued with the ETC/RMT system (Ricker *et al.* 1984; Teegarden *et al.* 1984, 1986) and the GMS (Pedersen 1987). In fact, the optical flash background rate is dominated by meteors and satellite glints (Schaefer *et al.* 1984b; Schaefer 1985; Schaefer *et al.* 1987a), the latter type of "light pollution" will become an even more severe problem for ground based observations in the future. As of this writing we must consider bright optical flashes accompanying γ-ray bursts as a possibility, but not an established fact. Simultaneous observations of γ-ray bursts and accompanying optical emission that are now in the planning stage are needed to provide evidence for or against this paradigm (see § 5).

Even with a small uncertain data base at hand, a variety of theoretical models have been suggested for this phenomenon. London and Cominsky (1983) investigated reprocessing of a small fraction of the hard radiation by a companion main sequence star, a degenerate dwarf, or an accretion disk (see also Epstein 1985). London and Cominsky (1983) concluded that the particular event 1928-OT/1978-GRB could not be explained within the *binary reprocessing scenario*. The problem stems from the fact that one has to explain both the bright optical display during outburst ($R_{\gamma o}$ is *only* of order 1000) and the extremely low emission during quiescent times, as inferred from the deep CCD surveys.

Rappaport and Joss (1985) reconsidered the work of London and Cominsky and concluded that this model might be viable for source distances less than about 100 pc. The new twist in their work was the consideration of nearby binary systems with H-rich secondaries of mass less than about 0.06 $M_\odot$ (dark dwarfs).

A detailed investigation of the reprocessing of hard photons in the atmospheres of such low-mass stars was consequently carried out by Melia, Rappaport, and Joss (1986), who found only marginal agreement between the observed and calculated ratios of γ-ray fluence to optical fluence at the earth for source distances as small as 25 pc. Similar calculations by Cominsky, London, and Klein (1987) again indicate that it is extremely difficult to find a candidate to match both the great brightness of the flash and the low quiescent luminosity if one confines the possible companion objects to low mass main sequence stars or white dwarfs. On the other hand, bright "optical" flashes might be generated if the reprocessing occurs in an extended accretion disk around the neutron star, as in the case of X-ray bursters (Melia 1987, 1988).

A different scenario involving a binary system has recently been suggested by Tremaine and Zytkow (1986). Reconsidering the model of Harwitt and Salpeter (1973) and Colgate and Petschek (1981) in which a comet impact on a neutron star causes the γ-ray burst, the authors propose that optical flashes might be generated by a comet impact on a white dwarf companion. In this case the optical flashes would not occur in coincidence with the γ-ray bursts. However, the observational constraints on the ratio of transient to quiescent optical flux appear to rule out white dwarfs as possible sites (Katz 1987).

Interpreting the observation that no γ-ray burst source is associated with a luminous stellar companion as an indication that they are in fact isolated objects, Katz (1985) proposed non-thermal processes in the neutron star magnetosphere as origin of the optical emission, but no detailed calculation was performed. Ruderman (1987) suggested that curvature radiation from electron-positron pairs that rapidly fill plasma starved regions in the magnetosphere could produce transient optical emission. Collective processes for burst emission at wavelengths other than γ-rays were invoked by Liang (1985), who considered emission in the EUV region resulting from enhanced plasma line radiation at ω_p and $2\omega_p$, and Sturrock (1986), who discussed infrared, optical, and UV emission resulting from plasma oscillations driven by a two stream instability as in a radio pulsar. Unfortunately, all the proposals of this paragraph are very qualitative at present.

Alternatively, the formation of optical flashes far above the surface of strongly magnetized neutron stars undergoing either a thermonuclear explosion of plasma accreted over long periods of time (*nuclear model*), or equivalently a sudden impact of a large amount of plasma (*gravity model*), be it a solid body or plasma from an accretion disk, was considered by Woosley and Arons (Woosley 1984b). Strong surface fields are indicated in γ-ray burst sources by the observed low-energy features in the spectra of a number of γ-ray bursts if these "lines" can in fact be interpreted as cyclotron resonance lines (see Mészáros, Bussard, and Hartmann

1986 for a discussion of this interpretation). In Woosley and Arons's model the mechanism for the optical emission is cyclotron radiation in a region far above ($r \sim 10^8$ cm) the surface of the neutron star. The radiation is self-absorbed for about the first 100 harmonics and the electrons are radiatively excited by Compton collisions with photons of the γ-ray burst. This so-called *thermal cyclotron reprocessing* (TCR) model, does not require the existence of a companion star, though it is a convenient source of donated plasma. Model spectra for a number of different neutron star magnetospheric conditions, notably surface field strength, temperature, and plasma density, were presented during the Taos meeting by Hartmann. The main results and predictions of the TCR model can be summarized as follows: (i) Luminous bursts in the IR to UV spectral range result if a γ-ray burst occurs inside or near a plasma filled magnetosphere around a strongly magnetized neutron star. (ii) The ratio of γ-ray to optical luminosity is expected to be of order 10^{4-5}. (iii) The "optical" spectrum provides a means to constrain the neutron star surface field strength. (iv) The model spectra are of characteristic shape, with detectable fluxes over a large spectral range. (v) The duration of the "optical" flash is essentially the same as that of the γ-ray burst, in fact, the time histories of both wavebands should be correlated down to time scales short in comparison to the burst duration. (vi) There is no significant delay between the reprocessed light and the γ-ray burst. (vii) Polarization should be detectable. (viii) The model does not require a luminous companion.

The predictions of the various theoretical models are very different from each other so that simultaneous $\lambda\lambda\lambda$ observations will most certainly provide distinguishing diagnostics. Figure 9 shows the expected "optical" display from some of the models described above in comparison with observations and the expected sensitivity of the HETE detector (see § 5).

4. Theoretical Issues

X-ray burst models

Type I X-ray bursts are currently interpreted as hydrogen and helium explosions on the surfaces of weakly magnetic, accreting neutron stars (Woosley and Taam 1976; Maraschi and Cavaliere 1977; Woosley 1982; Taam 1982; Ayasli and Joss 1982; Wallace, Woosley and Weaver 1982; Lewin and Joss 1983; Woosley and Weaver 1984; Melia and Joss 1986). The thermonuclear model matches remarkably well most of the observational features of X-ray bursts described in §2. In this model, the observed ratio of average persistent to burst luminosity reflects the ratio of the gravitational energy to the nuclear energy released per unit mass ($\sim$100 for He flash and larger than $\sim$ 20 if H-burning is important). Replacing

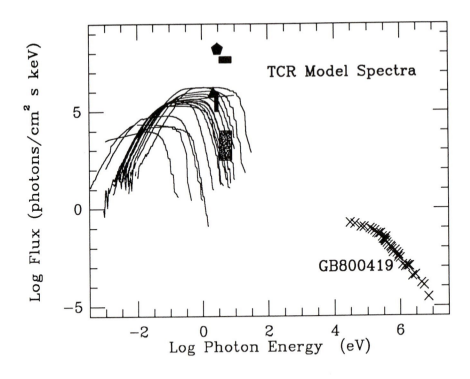

Figure 9 Optical flash spectra generated by thermal cyclotron reprocessing (TCR) of γ-rays in the neutron star magnetosphere (Hartmann, Woosley, and Arons 1987; 1988) (solid lines) in comparison with archival flashes reported by Schaefer (1981) (filled pentagon), observations obtained during optical monitoring of the GB790305b source region (Pedersen *et al.* 1984) (filled triangle), predictions of models involving reprocessing in a cold disk (Melia 1987) (filled rectangles), and the expected sensitivity of the proposed HETE mission (Ricker *et al.* 1987) (dotted rectangle). Also shown is the spectrum of the γ-ray burst GB800419 (crosses).

the neutron star by a white dwarf one obtains the current model for classical novae (*e.g.*, Starrfield, Sparks, and Truran 1985). However, owing to the much greater surface gravitational potential in the case of neutron stars, the explosions occur with a much smaller critical mass ($\sim 10^{-12}$ M$_\odot$) and reach much higher burning temperatures, T$\sim 1.5 \times 10^9$ K being typical, but with even larger values occurring in some models. Typical densities of the H/He plasma are of order 10^6g cm^{-3}. Nuclear processes occurring under those thermodynamic conditions have no parallel elsewhere in the universe (see Taam 1985 and Woosley 1986 for recent reviews). The energy released on time scales typically of order 1 minute is roughly 10^{39} ergs. In the thermonuclear explosion almost no matter escapes the deep gravitational well of the neutron star, so that X-ray burst sources, unlike novae (discussed by S. Starrfield elsewhere in this volume), do not contribute significantly to the chemical evolution of the galaxy.

An outstanding impediment to our current understanding of Type I X-ray bursts in the context of the thermonuclear model is the fact that a few sources have shown burst separations of only minutes (§2). The difficulty is that it is impossible to accrete the critical mass in such a short time. For typical accretion rates of order 10^{16}g s^{-1} the time interval between bursts is of the order 10^5 s. On the other hand, it is very difficult to get the burning to halt for ~ 10 minutes and then start again (Fryxell and Woosley 1982). Survival of some amount of unburnt H/He fuel has been proposed to explain the observations (Ayasli and Joss 1982; Hanawa, Sugimoto, and Hashimoto 1983; Woosley and Weaver 1984). How much hydrogen can survive the runaway? Unfortunately, uncertainties in the nuclear input physics along the relevant reaction path prevent even a qualitative answer to this question (Taam 1985; Woosley 1986). Based upon the recent EXOSAT observations of MXB1636-53 (Lewin *et al.* 1987a) a novel triggering mechanism for "premature bursts" was proposed (Fujimoto *et al.* 1987) that involves elemental mixing and dissipative heating associated with hydrodynamical instabilities caused by the accretion of angular momentum.

γ-ray burst models

Considering that we do not know the γ-ray burst distance scale to perhaps a factor of 100 and that no counterparts have been found, it does not come as a surprise that theoreticians still consider more than a handful of competing scenarios to explain this enigmatic phenomenon. For brevity we concentrate on the thermonuclear model of γ-ray bursts (for recent reviews of several alternative models see Verter 1982; Liang and Petrosian 1986). Successful in explaining the main features of X-ray bursts as thermonuclear runaways on the surface of rapidly accreting ($\sim 10^{-10}$ $M_\odot$ y^{-1}), weakly magnetized ($\sim 10^8$ Gauss) neutron stars the same mechanism can explain at least some γ-ray bursts as thermonuclear explosions on slowly accreting ($\sim 10^{-13}$ $M_\odot$ y^{-1}), strongly magnetized ($\sim 10^{12}$ gauss) neutron stars (Woosley and Taam 1976). Figure 10 illustrates the approximate locations of various high energy transients in accretion rate - magnetic field phase space.

Details of the thermonuclear model for γ-ray bursts have been discussed in a number of papers (Woosley and Taam 1976; Woosley and Wallace 1982; Fryxell and Woosley 1982; Woosley 1984a,b; Hameury *et al.* 1982, 1983a,b, 1984, 1985; Ergma 1984). Here we wish to summarize those characteristic properties and predictions of the model that can be tested by λλλ observations. The approximate ranges of allowed conditions of the thermonuclear model are given in Table 3.

The following predictions of the thermonuclear model allow relatively clear-cut λλλ observations to support or rule-out the model. Because the critical mass ($\sim 10^{20-21}$g) must be supplied by accretion the model requires an average (not

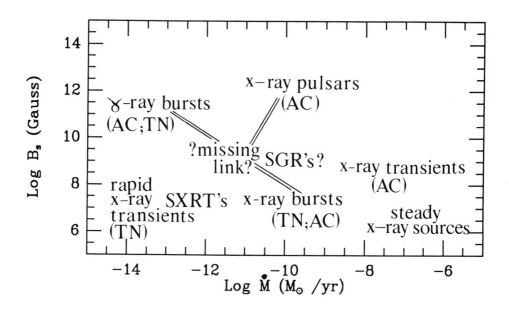

Figure 10 Location of several high energy transients in the mass accretion rate – neutron star magnetic field strengths phase space. The dichotomy between the high accretion rate bursters and pulsars is caused by the confining power of the magnetic field. The recently discovered Soft Gamma Repeaters (SGR) (Atteia *et al.* 1987; Laros *et al.* 1987; Kouveliotou *et al.* 1987) may be the missing link between X-ray pulsars, X-ray bursters and the low accretion rate γ-ray bursters, believed to involve strongly magnetized neutron stars.

Table 3 Ranges of allowed conditions of the thermonuclear model[a]

Level	Fuel	Ignition	M_{crit}	$\dot{M}$	E_{tot}	τ_{rec}
			g	g s^{-1}	erg	
Low	H/He	3α/rp	10^{19}	10^{-11}	10^{38}	few hours
High	He	3α	10^{24}	10^{-15}	10^{42}	days-10^6yrs
Typical	H/He	p(e$^-$,ν)n	10^{21}	10^{-13}	10^{40}	~1 - 10 yrs

[a] A fraction of 10% is assumed for the extend of the fuel over the neutron star surface.

neccessarily steady) luminosity

$$L \sim \alpha_{TN} \left(\frac{E_{burst}}{\tau_{recurr}} \right)$$

where the efficiency α_{TN} is bounded between ~20 and ~100. Typical mass accretion rates are of order 10^{-15} M$_\odot$y^{-1}. Upper limits on the X-ray flux obtained from *Einstein* and EXOSAT measurements are consistent with these rates if source distances are larger than ~ 500 pc (Boer *et al.* 1986, 1987; Pizzichini *et al.* 1986). Fuel could be provided by accretion from the interstellar medium (Higdon and Lingenfelter 1984) but is more likely from a companion of some sort. Deep CCD searches for optical/IR counterparts in a few γ-ray burst error boxes have already severely constrained possible companion properties. If accretion from the ISM is the source of fuel then one would expect to observe an apparent disk association because most of the fuel reservoir is concentrated in the galactic disk. However, neutron stars are born near the galactic plane with velocities of order of several 100 km s^{-1} (Woosley 1987), so that the accretion efficiency is reduced (Bondi rate scales as v^{-3}). The observations of radio pulsars provide the most complete sample of neutron star kinematic properties. From these data it is quite clear that neutron stars are at least a very thick disk if not a halo population. Reaching heights above the galactic plane of order of several kpc neutron stars are then slow enough to accrete efficiently but by then there is essentially no gas left to accrete from.

Therefore, if γ-ray bursts do originate on isolated neutron stars accreting galactic disk matter one may expect bursts to occur more frequently at some intermediate height above the plane, which could result in a non-isotropic source distribution of the order of the scale height of galactic H I. As is apparent from Figure 5 the observed γ-ray burst source distribution is isotropic (e.g., Atteia *et al.* 1987a; Hartmann and Epstein 1988).

About 1.5% of the presently known pulsars are binary pulsars (Manchester 1987), indicating that a small fraction of binary systems survive the high velocities given to them. A detailed Monte-Carlo simulation of radio pulsars and their progenitors (Dewey and Cordes 1987) showed that a model (their model C) in which supernova explosions impart a kick velocity to neutron star remnants with an average speed of 90 km s^{-1} is consistent with both the observed pulsar velocity distribution and the observed incidence of binary pulsars. In those systems matter for thermonuclear explosions is perhaps provided by the companion and the mechanism does not depend on accretion from the ISM and therefore no velocity induced anisotropy exists.

Models with internal energy sources such as starquakes and glitches (Fabian, Icke, and Pringle 1976; Epstein 1988 and references therein) work for isolated neutron stars as well as for neutron stars in binary systems. Models in which the γ-ray burst is fueled by external sources invoke accretion disk systems (Taam and Lin 1984; Epstein 1985), impacts of comets assuming that the neutron star birth process allowed a hypthesized "Oort cloud" to remain bound to the system (Harwitt and Salpeter 1973; Colgate and Petscheck 1981; Tremaine and Zytkow

1986; but see also Katz 1986 and Woosley 1987), or asteroid collisions with neutron stars (van Buren 1981; Newman and Cox 1980; Howard, Wilson, and Barton 1981).

The answer to the question of the neutron star membership in a binary system or the presence of an accretion disk can probably be given when more sensitive $\lambda\lambda\lambda$ observations of γ-ray bursts are performed. In particular, improved magnitude limits from the non-simultaneous IR study of γ-ray burst error boxes (Schaefer *et al.* 1987c) provide extremely strong constraints on the theoretical models. The AXAF observatory (Weisskopf 1987) could provide a similar improvement from bounds on or detections of quiescent X-ray flux. If γ-ray bursts in fact exhibit simultaneous long wavelength emission then missions such as HETE are likely to provide the answer to the binary membership puzzle (see below).

Another important prediction of the thermonuclear model is the existence of a substantial X-ray afterglow (Woosley 1982; Woosley and Wallace 1982), having a thermal spectrum with T $\sim$ 2 keV and declining over a period of about 100 s. Radiation having this characteristics has recently been detected by the *Ginga* satellite (§3). These observations lend strong support to the thermonuclear model for γ-ray bursts and provide strong evidence for their galactic origin. Here again, advances in our understanding are based upon simultaneous $\lambda\lambda\lambda$ observations (X-ray/γ-ray with *Ginga* and UV/X-ray/γ-ray with missions of HETE design). Refined theoretical models are needed to more accurately determine the expected properties of the X-ray afterglow.

The soft gamma repeaters

A longstanding problem is the search for a possible link (if it exists) between X-ray pulsars, X-ray bursts, and γ-ray bursters (Figure 10). The possible new class of soft repeaters may hold the key to this question. The observational properties of the three sources conjectured to constitute the new class have been summarized in §3. The observed properties of the soft repeaters are consistent with the idea that periodic comet accretion onto a strongly magnetized neutron star causes the bunched burst behavior (Livio and Taam 1987), perhaps orchestrated by a companion star as suggested for GB790305 (Rothschild and Lingenfelter 1984). Whether powered by transient accretion onto a magnetized neutron star or by thermonuclear explosion, the burst spectrum is expected to be very similar to that of an X-ray pulsar (see Figure 7). Unfortunately, the extremely short duration of the bursts seen from SGR 1806-20 render it virtually impossible to detect periodic modulations expected in this scenario if the neutron star rotates.

A possible model for gamma-ray bursts

The family of γ-ray transients has become so diverse that it is increasingly inappropriate, even misleading to speak of a single "γ-ray burst mechanism." How can one reconcile, for example, within the context of a single mechanism, the fact that some transients have a spectrum like that of an X-ray pulsar, others show spectral features indicative of cyclotron absorption in a field of greater than 10^{12} gauss and show gravitationally redshifted lines from pair annihilation, while many others produce γ-rays having energies in excess of ($\gtrsim 10$ MeV)? The first three properties require that the burst originate near the surface of a strongly magnetic neutron star while the last requires that it originate in a weak magnetic field (Matz *et al.* 1985). How too are we to understand the paucity of clear evidence for rotational periodicity within a given transient if the source is confined to a small region on the surface.

In part to address these questions, Woosley suggested at the Taos meeting that a "gamma-ray burst" should be considered as consisting of two components (see also Lasota and Belli 1983), a soft component, $kT \lesssim 30$ keV, and a hard one extending up to 100 MeV and perhaps beyond. The soft component originates near the surface and, in both radiative mechanism (inverse Compton) and appearance, resembles an X-ray pulsar. The ultimate energy source could be thermonuclear, transient accretion, quake, or any phenomenon that gives rise to magnetically confined, optically thick plasma near the surface of the neutron star. The well developed theory of X-ray pulsar spectra (Mészáros 1984; Mészáros and Nagel 1985a,b; Kirk 1986) could be brought to bear to describe the emissions from this region. Cyclotron lines, when present, would be an obvious diagnostic. This soft component might also give clearer evidence for rotation, in the longer bursts, than the bolometric luminosity. The X-ray afterglow reported by Murakami also originates from this region.

In the soft repeaters, for reasons that will be speculated upon below, we see only this soft component. The short rise times and durations suggest that transient accretion is the most likely energy source, but a thermonuclear explosion might also produce a similar spectrum. The resemblance of the spectrum and assumed similarity in radiative mechanism to those of X-ray pulsars also suggest that the luminosity of the soft repeaters should not greatly exceed that of the brightest known X-ray pulsars, LMC X-4 and SMC X-1, which have log L = 38.8 and 38.7 respectively. This would imply that the soft repeaters, including the March 5 event, are *not* to be found in the Galactic center or the LMC but less than 1 kpc away.

The *hard* component of γ-ray bursts, of course, is the mystery that has anguished theoreticians for years. In order to maintain the favored model of a strongly magnetic neutron star and avoid excessive beaming or opacity due to pair production, it is useful if the hard component does *not* originate near the surface of the neutron star but farther out, presumably in the magnetosphere (see also Lingenfelter and Hueter 1984; Katz 1983). Three possible sources of energy exist in the magnetosphere: rotation, the magnetic field, and the kinetic energy of any resident plasma. The magnetostatic energy by itself, ($B_s^2 R^3 \sim 10^{42}$ erg), though adequate to power many γ-ray bursts, will not spontaneously annihilate and the magnetic field more likely acts either as a containment vessel or as an intermediary for another energy source.

Rotation is the ultimate power source for radio pulsars, some of which are also known emitters of hard γ-radiation. Could it also be the basis for γ-ray bursts? At this and other meetings Woosley has suggested that it could (see also Ruderman 1987; Sturrock 1986). In Woosley's "model," scarcely more than dimensional analysis at this point, a large electric field develops parallel to the magnetic field because of the stresses associated with rotation. However, unlike for radio or γ-ray pulsars, the stresses do not occur near the light cylinder and the region of interest is not starved for plasma. Rather they occur closer to the neutron star and exist because the field is suddenly loaded with plasma. In short, the magnetic stress induced by the differential rotation of plasma that has inertia generates a twisted flux tube which in turn gives rise to a large parallel electric field. It is the resistive dissipation of the current that flows in response to this parallel field that produces the hard component of classical γ-ray bursts. This mechanism may generate luminosities on the order of $10^{36} P(s)^{-2}$ erg s^{-1} as has previously been discussed (Woosley 1984b). Periods are not known for γ-ray bursts as a class, but values between 1 s and 10 s are suggested by quasi-periodic structure in the time histories of a few bursts (see § 3).

An alternative mechanism for producing the hard component also involves plasma that is injected into the magnetosphere by accretion or by a thermonuclear explosion, but does not invoke a large parallel electric field. Most important is the plasma that finds itself on closed field lines (*i.e.*, those that bend over at distances ≲1000 km). Such plasma will be trapped and will be characterized by large velocity shear from flux tube to flux tube. It may, either by shocks (*e.g.*, at the opposite pole) or by two-stream instability, convert a large fraction of its streaming energy into hard radiation (Woosley 1982). A thermonuclear flash or accretion of a solid body would load the closed field lines with greater ease than accretion from a disk or companion.

Within the context of these models the soft repeaters would avoid producing the hard component either because they are powered, as are X-ray pulsars, by accretion along *open* field lines or, in the rotationally powered case, because they have relatively weak magnetic fields and/or slow rotation rates. Interestingly the only γ-ray transient for which we have an accurate rotation period, the March 5, 1979 event, was both a soft repeater and a slow rotator (8 s).

In the classical bursts, on the other hand, one might expect, as seems to be the case, that the hard component would be favored by high luminosity (hence larger amounts of plasma loaded into the magnetosphere) and would be strongest at the onset of the burst. The hard component would cease just as the temperature dropped below the Eddington value (2 keV) because plasma could no longer be maintained in the magnetosphere (at least in the thermonuclear model). The temperature of the X-ray afterglow observed by *Ginga* for some γ-ray bursts is consistent with this prediction.

5. Strategies for λλλ Observations of γ-ray Bursts

Our discussion in §2 demonstrates the importance of λλλ observations for the understanding of X-ray burst sources . Although more observations are still needed to work out details, the nature and general properties of X-ray burst sources appear to be relatively well understood. This success, however, was the result of their predictable bursting behavior and the existence of binary companions. Furthermore, their locations on the celestial sphere reveal their galactic origin. These favorable circumstances are not found in the case of the elusive γ-ray burst phenomenon. Unpredictable, non-repetitive, and devoid of luminous companions, γ-ray bursts require unique λλλ strategies to yield their secrets.

Although λλλ observations of γ-ray burst error boxes with different instruments at different sites and times are extremely useful and continue to be pursued, crucial λλλ data need to be obtained simultaneously, *i.e.*, when the source is "on." Ground- and space-based monitoring of specific γ-ray burst source locations is feasible only for sources that are known to repeat. In those few cases where a possible outburst date was predicted on observational and/or theoretical grounds, worldwide λλλ watch programs were organized. The generally inconclusive results obtained from these efforts have shown that a better coverage (more redundancy by a larger number of participating organizations and instruments) is essential for "transient watch programs." To this end the Taos workshop participants recommended the formation of a Multiwavelength Observing Review and Evaluation (MORE) Committee to facilitate a more effective time allocation of the involved observatories (see the chapter by F. Córdova).

Although unlikely, a serendipitous detection of bright optical flashes could result from a systematic deep optical monitoring program for several small γ-ray burst error boxes that was recently initiated at the Naval Observatory in Flagstaff (USNO) (Jennings and Vrba 1987, private communication). This program employs fully automated CCD data taking systems on a 40-inch Cassegrain reflector. In a complementary effort the Santa Barbara Astronomy Group (SBAG), a team of advanced amateur astronomers, patrols γ-ray burst error boxes with several photoelectric photometer equipped small telescopes (12-14 inch) operating in co-incidence over a 7 km baseline (Schwartz 1986). Both monitoring programs have not yet detected any optical flashes from known γ-ray burst positions. On the other hand, ground-based wide-FOV optical monitoring of a substantial fraction of the sky, such as the ETC/RMT system (Ricker *et al.* 1984; Teegarden *et al.* 1984) and the GMS at ESO (Pedersen 1987) suffer from the enormous background rates of transient events caused by meteors and reflections of sunlight off space debris in near earth orbits (Schaefer 1985; Maley 1987). Thus it appears that the most promising strategy is one in which different wide-FOV instruments (optical, UV, X, and γ band) observe simultaneously from a single platform in space. A space based $\lambda\lambda\lambda$ station operating with modest pointing accuracy is required and a wide FOV monitoring mission could be developed at a very nominal cost (see the discussion of the *High Energy Transient Experiment* (HETE) by G. Kriss).

What could be accomplished with a true $\lambda\lambda\lambda$ space mission dedicated to the study of transients? Most importantly, X-ray positions of fast transients (X-ray as well as γ-ray bursts) with accuracies better than 3 arcminutes would be available within minutes to hours after the burst (compare this to the usual delay of order of years for positions obtained from multi-satellite triangulation). A simultaneous UV detection would increase the accuracy to 3 arcseconds! If these positions are relayed instantaneously to participating $\lambda\lambda\lambda$ observatories (in space or on earth), then a possible $\lambda\lambda\lambda$ afterglow on time scales long compared to the burst duration could be detected. The thermonuclear model predicts a substantial X-ray tail that has in fact been detected by the Japanese satellite *Ginga*, and the H II region created by the ionizing burst of radiation may under favorable conditions be detectable for several days (Jennings 1983). Followup $\lambda\lambda\lambda$ observations are likely to solve the longstanding problem of their unknown distances. To make this work, it is neccessary to establish a fast (electronic) alert system between all participating observatories (see the chapter by F. Córdova). It would be desirable to include several dedicated small/medium scale telescopes in the response network.

Simultaneous γ-ray and X-ray data would provide constraints on the geometry of the high energy emission regions (Epstein 1986; Imamura and Epstein 1987). A soft, thermal "X-ray tail," such as reported by the *Ginga* team during the Taos workshop, provides an accurate determination of the γ-ray burst distance scale. Possible interstellar absorption lines could confirm the distances independently. Cyclotron lines at energy $\hbar\omega_c \sim 11.6\ B_{12}\ keV$ might be present in the soft X-ray range because there is no reason to assume that the neutron star magnetic field strengths is always of order 10^{12} gauss. Last, but not least, observational selection against those bursts that have a substantial fraction of their emission at soft X-ray energies (like the soft repeaters) could be avoided.

If simultaneous optical/UV flashes from γ-ray bursts do exist, a wealth of information could be gathered. In particular, the existence or non-existence of a binary companion of some sort could be revealed. If the neutron star is a member of a binary system the time history in different bands will tell us about the geometric dimensions of the system. Optical and UV photometry and spectroscopy would be desirable for future missions. Independent evidence for magnetic fields would come from polarimetry data. The ratio of optical to γ-ray luminosity constrains the size of the emission region and gives additional information on the magnetic field strength in case of cyclotron reprocessing (Hartmann, Woosley, and Arons 1987, 1988), or the solid angle of the γ-ray intercepting companion object.

In summary, if simultaneous λλλ emission accompanies and follows some or even most γ-ray bursts and if time–resolved λλλ data can be obtained, a break-through in our understanding of the enigmatic γ-ray burst phenomenon appears inevitable.

It is our pleasure to thank the organizers of the Taos workshop, F. Córdova and W. Priedhorsky and their LANL staff for providing us with a week of stimulating scientific exchange combined with a taste of New Mexican hospitality. This chapter is based on talks given by a number of workshop participants, but we also included results not presented at the Taos meeting. We have greatly benefited from extensive discussions during, and after, the Taos workshop. In particular, we would like to express our appreciation to Drs. L. Cominsky, F. Córdova, R. Epstein, E. Fenimore, J. Laros, W. H. G. Lewin, and G. Ricker. We thank J. Brodie, L. Cominsky, F. Córdova, R. Epstein, and K. Hurley for many valuable suggestions and a careful reading of this manuscript. This work was supported in part by the National Science Foundation under grant AST 84-18185, the California Space Institute under grant CS-69-85 and the IGPP at Los Alamos National Laboratory under project No. 142.

References

Alme, M. L., and Wilson, J. R. 1974, *Ap. J.*, **194**, 147.
Ambruster, C., Wood, K. S., Meekins, J. F., Yentis, D. J., Smathers, H. W., Byram, E. T., Chubb, T. A., and Friedman, H. 1983, *Ap. J.*, **269**, 779.
Ambruster, C., and Wood, K. S. 1986, *Ap. J.*, **311**, 258.
Apparao, K. M. V., and Allen, D. 1982, *Astr. Ap.*, **107**, L5.
Apparao, K. M. V. 1984, *Astr. Ap.*, **139**, 375.
Arons, J., and Lea, S. M. 1976, *Ap. J.*, **207**, 214.
—— 1980, *Ap. J.*, **235**, 1016.
Arons, J. 1987, in *Origin and Evolution of Neutron Stars*, Proc. IAU Symposium No. 125, ed. D. J. Helfand, (Reidel Publishing Co.), pg. 207.
Arons, J. Klein, R. I., and Lea, S. M. 1987, *Ap. J.*, **312**, 666.
Atteia, J.-L., *et al.* 1985, *Astr. Ap.*, **152**, 174.
Atteia, J.-L., and Hurley, K. 1986, *Adv. Space Res.*, **6**, (4), 39.
Atteia, J.-L., *et al.* 1987a, *Ap. J. Suppl.*, **64**, 305.
Atteia, J.-L., *et al.* 1987b, *Ap. J. (Letters)*, **320**, L105.
Atteia, J.-L. 1987, Ph. D. thesis, Universite Paul-Sabatier de Toulouse, France.
Ayasli, S., and Joss, P. C. 1982, *Ap. J.*, **256**, 637.
Babul, A., Paczynski, B., and Spergel, D. N. 1987 *Ap. J. (Letters)*, **316**, L49.
Babul, A., and Paczynski, B. 1987, *Ap. J.*, **323**, 582.

Baird, G. A., *et al.* 1975, *Ap. J. (Letters)*, **196**, L11.

Baird, G. A., Meikle, W. P. S., Jelly, J. V., Palumbo, G. G. C., and Partridge, R. B. 1976, *Ap. Space Sci.*, **42**, 69.

Barat, C., *et al.* 1984a, *Ap. J.*, **280**, 150. with Errata in *Ap. J.*, **299**, 1079 (1984) and *Ap. J.*, **288**, 833 (1985)

Barat, C., *et al.* 1984b, *Ap. J. (Letters)*, **286**, L5.

Basińska, E. M., Lewin, W. H. E., Sztajno, M., Cominsky, L., Marshall, F. J. 1984, *Ap. J.*, **281**, 337.

Belian, R. D., Conners, J. P., and Evans, W. D. 1976, *Ap. J. (Letters)*, **206**, L135.

Bhat, P. N., Gopalakrishnan, N. V., Gupta, S. K., Ramana Murthy, P. V., Sreekantan, B. V., and Tonwar, S. C. 1981, *Phil. Trans. R. Soc. London*, **A301**, 659.

Beurle, K., Bewick, A., Mills, J. S., and Quenby, J. J. 1981, *Ap. Space Sci.*, **77**, 201.

Bewick, A., Coe, M. J., Quenby, J. J., and Mills, L. S. 1975, *Nature*, **258**, 686.

Bisnovatyi-Kogan, G. S., and Illarionov, A. F. 1987, *Sov. Astron.*, **30**, (5), 582.

Boer, M., *et al.* 1986, *Adv. Space Res.*, **6**, (4), 65.

Boer, M., *et al.* 1987, *Astr. Ap.*, submitted.

Bradt, H. V. D., Doxey, and Jernigan 1979, in *X-Ray Astronomy*, ed. W. A. Baity and L. E. Petersen, (Oxford, Pergamon), p. 3.

Bradt, H. V. D., and McClintock, J. E. 1983, *Ann. Rev. Astron. Astrophys.*, **21**, 13.

Brown, S. E. 1987, reported in *Bull. Am. Astr. Soc.*, **19**, 571.

Brown, S. E., and Fried, R. E. 1987, reported in *Bull. Am. Astr. Soc.*, **19**, 571.

Canizares, C. R., McClintock, J. E., and Grindlay, J. E. 1979, *Astrophys. J.*, **234**, 556.

Canizzo, J. K., Ghosh, P., and Wheeler, J. C. 1983, in *Cataclysmic Variables and Low Mass X-ray Binaries*, ed. J. Patterson and D.Q. Lamb, Reidel.

Cavallo, G., and Jelly, J. V. 1975, *Ap. J. (Letters)*, **201**, L113.

Charles, P. A., *et al.* 1980, *Ap. J.*, **37**, 154.

Chester, T. J. 1978, *Ap. J.*, **222**, 652.

Chevalier, C., *et al.* 1981, *Astr. Ap.*, **100**, L1; and Erratum in *Astr. Ap.*, **103**, 428, (1981).

Chevalier, C., Ilovaisky, S. A., and Charles, P. A. 1985, *Astr. Ap.*, **147**, L3.

Chevalier, C., and Ilovaisky, S. A. 1987, *Astr. Ap.*, submitted.

Cline, T. L. 1980, *Comm. Astrophys.*, **9**, 13.

Cline, T. L. 1982, in *Gamma Ray Transients and Related Astrophysical Phenomena*, AIP Conf. Proc. **77**, 17.

Cline, T. L., Desai, U. D., Pizzichini, G., Spizzichino, A., Trainor, J., Klebesadel, R., Ricketts, M., and Helmken, H. 1979, *Ap. J. (Letters)*, **229**, L47.

Cline, T. L., *et al.* 1980, *Ap. J. (Letters)*, **237**, L1.

Cline, T. L., *et al.* 1981, *Ap. J. (Letters)*, **246**, L133.

Cline, T. L., *et al.* 1982, *Ap. J. (Letters)*, **255**, L45.

Cline, T. L., *et al.* 1984, *Ap. J. (Letters)*, **286**, L15.

Colgate, S. A., and Petschek, A. G. 1981, in *Ap. J.*, **248**, 771.

Cominsky, L. R., London, R. A., and Klein, R. I. 1987, *Ap. J.*, **315**, 162.

Cominsky, L. R., and Wood, K. S. 1984, *Ap. J.*, **283**, 765.

Connors, A., Serlemitsos, P. J., and Swank, J. H. 1986, *Ap. J.*, **303**, 769.

Corbet, R. H. D., Thorstensen, J. R., Charles, P. A., Menzies, J. W., Naylor, T., and Smale, A. P. 1986, *M. N. R. A. S.*, **222**, 15p.

Corso, G. J., Ringwald, F. A., and Harris, R. W. 1987, *Astr. Ap.*, **183**, L9.

Courvosier, T. J.-L., Parmar, A. N., Peacock, A., and Pakull, M. 1986, *Ap. J.*, **309**, 265.

Crampton, D., Cowley, A. P., Stauffer, J., Ianna, P., and Hutchings, J. B. 1986, *Ap. J.*, **306**, 599.

Czerny, M., and Sztajno, M. 1983, *Acta Astr.*, **33**, 213.

Czerny, M., Czerny, B., and Grindlay, J. E. 1987, *Ap. J.*, **312**, 122.

Dennis, B. R., Frost, K. J., Kiplinger, A. L., Orwig, L. E., Desai, U., and Cline, T. L. 1982, in *Gamma-Ray Transients and Related Astrophysical Phenomena*, AIP Conf. Proc. **77**, ed. R. E. Lingenfelter, H. S. Hudson, and D. M. Worral, pg. 153.

Dewey, R. J., and Cordes, J. M. 1987, *Ap. J.*, **321**, 780.

Ebisuzaki, T., Hanawa, T., and Sugimoto, D. 1983, *Publ. Astron. Soc. Japan*, **35**, 17.
— 1984, *Publ. Astron. Soc. Japan*, **36**, 551.
Ebisuzaki, T., and Nomoto, K., 1985, in *Galactic and Extragalactic Compact X-Ray Sources*, ed. Y. Tanaka and W. H. G. Lewin, (Tokyo: ISAS), pg. 101.
— 1986, *Ap. J. (Letters)*, **305**, L67.
Ebisuzaki, T. 1987, *Publ. Astron. Soc. Japan*, **39**, 287.
Ebisuzaki, T., and Nakamura, N. 1988, *Ap. J.*, in press.
Epstein, R. I. 1985, *Ap. J.*, **291**, 822.
Epstein, R. I. 1986, *Proceedings of IAU Coll.*, **89**, publ. in *Lecture Notes in Physics*, **255**, Springer-Verlag, ed. D. Mihalas and K.-H. A. Winkler, pg. 305.
Epstein, R. I. 1988, *Phys. Rep.*, in press.
Ergma, E. V. 1984, *Adv. Space Res.*, **3**, No. 10-12, 275.
Evans, W. D., *et al.* 1980, *Ap. J. (Letters)*, **237**, L7.
Fabian, A. C., Pringle, J., and Rees, M. 1975, *M. N. R. A. S.*, **172**, 15P.
Fabian, A. C., Icke, V., and Pringle, J. E. 1976, *Ap. Space Sci.*, **42**, 77.
Fenimore, E. E., Klebesadel, R. W., and Laros, J. G. 1983, *Adv. Space Res.*, **3**, 207.
Fishman, G. J., Duthie, J. G., and Dufor, R. J. 1981, *Ap. Space Sci.*, **75**, 135.
Flohrer, J., Greiner, J., D. Möhlmann, and Wenzel, W. 1986, *Adv. Space Res.*, **6**, (4), 55.
Foster, A. J., Ross, R. R., and Fabian, A. C. 1987, *M. N. R. A. S.*, **228**, 259.
Fryxell, B. A., and Woosley, S. E. 1982, *Ap. J.*, **258**, 733.
Fujimoto, M. Y., Sztajno, M., Lewin, W. H. G., and van Paradijs, J. 1987, *Ap. J.*, **319**, 902.
Fujimoto, M. Y., and Taam, R. E. 1986, *Ap. J.*, **305**, 246.
Garnavich, P., and Temple, S. 1987a, *Bull. Am. Astron. Soc.*, **19**, 645.
— 1987b, preprint.
Gilman, D., Metzger, A. E., Parker, R. H., Evans, L. G., and Trombka, J. I. 1980; *Ap. J.*, **236**, 951.
Goodman, J. 1986, *Ap. J. (Letters)*, **308**, L47.
Goodman, J., and Hut, P. 1985, *Dynamics of Star Clusters*, Proc. of 113th IAU Symp., Princeton, May 1984, Reidel, Dordrecht.
Gottwald, M., Haberl, F., Parmar, A. N., and White, N. E. 1987a, *Ap. J.*, **308**, 213.
Gottwald, M., Stella, L., White, N. E., and Barr, P. 1987b, *M. N. R. A. S.*, in press
Greiner, J., and Flohrer, J. 1985, *Inf. Bull. Var. Stars*, Budapest, No. 2765.
Greiner, J., Flohrer, J., Wenzel, W., and Lehmann, T. 1987, *Adv. Space Res.*, submitted.
Grindlay, J. E., Wright, E., and McCrosky, R. 1974, *Ap. J. Lettr.*, **192**, L113.
Grindlay, J. E., Gursky, H., Schnopper, H., Parsignault, D. R., Heise, J., Brinkman, A. C., and Schrijver, J. 1976, *Ap. J. (Letters)*, **205**, L127.
Grindlay, J. E., McClintock, J. E., Canizares, C. R., van Paradijs, J., Cominsky, L., Li, F. K., and Lewin, W. H. G. 1978, *Nature*, **274**, 567.
Grindlay, J. E., *et al.* 1982, *Nature*, **300**, 730.
Grindlay, J. E. 1984, in *High Energy Transients in Astrophysics*, AIP Conf. Proc. **115**, Santa Cruz, July 1983, ed. S. E. Woosley, p. 306.
— 1985, in Proc. of Japan–U.S. Seminar on *Galactic and Extragalactic Compact X-Ray Sources*, ed. Y. Tanaka, and W. H. G. Lewin, (Tokyo: ISIS), pg. 215.
— 1986, in *The Evolution of Galactic X-Ray Binaries*, ed. J. Trümper *et al.*, D. Reidel Publ., pg. 25.
— 1987, in *13th Texas Symp. on Relativ. Astrophys.*, Chicago, Ill, December 1986, ed. M. P. Ulmer, World Scientific, pg. 509.
Grindlay, J. E., and Hertz, P. 1985, in *Proc. 7th North American Workshop on Cataclysmic Variables and Low-Mass X-ray Binaries*, ed. D. Q. Lamb, and J. Patterson, D. Reidel, Dordrecht, pg. 79.
Grindlay, J. E., and Seaquist, E. R. 1985, *Ap. J.*, **310**, 172.
Hackwell, J. A., Grasdalen, G. L., Gehrz, R. D., van Paradijs, J., Cominsky, L., and Lewin W. H. G. 1979, Ap. J. (Letters), **233**, L115.
Halliday, I., Blackwell, A. T., and Griffin, A. A. 1978, *Roy. Astr. Soc. Can.*, **72**, 15.
Halliday I., Feldman, P. A., and Blackwell, A. T. 1987, *Ap. J. Lettr.*, **320**, L153.
Hameury, J. M. 1985, *Ann. Phys. France*, **10**, 369.

Hameury, J. M., Bonazzola, S., Heyvaerts, J., and Ventura, J. 1982, *Astr. Ap.*, **111**, 242.
Hameury, J. M., Heyvaerts, J., and Bonazzola, S. 1983a, *Astr. Ap.*, **121**, 259.
Hameury, J. M., Bonazzola, S., and Heyvaerts, J. and J. P. Lasota 1983b, *Astr. Ap.*, **128**, 369.
Hameury, J. M., Bonazzola, S., and Heyvaerts, J. and J. P. Lasota 1984, *Adv. Space Res.*, **3**, (10-12), 297.
Hameury, J. M., Lasota, J. P., Bonazzola, S., and Heyvaerts, J. 1985, *Ap. J.*, **293**, 56.
Hameury, J. M., King, A. R., and Lasota, J. P. 1986, *Astr. Ap.*, **162**, 71.
— 1987, *Astr. Ap.*, **171**, 140.
— 1988, *Astr. Ap.*, in press.
Hanawa, T., Sugimoto, D., and Hashimoto, M. 1983, *Publ. Astron. Soc. Japan*, **35**, 491.
Hartmann, D., and Pogge, R. W. 1987, *Ap. J.*, **318**, 363.
Hartmann, D., Woosley, S. E., and Arons, J. 1987, *Bull. Am. Astr. Soc.*, **18**, 928.
— 1988, *Ap. J.*, in press.
Hartmann, D., and Epstein, R. 1988, *Ap. J.*, submitted.
Hartmann, D., Pogge, R. W., Hurley, K., Vrba, F., and Jennings, M. 1988, *Nature*, submitted.
Harwitt, M., and Salpeter, E. E. 1973, *Ap. J. (Letters)*, **186**, L37.
Hatchet, S., Buff, J., and McCray, R. 1976, *Ap. J.*, **206**, 847.
Hayakawa, S. 1981, *Adv. Space Res.*, **1**, 149.
— 1985, *Phys. Rep.*, **121**, 317.
Helfand, D. J. 1983, in *Proceedings of IAU Symp. No. 101 on "Supernova Remnants And Their X-Ray Emission"*, ed. J. Danziger and P. Gorenstein, (D. Reidel), Dordrecht, pg. 471
Helfand, D. J., and Vrtilek, S. D. 1983, *Nature*, **304**, 41.
Helfand, D. J., and Long, K. S. 1979, *Nature*, **282**, 589.
Higdon, J. C., and Lingenfelter, R. E. 1984, in *High Energy Transients in Astrophysics*, AIP Conf. Proc. **115**, 568.
— 1986, *Ap. J.*, **307**, 197.
— 1988, in *Ann. Rev. Astr. Astrophys.*, **26**, in press.
Hjellming, R., and Ewald, S. 1981, *Ap. J. (Letters)*, **246**, L137.
Hoffman, J. A., Lewin, W. H. G., and Doty, J. 1977, *Ap. J. (Letters)*, **217**, L23.
Hoffman, J. A., Marshall, H. L., and Lewin, W. H. G. 1978, *Ap. J. (Letters)*, **221**, L57.
Hoffman, J. A., Cominsky, L., and Lewin, W. H. G. 1980, *Ap. J. (Letters)*, **240**, L27.
Hoshi, R. 1984, *Ap. J.*, **247**, 628.
Howard, W. M., Wilson, J. R., and Barton, R. T. 1981, *Ap. J.*, **249**, 302.
Hudec, R. *et al.* 1984, *Adv. Space Res.*, **3**, (10-12), 115.
Hudec, R. 1985, *Space Sci. Rev.*, **40**, 715.
Hudec, R. *et al.* 1986, *Adv. Sp. Res.*, **6**, 51.
Hudec, R. *et al.* 1987, *Astr. Ap.*, **175**, 71.
Hudec, R., Borovicka, J., Danis, S., Franc, V., Peresty, R., and Valnicek, B. 1988, *Adv. Space Res.*, submitted.
Hueter, G. J. 1984, in *High Energy Transients in Astrophysics*, AIP Conf. Proc. **115**, ed. S. E. Woosley, pg. 373.
Hurley, K. 1983, *Adv. Space Res.*, **3**, (4), 163.
— 1987, in *The Origin and Evolution of Neutron Stars*, IAU Symposium No. **125**, ed. D. J. Helfand, and J.-H. Huang, pg. 489.
Hurley, K., and Vedrenne, G. 1986, *Gamma-Ray Astronomy, Adv. Space Res.*, **6**, (4).
— 1985, *Nature*, **315**, 715.
— 1987a, *Astro. Lett. and Comm.*, **25**, 145.
— 1987b, in *The Origin and Evolution of Neutron Stars*, ed. D. J. Helfand, and J.-H. Huang, pg. 489.
Imamura, J. N., and Epstein R. I. 1987, *Ap. J.*, **313**, 711.
Imamura, J. N., Steiman-Cameron, T. Y., and Middleditch, J. 1987, *Ap. J. (Letters)*, **320**, L41.
Inoue, H. 1984, *Publ. Astr. Soc. Japan*, **36**, 855.
Inzani, P., *et al.* 1978, *Ap. Space Sci.*, **56**, 239.
— 1982, in *Gamma Ray Transients and Related Astrophysical Phenomena*, AIP Conf.

Proc. **77**, ed. R. E. Lingenfelter, H. S. Hudson, and D. M. Worral, pg. 79.

Jennings, M. C. 1983, *Ap. J.*, **273**, 309.

Joss, P. C. 1977, *Nature*, **270**, 310.

Joss, P. C., and Rappaport, S. A. 1984, *Ann. Rev. Astr. Ap.*, **22**, 537.

Kallman, T. R., and McCray, R. 1980, *Ap. J.*, **242**, 615.

—— 1982, *Ap. J. Suppl.*, **50**, 263.

Katoh, M., *et al.* 1984, in *High Energy Transients in Astrophysics*, AIP Conf. Proc. **115**, ed. S. E. Woosley, pg. 390.

Katz, J. I. 1983, in *Positron-Electron Pairs in Astrophysics*, AIP Conf. Proc. **101**, ed. M. L. Burns, A. K. Harding, and R. Ramaty., New York, pg. 65.

Katz, J. I. 1985, *Astrophys. Letters*, **24**, 183.

Katz, J. I. 1987, preprint.

Katz, B., *et al.* 1986, *Ap. J. Lett.*, **307**, L33.

Kirk, J. G. 1986, *Astr. Ap.*, **158**,305.

Klebesadel,R. W., Strong, I. B., and Olsen, R. A. 1973, *Ap. J. Lettr.*, **182**, L85.

Klemola, A. R. 1983, *Publ. Astron. Soc. Pac.*, **95**, 241.

Kouveliotou, C., *et al.* 1987, *Ap. J. (Letters)*, **322**, L21.

Kouveliotou, C., Desai, U. D., Cline, T. L., Dennis, B. R., Fenimore, E. E., Klebesadel, R. W., and Laros, J. G. 1988, *Ap. J.*, submitted.

Lamb, D. Q. 1984, in *Ann. N. Y. Acad. Sci.*, **422**, 237.

Langmeier, A., Sztajno, M., Hasinger, G., Trümper, J., and Gottwald, M. 1987, *Ap. J. (Letters)*, submitted.

Laros, J. G., *et al.* 1981, *Ap. J. (Letters)*, **245**, L63.

Laros, J. G., Evans, W. D., Fenimore, E. E., Klebesadel, R. W., Shulman, S., and Fritz, G. 1984, *Ap. J.*, **286**, 681.

Laros, J. G., *et al.* 1985, *Ap. J.*, **290**, 728.

Laros, J. G., *et al.* 1987, *Ap. J. (Letters)*, **320**, L111.

Laros, J. G., Fenimore, E. E., and Klebesadel, R. W. 1987, in *13th Texas Symp. on Relativ. Astrophys.*, Chicago, Ill, December 1986, ed. M. P. Ulmer, World Scientific, pg. 563.

Laros, J. G. 1988, *Nature*, submitted.

Lasota, J. P., and Belli, B. M. 1983, *Nature*, 304, 139.

Lawrence, A., *et al.* 1983, *Ap. J.*, **271**, 793.

Levine, A., Ma, C. P., McClintock, J., Rappaport, S., van der Klies, M., and Verbunt, F. 1987, *Ap. J.*, in press.

Lewin, W. H. G., *et al.* 1976, *M.N.R.A.S.*, **177**, P 83.

Lewin, W. H. G. 1977, *M.N.R.A.S.*, **179**, 43.

Lewin, W. H. G., Cominsky, L., and Oda, M. 1979, IAU *Circ.*, No. 3420.

Lewin, W. H. G., and Joss, P. C. 1981, *Space Sci. Rev.*, **28**, 3.

—— 1983, in *Accretion Driven Stellar X-Ray Sources*, ed. W. H. G. Lewin and E. P. J. van den Heuvel, Cambridge Univ. Press, pg. 41.

Lewin, W. H. G., Vacca, W. D., and Basinska, E. M. 1984, *Ap. J. (Letters)*, **277**, L57.

Lewin, W. H. G. 1985, in *Galactic and Extragalactic Compact X-Ray Sources*, ed. Y. Tanaka and W. H. G. Lewin, (Tokyo: ISAS), pg. 89.

Lewin, W. H. G., Pennix, W., van Paradijs, J., Damen, E., Sztajno, M., Trümper, J., and van der Klies, M. 1987a, *Ap. J.*, **319**, 893.

Lewin, W. H. G., van Paradijs, J., Damen, E., Jansen, F., McCall, M. L., Feldman, P. A., and Tapping, K. F. 1987b, *A. J.*, **94**, 429.

Liang, E. P. 1985, *Nature*, **313**, 202.

—— 1987, *Comments Astrophys.*, **12**, 35.

Liang, E. P., and Petrosian, V. 1986, *Gamma-Ray Bursts*, AIP Conf. Proc. **141**.

Lin, D. N. C., and Taam, R. E. 1984, in *High Energy Transients in Astrophysics*, AIP Conf. Proc. **115**, ed. S. E. Woosley, pg. 83.

Lingenfelter, R. E., Hudson, H. S., and Worrall, D. M. 1982, *Gamma Ray Transients and Related Astrophysical Phenomena*, AIP Conf. Proc. **77**.

Lingenfelter, R. E., and Hueter, G. J. 1984, in *High Energy Transients in Astrophysics*, AIP Conf. Proc. **115**, ed. S. E. Woosley, Santa Cruz 1983, pg. 558.

Livio, M., and Taam, R. E. 1987, *Nature*, **327**, 398.

London, R. A., Taam, R. E., and Howard, D. M. 1984, *Ap. J. (Letters)*, **287**, L27.

—— 1986, *Ap. J. (Letters)*, **306**, 170.

London, R. A., and Cominsky, L. R. 1982, *Bull. Am. Astr. Soc.*, **14**, 867.

London, R. A., and Cominsky, L. R. 1983, *Ap. J. (Letters)*, **275**, L59.

London, R. A. 1984, in *High Energy Transient in Astrophysics*, AIP Conf. Proc. **115**, ed. S. E. Woosley, pg. 581.

MacRobert, A. 1985a, *Sky and Telescope*, **69**, 148.

MacRobert, A. 1985b, *Sky and Telescope*, **70**, 54.

Makino, F., *et al.* 1987, *Astro. Lett. Comm.*, **25**, 223.

Maley, P. D. 1987, *Ap. J. (Letters)*, **317**, L39.

Manchester, R. N. 1987, in *The Origin and Evolution of Neutron Stars*, IAU Symposium No. **125**, ed. D. J. Helfand, and J.-H. Huang, pg. 3.

Mandolesi, N., Morigi, G., Inzani, P., Sironi, G., Delli Santi, F. S., Delpino, F., and Petessi, M. 1977, *Nature*, **266**, 427.

Maraschi, L. , and Cavaliere, A. 1977, in *Highlights of Astronomy*, **4**, 127.

Matsuoka, M., *et al.* 1984, *Ap. J.*, **283**, 774.

Matsuoka, M. 1985, in *Galactic and Extragalactic Compact X-Ray Sources*, ed. Y. Tanaka and Lewin, W. H. G. (Tokyo: ISAS), p.45.

Matz, S. M., Forrest, D. J., Vestrand, W. T., Chupp, E. L., Share, G. H., and Rieger, E. 1985, *Ap. J. (Letters)*, **288**, L37.

Mazets, E. P., Golenetskii, S. V., Ilinskii, V. N., Aptekar', R. L., and Guryan, Yu. A. 1979, *Nature*, **282**, 587.

Mazets, E. P., *et al.* 1980, *Sov. Astr. Lett.*, **5**, 163.

Mazets, E. P., Golenetskii, S. V., Guryan, Yu. A., and Ilynskii, V. N. 1982, *Ap. Space Sci.*, **84**, 173.

Mazets, E. P., Golenetskii, S. V., Petrov, G. G., Savvin, A. V., Prilutskii, O. F., and Rodin, V. G. 1987, Proc. 20th Int. Cos. Ray Conf., Moscow 1987, paper OG 9.1-1.

McClintock, J. E., Canizares, C. R., and Tarter, C. B. 1975, *Ap. J.*, **198**, 641.

McClintock, J. E., Canizares, R. C., van Paradijs, J., Cominsky, L., Li, F. K., and Lewin, W. H. G. 1979, *Nature*, **279**, 47.

Meegan, C. A., Fishman, G. J., and Wilson, R. B. 1985, *Ap. J.*, **291**, 479.

Melia, F., Rappaport, S., and Joss, P. C. 1986, *Ap. J. (Letters)*, **305**, L51.

Melia, F., and Joss, P. C. 1986, in *Radiation Hydrodynamics in Stars and Compact Objects*, Proc. of IAU Colloquium No. 89, Copenhagen, June 1985, ed. D. Mihalas and K.-H. A. Winkler, Spronger-Verlag, pg. 283.

Melia, F. 1987, *Fund. Cosmic Phys.*, **12**, 97.

—— 1988, *Ap. J. (Letters)*, **324**, L21.

Mészáros, P. 1984, *Space Sci. Rev.*, **38**, 325.

Mészáros, P., and Nagel, W. 1985a, *Ap. J.*, **298**, 147.

—— 1985b, *Ap. J.*, **299**, 138.

Mészáros, P., Bussard, R. W., and Hartmann, D. 1986, in *Gamma-Ray Bursts*, ed. E. P. Liang, and V. Petrosian, AIP Conf. Proc. **141** p. 121.

Metzger, A. E. *et al.* 1974, *Ap. J. (Letters)*, **194**, L19.

Michel, F. C. 1985, *Ap. J.*, **290**, 721.

Milgrom, M. 1978, *Astr. Ap.*, **67**, L25.

Moskalenko, E. I., *et al.* 1988, *Adv. Space Res.*, submitted.

Motch, C., Pedersen, H., Ilovaisky, S. A., Chevalier, C., Hurley, K., and Pizzichini, G. 1984, *Astron. Ap.*, **145**, 201.

Murakami, T. *et al.* 1980a, *Publ. Astr. Soc. Japan*, **32**, 513.

Murakami, T. *et al.* 1980b, *Publ. Astr. Soc. Japan*, **32**, 543.

Murakami, T. *et al.* 1983, *Publ. Astr. Soc. Japan*, **35**, 531.

Newman, M. J., and Cox, A. N. 1980, *Ap. J.*, **242**, 319.

Nolan, *et al.* 1984, in *High Energy Transients in Astrophysics*, AIP Conf. Proc. **115**, ed. S. E. Woosley, pg. 399.

Osaki, Y. 1985, *Astr. Ap.*, **144**, 369.

Paczyński, B. 1986, *Ap. J. (Letters)*, **308**, L43.
Paczyński, B., and Prószyński, M. 1986, *Ap. J.*, **302**, 519.
Parmar, A. N., White, N. E., Giommi, P., and Gottwald, M. 1986, *Ap. J.*, **308**, 199.
Pedersen, H., *et al.* 1982a, *Ap. J.*, **263**, 325.
— 1982b, *Ap. J.*, **263**, 340.
Pedersen, H., Motch, C., Tarenghi, M., Danziger, J., Pizzichini, G., and Lewin, W. H. G. 1983, *Ap. J. (Letters)*, **270**, L43.
Pedersen, H., *et al.* 1984, *Nature*, **312**, 46.
Pedersen, H. 1987, *Adv. Space Res.*, **6**, (4), 61.
Pennix, W., van Paradijs, J., and Lewin, W. H. G. 1987, *Ap. J. (Letters)*, **321**, L67.
Pizzichini, G. *et al.* 1981, *Space Sci. Rev.*, **30**, 467.
Pizzichini, G. *et al.* 1986, *Ap. J.*, **301**, 641.
Priedhorsky, W. C. 1986, *Ap. Space Sci.*, **126**, 89.
Priedhorsky, W. C., and Holt, S. S. 1987, *Space Sci. Rev.*, **45**, 291.
Pye, J. P., and McHardy, I. M. 1983, *M. N. R. A. S.*, **205**, 875.
Rappaport, S. A., and Joss, P. C. 1983, in *Accretion-driven stellar X-ray sources*, ed. W. H. G. Lewin, and E. P. J. van den Heuvel, Cambridge Univ. Press, pg. 1.
Rappaport, S. A., and Joss, P. C. 1985, *Nature*, **314**, 242.
Ricker, G., Doty, J. P., Vallerga, J. V., and Vanderspek, R. K. 1984, in *High Energy Transients in Astrophysics*, AIP Conf. Proc. **115**, ed. S. E. Woosley, pg. 669.
Ricker, G. R., Vanderspek, R. K., and Ajhar, E. A. 1986, *Adv. Space Res.*, **6**, (4), 75.
Ricker, G. R., *et al.* 1987, *High Energy Transient Explorer*, A proposal submitted to NASA.
Romani, R. W. 1987, preprint.
Rothschild, R. E., and Lingenfelter, R. E. 1984, *Nature*, **312**, 737.
Ruderman, M. 1987, in *13th Texas Symp. on Relativ. Astrophys.*, Chicago, Ill, December 1986, ed. M. P. Ulmer, World Scientific, pg. 448.
Schaefer, B. E. 1981, *Nature*, **294**, 722.
— 1983, *Publ. Astr. Soc. Pac.*, **95**, 1019.
— 1985, *A. J.*, **90**, 1363.
— 1986, *Adv. Space Res.*, **6**, (4), 47.
Schaefer, B. E., and Ricker, G. R. 1983, *Nature*, **302**, 43.
Schaefer, B. E., Seitzer, P., and Bradt, H. V. 1983, *Ap. J. (Letters)*, **270**, L49.
Schaefer, B. E., and Cline, T. 1985, *Astrophys. J.*, **289**, 490.
Schaefer, B. E., *et al.* 1984a, *Ap. J. (Letters)*, **286**, L1.
Schaefer, B. E., Vanderspek, R., Bradt, H. V., and Ricker, G. R. 1984b, *Ap. J.*, **283**, 887.
Schaefer, B. E., Pedersen, H., Gouiffes, C., Poulsen, J. M., and Pizzichini, G. 1987a, *Astr. Ap.*, **174**, 338.
Schaefer, B. E., *et al.* 1987b, *Ap. J.*, **320**, 398.
Schaefer, B. E., *et al.* 1987c, *Ap. J.*, **313**, 226.
Share, G. H., Matz, S. M., Messina, D. C., Nolan, P. L., Chupp, E. L., Forrest, D. J., and Cooper, J. F. 1986, *Adv. Space Res.*, **6**, No. 4, 15.
Scholz, M. 1984, *Inf. Bull. Var. Stars*, Budapest, No. 2615.
Schwartz, R. 1986, *Sky & Telescope*, December 1986, pg. 560.
Smith, A. 1987, Proc. of NATO ASI Conf. on *Hot Thin Plasmas in Astrophysics*, Cargese, 1987.
Starrfield, S. G., Sparks, W. M., and Truran, J. W. 1985, *Ap. J.*, **291**, 136.
Stella, L., Priedhorsky, W. C., and White, N. E. 1987, *Ap. J. (Letters)*, **312**, L17.
Sturrok, P. A. 1986, *Nature*, **321**, 47.
Sugimoto, D., Ebisuzaki, T., and Hanawa, T. 1984, *Publ. Astr. Soc. Jap.*, **36**, 839.
Swank, J. H., Becker, R. H., Boldt, E. A., Holt, S. S., Pravdo, S. H., and Serlemitsos, P. J. 1977, *Ap. J. (Letters)*, **212**, L73.
Sztajno, M., van Paradijs, J., Lewin, W. H. G., Trümper, J., Stollman, G., Pietsch, W., and van der Klies, M. 1985, *Ap. J.*, **299**, 487.
Sztajno, M., van Paradijs, J., Lewin, W. H. G., Langmeier, A., Trümper, J., and Pietsch, W. 1986, *M. N. R. A. S.*, **222**, 499.
Taam, R. E. 1982, *Ap. J.*, **258**, 761.

Taam, R. E., and Lin, D. N. C. 1984, *Ap. J.*, **287**, 761.
Taam, R. E. 1985, *Ann. Rev. Nucl. Part. Sci.*, **35**, 1.
—— 1987, in *13th Texas Symp. on Relativ. Astrophys.*, Chicago, Ill, December 1986, ed. M. P. Ulmer, World Scientific, pg. 546.
Taam, R. E., and van den Heuvel, E. P. J. 1986, *Ap. J.*, **305**, 235.
Tawara, Y., Hayakawa, S., and Kii, T. 1984a, *Publ. Astron. Soc. Japan*, **36**, 845.
Tawara, Y., Hirano, T., Kii, T., Matsuoka, M., and Murakami, T. 1984a, *Publ. Astr. Soc. Japan*, **36**, 861.
Tawara, Y. *et al.* 1984b, *Ap. J.*, (Letters), **276**, L41.
Teegarden, B. J. *et al.* 1984, in *High Energy Transients in Astrophysics*, AIP Conf. Proc. **115**, ed. S. E. Woosley, Santa Cruz 1983, pg. 687.
—— 1986, *Adv. Space Res.*, **6**, (4), 93.
Terrell, J., Fenimore, E. E., Klebesadel, R. W., and Desai, U. D. 1982, *Ap. J.*, **254**, 279; Erratum: *Ap. J.*, **269**, 806 (1983).
Thorstensen, J., Charles, P., and Bowyer, S. 1986, *Ap. J. (Letters)*, **220**, L131.
Tremaine, S., and Zytkow, A. N. 1986, *Ap. J.*, **301**, 155.
Trümper, J., van Paradijs, J., Sztajno, M., Lewin, W. H. G., Pietsch, W., Krautter, J., Stollman, G., and van der Klies, M. 1985, *Space Sci. Rev.*, **40**, 255.
Tsuruta, S. 1985, *Neutron stars: current cooling theories and observational results*, MPA report **183**.
Turner, M. J. L., *et al.* 1985, *Space Sci. Rev.*, **40**, 249.
Turner, M. J. L., and Breedon, L. M. 1984, *M. N. R. A. S.*, **208**, 29p.
van Amerongen, S., Pedersen, H., and van Paradijs, J. 1987, *Astr. Ap.*, **185**, 147.
van Buren, D. 1981, *Ap. J.*, **249**, 297.
van den Heuvel, E. P. J., van Paradijs, J. A., and Taam, R. E. 1986, *Nature*, **322**, 153.
Vanderspek, R. K., Zachary, D. S., and Ricker, G. R. 1986, *Adv. Space Res.*, **6**, (4), 69.
van Paradijs, J., Verbunt, F., van der Linden, T., Pedersen, H., and Wamsteker, W. 1980, *Ap. J. (Letters)*, **241**, L161.
van Paradijs, J. 1981, *Astr. Ap.*, **103**, 140.
van Paradijs, J. 1982, *Astr. Ap.*, **107**, 51.
van Paradijs, J. 1983, in *Accretion-driven Stellar X-Ray Sources*, ed. W. H. G. Lewin and E. P. J. van den Heuvel, Cambridge Univ. Press, pg. 189.
van Paradijs, J., and Verbunt, F. 1984, in *High Energy Transients in Astrophysics*, ed. S. E. Woosley, AIP Conf. Proc. No. 115, p. 49.
van Paradijs, J., *et al.* 1980, *Astrophys. J. Lettr.*, **241**, L161.
van Paradijs, J., and Lewin, W. H. G. 1985, *Astr. Ap.*, **142**, 361.
—— 1987, *Astr. Ap.*, **172**, L20.
Vanderspek, R. K., Zachary, D. S., and Ricker, G. R. 1987, *Adv. Space Res.*, **6**, in press.
Vedrenne, G., and Hurley, K. 1983, *Gamma-Ray Astronomy in Perspective of Future Space Experiments, Adv. Space Res.*, **3**, (4).
Verbunt, F., van Paradijs, J., and Elson, R. 1984, *M.N.R.A.S.*, **210**, 899.
Verter, F. 1982, *Phys. Rep.*, **81**, 294.
Voges, W. 1985, *MPE Report*, **191**, pp 1-123.
Waki, I., *et al.* 1984, *Publ. Astron. Soc. Japan*, **36**, 819.
Wallace, R. K., Woosley, S. E., and Weaver, T. A. 1982, *Ap. J.*, **258**, 696.
Walter, F. M., Bowyer, S., Mason, K. O., Clarke, J. T., Henry, J. P., Halpern, J., and Grindlay, J. E. 1982, *Ap. J. (Letters)*, **253**, L67.
Warner, B. 1986, *Ap. Space Sci.*, **122**, 397.
Wdowiak, T. J., and Clifton, K. S. 1985, *Ap. J.*, **295**, 171.
Weisskopf, M. C. 1987, *Astro. Lett. Comm.*, **26**, 1.
Wheaton, Wm. A., *et al.* 1973, *Ap. J. (Letters)*, **185**, L57.
White, N. E., and Swank, J. H. 1982, *Ap. J. (Letters)*, **253**, L61.
White, N. E., Swank, J. H., and Holt, S. S. 1983, *Ap. J.*, **270**, 711.
White, N. E., Kaluzienski, J. L., and Swank, J. H. 1984, in *High Energy Transients in Astrophysics*, ed. S. E. Woosley, AIP Conf. Proc. No. 115, p. 31.
White, N. E., and Mason, K. O. 1985, *Space Sci. Rev.*, **40**, 167.

White, N. E., Peacock, A., Hasinger, G., Mason, K. O., Manzo, G., Taylor, B. G., and Branduardi-Raymont, G. 1986, *M.N.R.A.S*, **218**, 129.
White, N. E. 1986, *in The Evolution of Galactic X-Ray Binaries*, ed. J. Trümper, W. H. G. Lewin, and W. Brinkmann, D. Reidel Publ. Comp., 227.
White, N. E. 1987, in *The Origin and Evolution of Neutron Stars*, IAU Symposium No. **125**, ed. D. J. Helfand, and J.-H. Huang, pg. 135.
White, N. E., Stella, L., and Parmar, A. N. 1987, *Astrophys. J.*, submitted.
Wilson, A. S. 1987, *Astro. Lett. Comm.*, **26**, 99.
Wood, K. S., Byram, E. T., Chubb, T. A., Friedman, H., Meekins, J. F., Share, G. H., and Yentis, D. J. 1981, *Ap. J.*, **247**, 632.
Woosley, S. E., and Taam, R. E. 1976, *Nature*, **263**, 101.
Woosley, S. E., and Wallace, R. K. 1982, *Astrophys. J.*, **258**, 716.
Woosley, S. E. 1982, in *Accreting Neutron Stars*, ed. W. Brinkmann, J. Trümper, MPE Report **177**, p. 189.
— 1984a, *High Energy Transients in Astrophysics*, AIP Conf. Proc. **115**, Santa Cruz 1983, ed. S. E. Woosley.
— 1984b, in *High Energy Transients in Astrophysics*, AIP Conf. Proc. **115**, Santa Cruz 1983, ed. S. E. Woosley. pg. 485.
Woosley, S. E., and Weaver, T. A. 1984, in *High Energy Transients in Astrophysics*, AIP Conf. Proc. **115**, Santa Cruz 1983, ed. S. E. Woosley. pg. 273.
Woosley, S. E. 1986, *Nucleosynthesis and Chemical Evolution*, 16th Advanced Course, Saas-Fee 1986, pp.1-195.
Woosley, S. E. 1987, in *The Origin and Evolution of Neutron Stars*, IAU Symp. No. 125, Nanjing, China May 1986, ed. D. J. Helfand and J.-H. Huang, D. Reidel Publ., pg. 255.

Supernovae and Supernova Remnants

Roger A. Chevalier

Department of Astronomy, University of Virginia

Frederick D. Seward

Harvard-Smithsonian Center for Astrophysics

1. Introduction

Early observations of supernovae showed that they could be divided into two major types defined by whether hydrogen lines are absent (Type I, hereafter SNI) or present (Type II, SNII) in their spectra. Spectrophotometry of supernovae (Kirshner *et al.* 1973) showed that in both cases, most of the light near maximum is emitted in a cooling continuum. This information led to the basic models for the light curves. Type II supernovae were interpreted as instantaneous explosions in massive red supergiant stars at the ends of their lives and Type I events as explosions of white dwarfs (see Woosley and Weaver 1986 for a review). Power from radioactive decay is crucial for the total radiation from SNI and the late emission from SNII.

More recently, it has been recognized that there are two major types of Type I supernovae; they have been designated Type Ia and Type Ib. While Type Ia events are associated with an old stellar population, the Type Ib's are associated with a very young population. Multiwavelength observations have been crucial in determining the differences between these subtypes. Here we take the point of view that SN Ib have a massive star origin, although this is still controversial (e.g. Branch and Nomoto 1986).

Our knowledge of supernovae comes from observations of extragalactic events because there has not been an observable galactic supernova since the development of astronomical instrumentation. Supernova remnants, with ages ≥ 300 years, are observed in our galaxy and their proximity allows detailed imaging studies. Multiwavelength images are presented here for a number of remnants.

In Section 2, we describe multiwavelength observations of the three major types of supernovae. For each case, we concentrate on a small number of supernovae that are particularly good examples of the type under consideration. For Type II supernovae, two major subtypes are represented by SN1979C and SN1987A. New types of multiwavelength observations of SN1987A have been possible because of its proximity and are continuing at the time of writing (1988 January). In section 3, we interpret the observations in terms of the initial energy deposition, radioactive energy input, interaction with the circumstellar medium, and pulsar energy input. Future prospects for multiwavelength observations are discussed. In section 4, we discuss observations of supernova remnants, with an emphasis on a small number of well-studied galactic remnants.

2. Observations of Supernovae

Type Ia Supernovae

Type Ia supernovae have a characteristic optical light curve with a rounded peak and late exponential decay. The long tail, observed for >700 days in the case of SN1972E (Kirshner and Oke 1975), is one of the indications for energy input from radioactive decay. At infrared wavelengths, the light curve shows two maxima separated by tens of days (Fig. 1); this structure is presumably due to evolving spectral features. The infrared light curves show a strong degree of uniformity among objects (Elias *et al.* 1985).

One of the brightest recent Type Ia supernovae was SN1981B in NGC 4536. Fig. 2 shows the spectrum of the supernova near maximum light at ultraviolet through infrared wavelengths (Panagia 1985). The ultraviolet observations were obtained with the *IUE* (International Ultraviolet Explorer) and show a relatively small flux, presumably due to line blanketing. The infrared emission is also below an extrapolation of a blackbody fit to the optical data. The optical spectrum near maximum light shows broad features that can be identified with lines of intermediate weight elements like Mg, Si, S, and Ca (Branch *et al.* 1985). At late times (t > 200 days), the character of the optical spectrum changes and complexes of Fe II and Fe III lines dominate (Fig. 3). Figure 3 shows data on SN1972E from Kirshner and Oke (1975); SN1981B had a similar spectrum at this late phase (Branch 1984).

Type Ia supernovae have not been observed as radio sources. The flux limit on SN1981B is 0.2 mJy at an age of 1.6 years (Weiler *et al.* 1986), which translates into a luminosity of 13 times that of Cassiopeia A for a distance of 20 Mpc.

Type Ib Supernovae

The optical light curves of Type Ib supernovae are sufficiently close to those of Type Ia that they were not distinguished from each other. This is not true at infrared wavelengths where the two types of events are quite distinctive (Elias *et al.* 1985; see Fig. 1). The two maxima that are present in the curves for Type Ia events are not present.

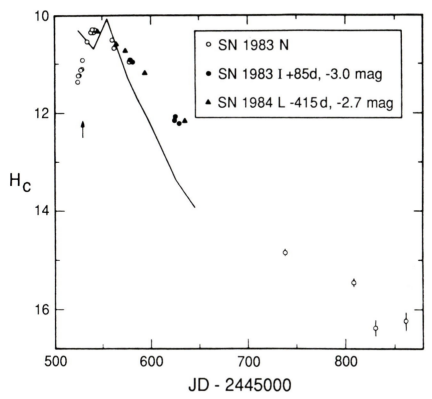

Figure 1 The infrared H-band (1.65 μm) light curves for Type I supernovae. The solid line is an average for the curves of the Type Ia supernovae 1980B, 1981B, and 1981D. The data points are for three Type Ib supernovae (from Elias *et al.* 1985).

Figure 4 shows the overall spectrum of the Type Ib supernova SN1983N in M83 near maximum light. Decreased emission in the ultraviolet is again evident, probably the result of line blanketing. The infrared part of the spectrum is also depleted relative to a blackbody fit at optical wavelengths.

The spectra of SN Ib near maximum light show many features in common with spectra of SN Ia, but with some clear differences (Porter and Filippenko 1987). Most notable is the absence of an absorption feature at $\lambda6150$ Å that is usually attributed to Si II $\lambda6355$. Also, the spectrum of a SN Ib at a given phase of the light curve corresponds to that of a SN Ia at a later phase of its light curve. The spectra of some SN Ib appear to show a series of features attributable to He I (Harkness *et al.* 1987). In the spectra of SN1983N and SN1984L there is an absorption minimum at $\lambda5700$ Å that is probably due to He I $\lambda5876$. The variation of this feature among SN Ib suggests that these objects have a range of He abundances.

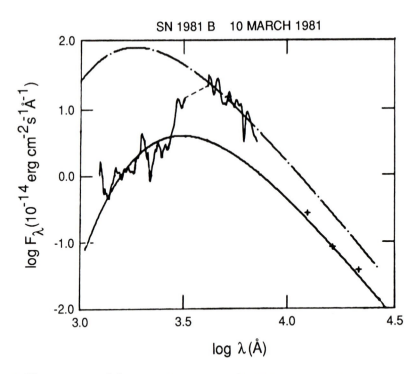

Figure 2 The spectrum of the Type Ia supernova SN 1981B near maximum light, dereddened with E(B-V) = 0.24. The upper curve is a blackbody at 15800K which fits the optical spectrum. The lower curve is a blackbody at 9400K which connects the infrared and ultraviolet spectra (from Panagia 1985).

While the spectra of SN Ib and SN Ia resemble each other near maximum light, they are completely different at late times when the supernovae have entered the nebular, emission line stage. Fig. 5 shows the spectrum of SN 1983N at an age of 236 days, when lines of [OI] λλ6300, 6363, Mg I] λ4562, [CaII] λλ7291, 7332, and NaI λλ5890, 5896 were especially prominent. The difference in late spectra suggests different abundances and a different supernova model for SN Ia and SN Ib. An infrared spectrum of SN 1983N by Graham *et al.* (1986) did show an emission line at 1.65 μm which they interpreted as due to [FeII]. However, SiI is another plausible identification (Oliva 1987).

Another difference with the SN Ia is that SN Ib have been detected as radio sources. Fig. 6 shows the radio light curve of SN 1983N. Radio emission was first detected 11 days before optical maximum when the supernova was only a few days old. At late times the supernova showed a clear power law decay of the form $F_\nu \propto t^{-1.6}$, where t is the supernova age, and the spectral index between 6 and 20 cm was -1 (Sramek, Panagia and Weiler 1984; Weiler *et al.* 1986).

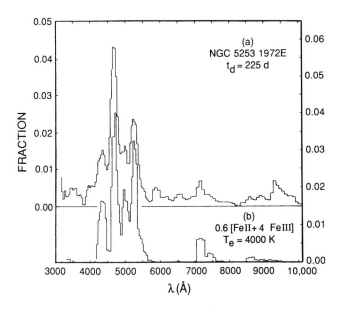

Figure 3 (a) The observed spectrum of the Type Ia supernova SN 1972E at an age of 255 days. (b) A synthetic spectrum of Fe II + 4 Fe III at an electron temperature of 4000K (from Meyerott 1980).

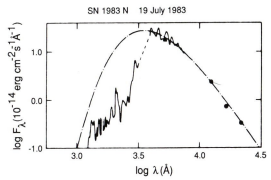

Figure 4 The spectrum of the Type Ib supernova SN 1983N near maximum light, dereddened with E(B-V) = 0.16. The dashed-dotted curve is a blackbody curve at T = 8300K (from Panagia 1985).

Type II Supernovae

SN 1987A, a SN II, has rapidly become the best observed supernova. However, it showed some unusual properties which can be traced to its progenitor star being a blue supergiant, Sk −69°202. Other well-observed SN II are inferred to have red supergiant progenitors and we use SN 1979C as a primary example of this kind of event. The optical light curves of several recent SN II are shown in Fig. 7. SN

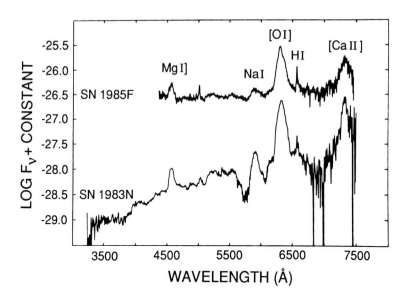

Figure 5 Spectra of the Type Ib supernova SN 1985F and SN 1983N at an age of about 250 days. Both spectra are presented in the rest frame of the host galaxy. The flux scale applies to SN 1985F; the spectrum of SN 1983N has been shifted down by a factor 10^2 (from Gaskell *et al.* 1986).

1979C showed a linear decline in magnitude with time. Near maximum light, the optical spectrum was characterized by weak broad features of Balmer hydrogen and Na I lines superposed on a continuum with temperature of about 10,000K (Branch *et al.* 1981). At late times, the spectrum showed a broad asymmetric Hα line, with a feature on the blue side of the line which is probably [OI] $\lambda\lambda$6300, 6363 (Branch *et al.* 1981).

Unlike the SNI spectra, the ultraviolet spectrum of SN 1979C did not have a deficit of ultraviolet continuum emission, but in fact had an excess over a blackbody curve fit through the optical photometry (Panagia *et al.* 1980). In addition, broad emission lines of highly ionized atoms appeared in the ultraviolet. The lines included NV λ1240, SiIV λ1400, NIV] λ1550, HeII λ1640, and NIII] λ1750 and they covered a velocity range up to at least 8400 km s^{-1} (Fransson *et al.* 1984), which was close to the photospheric velocity deduced from optical observations. These lines were observed at a time when the photospheric temperature from optical observations was $\leq$ 11,000K, which is too small to ionize the observed atoms.

At late times (after 8 months), SN 1979C showed a strong infrared excess (Merrill 1980; Dwek 1983). This feature completely dominated the luminosity of the supernova. Only JHKL photometry is available and it shows a peak at 3.5 μm at an age of 259 days with a distribution that is broader than a blackbody distribution.

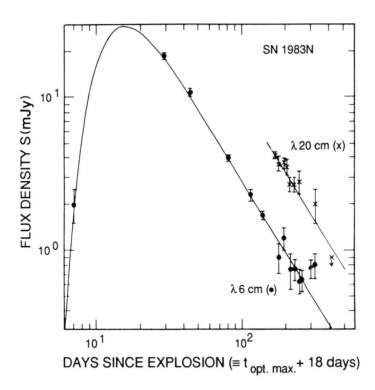

Figure 6 Radio light curves for the Type Ib supernova SN 1983N. The two wavelengths 20 cm (crosses) and 6 cm (filled circles) are shown together. The age of the supernova is measured in days from the estimated date of explosion on June 29, 1983. The solid lines represent the best-fit model light curves (from Weiler *et al.* 1986).

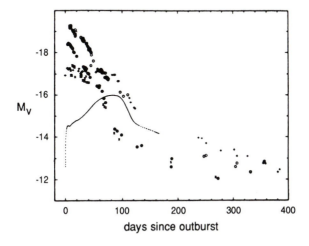

Figure 7 Absolute visual magnitudes for a sample of Type II supernovae with light curves that extend past 100 days. The symbols are: light circles, 1979C; dark circles, 1980K; pluses, 1969L; and crosses, 1970G. The solid curve is the data for SN 1987A (from Hamuy *et al.* 1988).

SN 1979C became a spectacular radio source, although it was first detected about one year after optical maximum (Weiler *et al.* 1986). The sharp rise of the radio light was well observed in this case. After peaking, the radio emission approached a nonthermal spectrum of the form $\nu^{-0.7}$. The early radio spectrum had a sharp drop at low frequencies. At peak radio light, the luminosity was approximately 200 times that of the bright supernova remnant Cassiopeia A. The supernova flux was sufficient to make VLBI (Very Long Baseline Interferometry) observations possible (Bartel *et al.* 1985; Bartel 1986). Observations at several times clearly showed the expansion of the radio supernova. If the supernova is assumed to be in the Virgo Cluster at a distance of 20 Mpc, the velocity of expansion is about 10^4 km s^{-1}.

SN 1979C was not detected as an X-ray source by the *Einstein Observatory* 63, 73, and 239 days past optical maximum; the upper limit on the luminosity was about 10^{40} erg s^{-1} (Palumbo *et al.* 1981). Another SN II, SN 1980K, which resembled SN 1979c in its optical, infrared, and radio properties, was observed as an X-ray source (Canizares, Kriss, and Feigelson 1982). It was first detected 35 days after maximum light at a luminosity of 2×10^{39} erg s^{-1} for a distance of 10 Mpc and subsequently dropped in luminosity by a factor of two or more 50 days later.

Figure 7 shows that the optical light curves of SN 1979C and SN 1980K were similar in character whereas the light curve of SN 1987A was very different during the first 100 days. SN 1987A started out relatively faint and gradually increased in brightness to become comparable to the other SN II. Another difference is that SN 1987A had more rapid color evolution than that of the other SN II (Hamuy *et al.* 1988). The blackbody fits to the observed colors evolved from a temperature of 14,000K at an age of two days to 7000K at an age of six days, where the ages are relative to the time of the neutrino burst observed by the Kamioka (Hirata *et al.* 1987) and Irvine-Michigan-Brookhaven (Bionta *et al.* 1987) experiments. The other SN II had a longer evolution time scale by nearly an order of magnitude. This rapid evolution was particularly evident in ultraviolet spectroscopy; the flux at 1500–2000 Å dropped by a factor of 10^3 over a period of days (Wamsteker *et al.* 1987). At later times, the supernova luminosity was dominated by optical emission. After an age of 120 days, the luminosity decayed exponentially with a time constant in the range 110–120 days (Feast 1988).

Optical spectroscopy of SN 1987A showed P-Cygni line profiles of the hydrogen lines and lines of low ionization heavy atoms. The maximum velocities observed in the Hα line were unusually large, greater than 30,000 km s^{-1} (Hanuschik and Dachs 1987). One interesting line identification in the optical spectrum was Ba II absorption lines (Williams 1987). Detailed model atmosphere calculations are needed to determine whether a significant overabundance of this heavy element is implied. In the months after maximum light, the optical spectrum became dominated by emission lines of Hα, Ca II] and [OI] (Danziger *et al.* 1987), which are the same lines observed in other SN II.

Because of the proximity of SN 1987A, it has been possible to identify the likely progenitor star from presupernova photographic plates. To better that 0.1 arcsec, the position of the supernova is coincident with the presupernova position of the B3 I star Sk −69° 202 (West *et al.* 1987; Walborn *et al.* 1987). This identification received support from ultraviolet observations where the supernova flux rapidly decayed to the point where it was possible to conclude that Sk −69° 202 is no longer present (Gilmozzi *et al.* 1987; Sonneborn, Altner, and Kirshner 1987). The optical and ultraviolet observations did show the presence of two early B type stars within a few arcsec of Sk −69° 202. The early ultraviolet observations showed an ultraviolet deficit compared to a blackbody fit to the optical data (Cassatella *et al.* 1987). In fact, the ultraviolet spectrum more closely resembled that of the SN I than that of SN 1979C. It is likely that line blanketing played an important role in the SN 1987A and the SN I spectra. While SN 1987A has been faint at ultraviolet wavelengths since an age of about 1 week, narrow ultraviolet emission lines of highly ionized atoms have been observed beginning in late May 1987 (Wamsteker, Gilmozzi, and Cassatella 1987; Kirshner *et al.* 1987). The lines include NV λ1240, NIV] λ1484, NIII] λ1750, HeII λ1640, OIII] λ1663, and CIII] λ1909. The relative strengths of the N and C lines imply that the N/C ratio is higher than the cosmic value by about a factor of 60 (Cassatella 1987). High resolution observations with the *IUE* in November 1987 showed that the N III] and CIII] lines have a width less than 30 km s^{-1} and that the ratio of the CIII] doublet lines implies an electron density of 2×10^4 cm^{-3} (Panagia *et al.* 1987). These observations clearly place the emitting region outside the expanding supernova gas.

Infrared spectroscopy of SN 1987A beyond 1 μm has been particularly rewarding. While an infrared excess was present quite early, it was not until an age of four months and later that a rich spectrum of emission features became evident. Observations have been carried out with ground-based telescopes in wavelength windows up to 13 μm and with the KAO (Kuiper Airborne Observatory). In June 1987, emission features due to CO bands at 2.3 μm and 4.8 μm were detected (Danziger *et al.* 1988). The width of the band heads made it clear that the CO had formed within the expanding ejecta; a temperature of about 2000 K was indicated. Ground-based observations and KAO observations in November 1987 showed a number of fine structure lines due to heavy elements as well as many hydrogen transitions (Cohen *et al.* 1987; Moseley *et al.* 1987). These data should be very useful in analyzing the physical conditions in the ejecta. In addition, the

KAO observations showed continuum radiation extending out to 100 μm (Harvey, Lester, and Joy 1987). The flux, F_ν, was approximately constant with wavelength, which suggests that free-free radiation is the emission mechanism.

The radio flux from SN 1987A peaked within days of the explosion and subsequently underwent a power law decline as $t^{-1.3}$ (Turtle *et al.* 1987). The peak radio luminosity was a factor 104 below that of SN 1979C. The later radio spectrum was not well defined by the observations, but appeared to be of a steep power law form. VLBI observations between Australia and South Africa were undertaken five days after the explosion (Shapiro *et al.* 1987). The radio emission was resolved, implying a lower limit to the angular size of 1.9 milliarcsec, or a minimum velocity of the radio emitting region of 16,000 km s^{-1}. This velocity was greater than the estimated photospheric velocity at the time of the observation.

Initial X-ray observations of SN 1987A with the *Ginga* satellite did not detect the supernova (Makino 1987). However, hard X-rays were detected with *Ginga* in the 10–30 keV range (Dotani *et al.* 1987) and with the Röntgen experiments on the Mir space station in the 20–300 keV range (Sunyaev *et al.* 1987) beginning in August 1987. The flux rose for several weeks before leveling off. The spectrum was remarkably hard, even harder than the spectrum of the Crab Nebula.

The expectation of γ-ray lines from radioactive isotopes synthesized in the explosion of SN 1987A has stimulated an extensive program of γ-ray observations. The means of detection include the SMM (Solar Maximum Mission) and a set of detectors flown by balloon from Australia. Detection of the ^{56}Co lines at 847 and 1238 keV was reported by Matz *et al.* (1987) using the SMM satellite. The lines were observed over the period August to October 1987 at flux levels of $(1.0 \pm 0.25) \times 10^{-3}$ and $(6 \pm 2) \times 10^{-4}$ photons cm^{-2} s^{-1}, respectively. The ratio of the line intensities was approximately that expected for an optically thin source. The lines were not increasing in flux during the time that they were observed and may have been decreasing. Detection of the 847 keV line was also reported by Sandie *et al.* (1987) from a balloon flight in November 1987. The line flux was 5×10^{-4} photons cm^{-2} s^{-1}.

A remarkable observation of SN 1987A was the detection of a bright source separated from the supernova by approximately 60 milliarcsec with a magnitude difference of 2.7 at the wavelength of Hα (Nisenson *et al.* 1987). The speckle imaging observations were made on March 25 and April 2, 1987 and were confirmed by independent speckle observations by Meikle, Matcher, and Morgan (1987). The brightness of this companion source implies that it is a phenomenon related to the supernova. The nature of this unexpected source remains obscure.

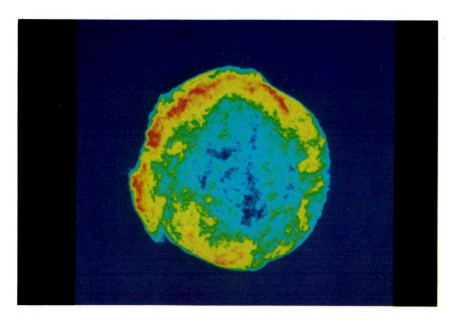

C1 Radio observation of Tycho's SNR by Green and Gull (1983) at 11cm with the Cambridge 5km telescope.

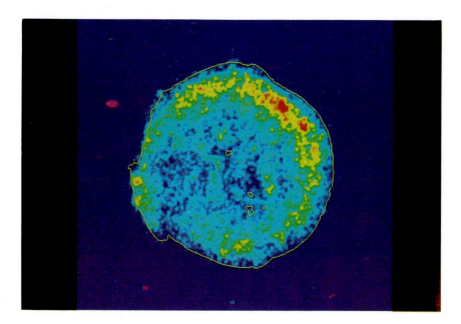

C2 Einstein HRI X-ray observation of Tycho's SNR by Seward, Gorenstein and Tucker (1983). The energy range is 0.3 to 3.5 keV and the contour shows the envelope of the radio emission.

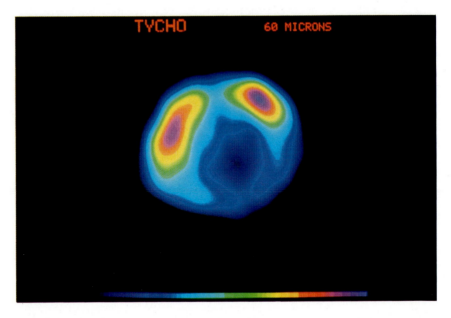

C3 IRAS infrared emission from Tycho's SNR at 60 microns. Picture courtesy of R. Braun.

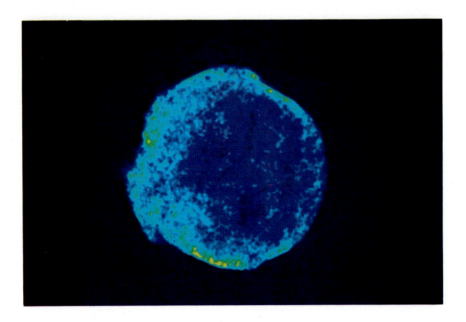

C4 The ratio of radio to X-ray emission from Tycho's SNR. This is brightest at the limb, illustrating a high ratio of radio to X-ray emission from the outer shock.

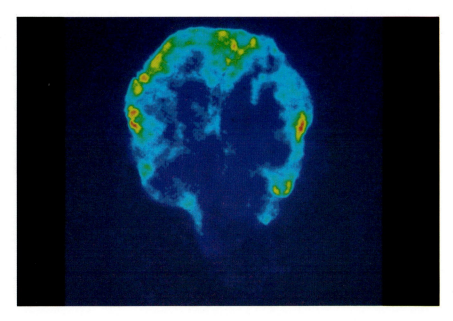

C5 The Cygnus Loop as observed with the Einstein IPC. Energy range is 0.2 to 0.6 keV. This is a mosaic of 59 fields and there are some minor artifacts in the picture.

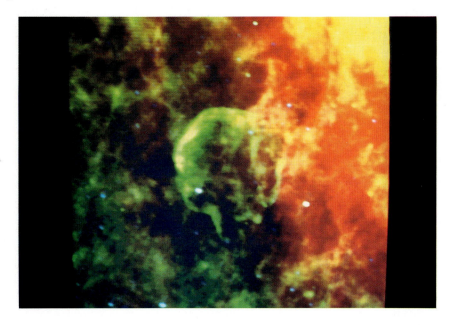

C6 IRAS map of infrared emission from the region of the Cygnus Loop. The color indicates the IR spectrum: red is 100 micron, green is 60 micron and blue is 25 micron emission. Note that the Cygnus Loop is warmer than the gas in the galactic plane to the west. Picture courtesy of E. Dwek.

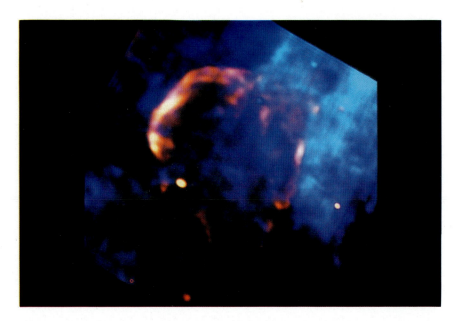

C7 Emission from warm (red) and cool (blue) dust components as determined by R. Braun. The warm component follows the X-ray morphology of the Cygnus Loop. Picture courtesy of R. Braun.

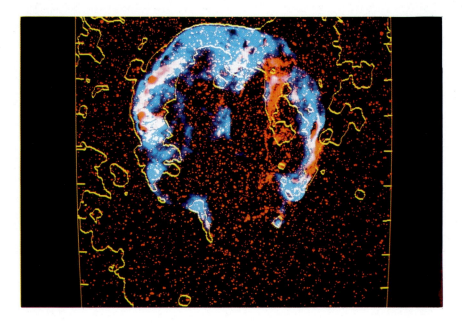

C8 The Cygnus Loop at 3 wavelengths. Blue indicates X-ray, the red, optical, and the contour shows 60 micron infrared emission. Picture courtesy of E. Dwek.

3. Interpretation and Future Prospects

The previous section shows that multiwavelength observations of supernovae are now a mature subject. Detailed physical aspects of supernova explosions can be modeled with the aid of these observations. Once models have been developed, they can be used to predict future multiwavelength observations. In this section, we describe some of the physical situations thought to be responsible for the supernova light.

Shock Heating by the Initial Explosion

For both SN I and SN II, it is thought that most of the kinetic energy of the explosion is deposited in the star on a time scale less than a few seconds. In the case of SN Ia, the energy source is thermonuclear burning and there is no compact remnant. For SN II and probably for SN Ib, the primary energy source is gravitational and a compact remnant, either a neutron star or a black hole, is expected. In both cases, the energy deposition results in strong heating and expansion of the star. The total energy is about 10^{51} ergs.

While a large fraction of the energy is initially in the form of internal energy, the adiabatic expansion of the star converts this energy to kinetic energy. The total radiative energy, which is likely to dominate the internal energy drops as R^{-1} where R is the supernova radius. Thus the observable effects of the initial energy deposition depend sensitively on the initial stellar radius. In order to achieve a luminosity comparable to that of a supernova near maximum light, an initial stellar radius of order 3×10^{13} cm is required (Chevalier 1981; Woosley and Weaver 1986). This is the radius of a red supergiant star and is the expected radius of a massive star after it has evolved off the main sequence. Models for the explosion of such a star have been successful in reproducing the basic features of SN II light curves, other than that of SN 1987A, over their first 100 days. The photospheric temperature peaks when the shock source first breaks out of the star and the temperature is a few 10^5K. The temperature then declines due to expansion and radiative losses and the rate of decline below 20,000 K is consistent with the observations. The photospheric temperature does not appear to cool below 5000K because the gas opacity drops sharply below this temperature due to recombination.

SN 1987A is a special case because it is known that the likely progenitor was not a red supergiant, but was a B3 I star with a radius of about 3×10^{12} cm. The additional expansion cooling is expected to lead to a low initial luminosity and this was observed (see Fig.7). The fact that a given temperature is reached at a smaller radius than is typical of other SN II gives a faster evolutionary time (radius/velocity), as was also observed. Finally, the first optical observations of the supernova within a few hours of the neutrino burst require that the stellar radius be considerably less than that of a red supergiant (Arnett 1987; Woosley 1988). While the initial energy deposition provides a very satisfactory explanation of the properties of SN 1987A over the first ten days, it does not account for the later luminosity rise.

The expected hard radiation burst from a SN II has never been directly observed. Klein and Chevalier (1978) noted that since there is a scattering atmosphere, the emitted radiation is shifted to the blue relative to a blackbody, so that a soft X-ray burst is predicted with a duration of about 30 min. No X-ray satellite has yet had the ability to effectively search for such a burst; a sensitive, soft X-ray all-sky monitor is needed. However, the effects of the ionizing radiation on the surrounding medium can be observable; the emission from highly ionized atoms from SN 1987A can probably be explained in this way (see section IIIc).

The initial energy deposition probably plays even less of a role in the light curves of SN I. Wolf-Rayet stars, which are plausible progenitors of SN Ib, have radii less than 10^{12} cm and 1.4 $M_\odot$ white dwarfs, which are plausible progenitors of SN Ia, have radii less than 10^9 cm. Although their initial radii are small, there is still the possibility of an energetic burst at the time of shock break-out. For example, Colgate (1975) has suggested that matter accelerated to relativistic velocities in a SN Ia explosion interacts with the magnetic field of the progenitor white dwarf and produces a radio pulse. This has not yet been observed.

Radioactive Energy Input

The relatively small amount of radiated energy from the instantaneous explosion of a small radius star led Colgate and McKee (1969) to suggest that energy from the radioactive decay chain ^{56}Ni $\rightarrow ^{56}$Co $\rightarrow ^{56}$Fe is important for the light from supernovae. The first part of the chain has a half-life of six days and the second a half-life of 77 days. The energy is deposited in the form of γ-rays and positrons with energies of about 1 MeV. It is now thought that this energy source is responsible for the optical light from most SN II after an age of about 100 days, for the light from SN 1987A after about 40 days, and for all the light from SN I (Woosley and Weaver 1986; Woosley 1988). The required amount of ^{56}Ni that must be synthesized in the explosion varies from 0.07 $M_\odot$ for SN 1987A to 0.5–1.0 $M_\odot$ for SN Ia. These amounts are consistent with the nucleosynthesis that is expected in the explosion process.

Initially, the supernova is optically thick to both the γ-rays and the optical photons, and there is an approximately blackbody photosphere. While the diffusion time for the optical photons is much longer than the age, much of the internal energy is lost to adiabatic expansion, but eventually the supernova luminosity can directly reflect the radioactive energy input. Since late June 1987, the luminosity of SN 1987A has followed the exponential decay expected for ^{56}Co, giving strong evidence for this energy source. As the gas becomes optically thin to the γ-rays, the rate of decline is faster than that of the ^{56}Co input; this is observed in the light curves of SN Ia. Decreasing continuum optical depth results in the emergence of line emission. This is the nebular phase, which is observed in all types of supernovae. Models for the nebular phase involve the calculation of heating and ionization of a gas by γ-rays. Successful models have been computed for the late spectra of SN Ia (Meyerott 1980; Axelrod 1980) which support the carbon deflagration model. Models for the late spectra of SN Ib support the Wolf-Rayet progenitor model (Fransson and Chevalier 1988).

One prediction of the late emission models is that there is eventually a thermal instability where the temperature drops from a few 10^3K to a few 10^2K. When this occurs, at an age of about two years, most of the low energy supernova emission appears in infrared fine structure lines. This transition has not yet been observed, but it may eventually be observable in SN 1987A.

The γ-rays from radioactivity initially lose energy by electron scattering and eventually are photoabsorbed at hard X-ray energies. The X-rays can eventually escape from the supernova when the Thompson optical depth is of order ten (McCray, Shull, and Sutherland 1987). This effect was predicted for SN 1987A and in fact gives a convincing description of the observed hard X-rays (Sunyaev *et al.* 1987: Itoh *et al.* 1987; Pinto and Woosley 1988), although minor modifications of the models, such as mixing, appear to be required. As the Thompson depth drops, the photons can no longer reach the lower X-ray energies and eventually the γ-rays escape directly. The detection of the ^{56}Co γ-ray lines from SN 1987A gives definitive support for the radioactive energy source model. SN 1987A provided the first detection of γ-ray lines from a supernova because of its proximity. SN Ia are in fact expected to be more luminous in γ-ray lines by orders of magnitude because of their relatively small total mass and large ^{56}Ni mass.

Circumstellar Interaction

The degree to which circumstellar interaction is an important phenomenon is expected to be related to the nature of the progenitor star. Progenitors with dense winds should have the largest effects. Red supergiant stars in our Galaxy are known to have slow dense winds. These are thought to be the progenitors of most SN II and these supernovae do show the best evidence for circumstellar interaction. Multiwavelength observations have played a crucial role in interpreting these events.

Circumstellar interaction can occur both through the high velocity motion of the outer parts of the supernova and the light from the supernova. The outer region of a supernova rapidly approaches a velocity profile proportional to radius. Hydrodynamic simulations of the explosion of a massive star show that the outer density profile approaches a steep power law with radius (Jones, Smith, and Straka 1981). Similarity solutions for the expansion of an initially exponential atmosphere shows that a power law form $r^{-9.58}$ is attained (Chevalier 1987a). The interaction of a steep power law with a circumstellar medium which has a density proportional to r^{-2} can also be described by self-similar solutions (Chevalier 1982a). Two regions of hot gas are formed bounded by shock fronts. The inner shocked supernova gas is at a higher density and lower temperature than the outer shocked circumstellar gas. The density ratio between these two regions is higher for a steeper supernova density profile.

The light from the supernova can heat and ionize circumstellar gas and dust. Observations of the light from a particular supernova are generally incomplete, especially at early times. Since the energetic radiation emitted at the time of shock break-out (see section IIIa) can have unique effects, theoretical models must be used to estimate the properties of this burst. The light travel time across the circumstellar region can be comparable to the age of the supernova so that light travel time effects must be taken into account when modeling the radiative interaction. The observed emission is then referred to as a light echo.

Observations of SN 1979C at radio, ultraviolet, and infrared wavelengths (see section IIc) can be interpreted in terms of circumstellar interaction. It is plausible that relativistic electrons are produced in the shocked region between the supernova and the ambient matter. Observations of supernova remnants show that 0.1–1% of the postshock thermal energy goes into relativistic electrons and magnetic fields. An efficiency on this order can explain the radio luminosity of SN 1979C. The early low frequency turnover can be attributed to free-free absorption and the amount of absorption gives an estimate of the density of the circumstellar wind (Chevalier 1982b). Lundqvist and Fransson (1988) have shown that detailed features of the radio light curve can be understood if one takes into account the temperature and ionization of the circumstellar medium. They deduce a presupernova mass loss rate of 1×10^{-4} $M_\odot$ yr^{-1} for a wind velocity of 10 km s^{-1}. The apparent size of the radio source from VLBI observations compared to the highest velocities observed in the supernova, about 10,000 km s^{-1}, is entirely consistent with the circumstellar interaction model.

The presence of the hot gas outside the photosphere can affect the ionization in the outer parts of the supernova; both through direct emission from the hot gas and through photospheric photons that are scattered up in energy by their interaction with the hot gas (Fransson 1984). Fransson (1984) showed that the latter process is of particular importance for SN 1979C and that the required supernova mass loss rate is consistent with that deduced from the radio observations. The scattered photons can ionize the supernova gas, producing the highly ionized atoms that are observed with the *IUE*. In addition, the photon shift to high frequency can produce the ultraviolet excess that was observed. Finally, the generally higher level of ionization in the supernova atmosphere may prevent line blanketing from being an important effect (Fransson *et al.* 1987). A strong ultraviolet deficit was not observed in SN 1979C, but was observed in the Type II supernova SN 1987A.

Red supergiant stars are known to have dust in their winds and the supernova light is expected to illuminate the dust. The radiation from the heated dust can give an infrared echo (Dwek 1983). The strong initial radiation burst from the supernova evaporates dust out to some radius, r_v. The infrared echo has a plateau phase until a time $2r_v/c$ after which the light declines. Near the time $2r_w/c$, where r_w is the total extent of the dusty wind envelope, the flux declines more sharply. The late infrared observations of SN 1979C are approximately reproduced with a mass loss rate consistent with the radio observations and $r_w = 1 \times 10^{18}$ cm (Dwek 1983). SN 1979C was not observed as an X-ray source, although thermal emission is expected from the hot shocked supernova gas. The X-ray emission that was observed from SN 1980K can be plausibly attributed to this mechanism.

Circumstellar interaction thus appears to be relevant to a wide range of multiwavelength observations and there are implications for future observations of normal SN II. At radio wavelengths, VLBI observations are expected to reveal a shell structure, although it is likely to be irregular. The fact that relativistic electrons are produced suggests that relativistic nucleons are also present. For a nearby supernova, these might give rise to a detectable flux of high energy γ-rays from the decay of pions produced by nuclear interactions. Spectroscopy of the grain emission can give information on the grain composition and should be able to distinguish between graphite and silicate grains (Dwek 1985). The circumstellar grains can scatter the supernova light as well as absorb it and a scattered light echo is expected. Chevalier (1986) checked for such an echo in the spectra of SN 1979C, but did not find evidence for the effect.

The situation with SN 1987A is different from that of normal SN II because it had a blue supergiant progenitor star. A B3 I star is expected to have a mass loss rate of a few time 10^{-6} M$_\odot$ yr^{-1} and a wind velocity of 550 km s^{-1}, which leads to a much lower wind density than that thought to be present around a red supergiant. Direct evidence for this wind comes from the observed radio emission.

The early radio turn-on and the low radio luminosity compared to normal SN II are consistent with the low wind density (Chevalier and Fransson 1987). The low density also explains the non-detection of early thermal X-ray emission.

Ultraviolet spectra of SN 1987A show emission from some of the same highly ionized atoms that were observed in SN 1979C. However the narrowness of the lines and the different time evolution suggest a different origin for the lines. Both the presence of red supergiants in the Large Magellanic Cloud and the theoretical studies of the evolution of Sk - 69° 202 (Woosley 1988) suggest that the progenitor star had a red supergiant phase. Dense mass loss from this phase would be swept into a shell on its inner edge by the blue supergiant wind. The ionization of this shell by the initial ultraviolet burst at the time of shock break-out can produce the required ionization; the evolution of the emission is determined by recombination of the gas and by light travel time effects (Lundqvist and Fransson 1987; Chevalier 1988). The evolution of the line fluxes, widths and velocities, as determined by high-resolution observations, should yield the basic shell properties; i.e. its velocity and radius. An important aspect of the line fluxes is that they imply a large N/C ratio. This overabundance suggests significant presupernova mass loss so that the N-rich layers were revealed (Maeder 1987).

The ultraviolet observations give the best evidence for the presence of dense gas close to SN 1987A. This gas is probably at some distance, about 10^{18} cm, from the supernova so that it may be 10-30 years before the supernova shock front reaches the gas. At that time, a brightening at radio and X-ray wavelengths is expected.

The radio emission observed from SN 1983N suggests that circumstellar interaction is a significant factor for SN Ib (Chevalier 1984; Sramek, Panagia, and Weiler 1984). The relatively early turn on implies a circumstellar density intermediate between normal SN II and SN 1987A. The circumstellar density is somewhat above that typical of a Wolf-Rayet star, but this may be an acceptable interpretation, considering the uncertainties. The radio emission from SN Ib is the only evidence for circumstellar interaction in SN I of both types. The lack of radio emission from SN 1981B rules out a dense circumstellar medium like that around SN 1979C for this SN Ia, but allows the presence of moderately dense circumstellar matter.

Pulsar Energy Input

While pulsars have been suggested as the primary energy sources for supernovae, the models have not been successful in reproducing detailed supernovae observations (e.g. Chevalier 1981). There is no indisputable evidence for pulsar activity in any supernova observations. Pacini and Salvati (1981) and Bandiera,

Pacini, and Salvati (1984) have suggested that the radio emission from SN 1979C and other normal SN II is from a pulsar nebula. This model requires the supernova envelope break up into filaments at an early age so that free-free absorption does not prevent the radio emission from escaping. The details of how this occurs and the likely effects on optical spectroscopy are not yet understood. While the pulsar model for the radio emission cannot be definitely ruled out, it does have the drawback that it does not relate to observations at other wavelengths. The picture of circumstellar interaction, in contrast, is supported by a range of multiwavelength observations. In addition, the early radio emission observed from SN 1983N and SN 1987A cannot be plausibly attributed to pulsar emission. The signature of pulsar energy input is expected to be a luminosity curve with a decay time that is considerably longer than that of ^{56}Co decay. Unfortunately, there are very few supernova observations outside of radio wavelengths beyond an age of two years. SN 1986J, which was discovered by radio observations, has shown such a slow decay of its optical light (Rupen *et al.* 1987). The radio emission does give evidence for strong circumstellar interaction but this does not seem to be responsible for the optical emission because the gas velocities deduced from line widths are relatively low. While pulsar input is suggested for the optical emission (Chevalier 1987b), it is difficult to confirm this hypothesis.

The neutrino burst observed from SN 1987A is strongly suggestive of neutron star formation in this object. However, at the present time there is no clear evidence for pulsar activity. Considering the exponential decay of the radioactive energy input, either pulsar effects will eventually be revealed, or it will be possible to set strong limits on pulsar energy input.

4. Supernova Remnants

We study SNR to learn the mechanism of the SN explosion itself and the nature of the precursor star. For about 200 years after the explosion the remnant is expected to be "young". About 1 $M_\odot$ of material is ejected and expands freely into the surrounding medium. During this time, the characteristics of the remnant are determined by the nature of the explosion, and with suitable observations, we should be able to determine the mass of ejected material, its elemental composition, its kinetic and thermal energy, and its distribution in space.

The material surrounding a Type II SN is likely to be from the star itself. A massive star can shed several solar masses as an energetic stellar wind before conditions in the core become catastrophic and result in a SN. A strong stellar wind over a million years might form a hot bubble with radius 10-30 pc around the star, and the structure of this bubble will reflect the early history of the star. Outside the bubble, there is the interstellar medium from which the star was born.

During the next phase of the SNR, sometimes called "adolescence", the expanding ejecta has swept up an equivalent or greater mass of circumstellar material. The material which radiates is a mixture of ejecta and circumstellar material, and as time goes on, the ejecta become more difficult to distinguish. The properties of the remnant become determined not by the explosion, but by the properties of the circumstellar medium.

During the adolescent phase, a remnant should expand adiabatically. The original kinetic energy of the ejecta is transformed slowly into thermal energy as the remnant sweeps up more and more material. Then, after a few thousand years, the outer shell of the remnant is expected to be dense enough so that much of the internal energy is radiated. At the end of this radiative phase at an age of 30-100 thousand years, the SN energy has been radiated away and the remnant is difficult to distinguish from the surrounding environment.

We present here data from several shell-like remnants representing different phases. Two, in particular, are emphasized: Tycho's SNR, a young remnant; and the Cygnus Loop which is certainly mature, if not middle-aged. It is dangerous to call these two typical because most remnants have strong individual personalities. Nevertheless, these do have some interesting features and both have been well observed over a variety of wavelengths.

Tycho's Remnant (2.5 Kpc distant, 415 years old)

A young remnant is expected to have a definite spatial structure. The expanding ejecta should drive a shock into the circumstellar material and the resistance to the expansion in turn drives a shock in the other direction into the ejecta. The ejecta cools quite rapidly through expansion and without this reverse shock would not be hot enough to radiate X-rays.

This remnant was first mapped in the radio band. A recent radio map, in Figure C1, obtained by Green and Gull (1983) with the Cambridge 5 km interferometer, shows the remnant to be a limb-brightened shell, with the outer boundary sharp, smooth, and almost circular over large areas, convincing evidence for a shock wave in circumstellar material. The origin of the radio emission is nonthermal. The shock compresses the magnetic field in the circumstellar material and electrons are also accelerated in this environment. These energetic electrons moving in the magnetic field produce synchrotron radiation, and the average direction of the magnetic field can be determined by measurement of polarization of the radio waves. Strom, Goss, and Shaver (1982), using 2 high-resolution radio observations taken eight years apart, have measured an expansion velocity of 3000 km s^{-1}. The average expansion velocity since the SN is 6400 km s^{-1} so the material has decelerated appreciably.

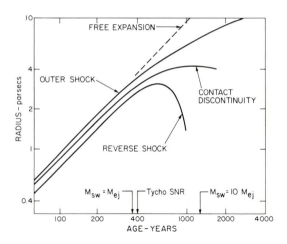

Figure 8 The calculated position of the shocks as function of time (Rosenberg and Scheuer 1973, Chevalier 1982c). The explosion is assumed to be spherically symmetric and expanding into a uniform medium. The boundary between the ejecta and the ISM, the contact discontinuity, expands continuously outward. The outer shock is shown propagating at constant velocity until the swept up mass becomes comparable to the mass of the ejecta, at which time, the velocity slows appreciably. [This is only true if the ejecta act as a piston traveling at uniform velocity. Since the ejected envelope is expected to have a complex density and velocity distribution the "free expansion" is an oversimplification of the position of the main shock.] The reverse shock, as viewed by someone riding on the contact discontinuity is always propagating back to the origin of the explosion. As viewed by an outside observer, however, the reverse shock appears to travel outward from the origin until the expansion slows, and it then proceeds back to the origin. The position of Tycho's remnant is indicated by a tick mark.

Figure C2 shows the X-ray picture obtained with the *Einstein Observatory* High Resolution Imager (HRI). The remnant is again a limb-brightened shell, very similar to the radio remnant but not identical. The X-ray emission is thermal, the radiation is a combination of continuum and line emission from hot gas. The X-ray luminosity, together with assumptions about the geometry of the emitting region, can be used to derive the mass of hot gas and if the outer shock can be distinguished from the ejected material, the mass of the ejecta can be determined. This has been done for Tycho's remnant by Seward, Gorenstein, Tucker (1983) although Dickel and Jones (1985) have offered a slightly different interpretation. The elemental composition and temperature of the emitting material are given by the X-ray spectrum.

The X-ray luminosity was calculated assuming a particular balance between the kinetic energy of the positive ions, the kinetic energy of the free electrons, and the ionization state of the heavy elements. The calculation was accomplished assuming that the outer shock is spatially isolated in two places and that the ejecta lies behind this in the form of clumps of dimension $\sim 24''$ or 0.2 parsecs. The mass of the ejecta, greatly enriched in silicon-group elements, is 1.2 $M_\odot$. The swept up mass is 1.4 M$_\odot$. The velocity of the main shock is 5000 km s^{-1} and the velocity of the ejecta is 4000 km s^{-1}, higher velocities than those derived from the radio observations.

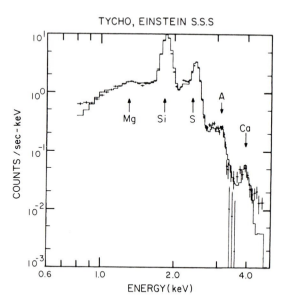

Figure 9 The Einstein Solid State Spectrometer (SSS) spectrum of Tycho's SNR (Becker et al 1980), is remarkable because half the emitted X-rays are in strong lines from silicon and sulphur. This spectrum illustrates that the abundances of these elements is probably higher than expected from cosmic material, (Here is material that has been generated in the SN explosion) and that the equilibrium in the remnant, or rather the lack of equilibrium, is an important problem. The ionization state of the heavy elements is expected to lag behind the temperature of the electrons which ionize them.

Tycho's SN is always assumed to have been Type I, since the light curve was accurately determined and the decay followed that observed for other Type I Supernovae. Type I SN show no evidence of H in their optical spectra and are believed to originate in stars which have used or lost all their H. One popular model for the explosion is the deflagration of a 1.4 solar mass white dwarf. The X-ray determined ejecta mass and the X-ray spectrum are compatible with this (Hamilton et al 1985).

Optical emission from Tycho's remnant is very faint, since most of the material is too hot to emit radiation in the optical band. Very faint filaments are seen on the northwestern limb of the remnant and there is one knot in close proximity to the radio knots on the eastern limb. This has been studied spectroscopically by Kirshner, Winkler, and Chevalier (1987). They find a narrow Hα line with broad wings. The narrow line indicates the existence of neutral hydrogen, outside the remnant and not ionized by the initial radiation of the SN. The broad wings are interpreted as shock-heated material and imply a shock velocity of 2,000 km s^{-1}. This number is lower than other results but the X-ray and radio results are global and concern the average properties of the remnant. The optical knot is probably a dense region, overrun by the shock and heated, but to a lower temperature than the material just behind the shock.

The latest morphological observations come from *IRAS* (the *Infrared Astronomy Satellite*). Most young remnants which are strong radio and X-ray sources are also bright at infra-red wavelengths. The infrared comes from imbedded dust heated by the hot X-ray emitting gas. Figure C3 shows an infrared picture of Tycho's remnant by Robert Braun (1985). This figure shows that the intensity at 60 microns stands out clearly and looks similar to the X-ray and radio pictures: a limb-brightened shell, brightest in the north. Since the *IRAS* resolution was $\sim 2'$, we have not directly compared this with the $5''$ resolution X-ray and radio images.

Dwek, Symkowiak, Petre, and Rice (1987) have shown that the infrared luminosity of Tycho's remnant is 8 times the X-ray luminosity. Therefore, the infrared radiation might be important in the evolution of remnants, although they will probably not cool and reach the radiative phase faster than previously expected. There is also some uncertainty over the origin of the infrared emission. There should be some emission in spectral lines from the heavy atoms. Estimates of this fraction in lines range from 0 to 40%, but the *IRAS* detectors were not capable of resolving line emission.

Comparison of the Tycho radio and X-ray images may indicate a difference in spectral signature between the main shock and the ejecta. The main shock is relatively brighter at radio wavelengths. Figure C4 shows the radio brightness divided by the X-ray brightness, which emphasizes regions where radio emission is relatively strong. Notice this image is brightest at the edges of the remnant. Several authors have noted that high resolution radio images of young SNRs tend to be a bit larger than the X-ray images. This is an illustration of that phenomenon.

Kepler's Remnant (4 Kpc distant, 383 years old)

This remnant is rather similar to Tycho's remnant, not surprising since Kepler's light curve for the SN also indicates a Type I origin. Both radio and X-ray maps show a limb-brightened shell, but with asymmetries greater than in Tycho's remnant. The shell is much brighter in the north than in the south. These asymmetries are not understood and may be characteristic of the explosion, perhaps in the mass or composition of the ejected material.

Matsui et al. (1984) have compared the HRI X-ray map with radio maps at 6 and at 20 cm. The X-ray brightness gives an average density of 7 electrons cm^3

in the bright part of the remnant. Faraday rotation and depolarization measured from the 2 radio maps indicates an average value of 7×10^{-5} gauss for the magnetic field. The energy in relativistic electrons within the remnant is about that of the magnetic field, and the flux is about 10^4 times that of the cosmic-ray electron flux in the vicinity of Earth.

A 4% linear polarization of the 6 cm radio emission shows the magnetic field to be radial in the outer parts of the remnant. A similar result was obtained by Duin and Strom (1975) for Tycho's remnant.

If the origin of a SN is the gravitational collapse of the core of a massive star, as is believed to be the case for Type Ib and Type II SN, the ejecta are the layered material just outside the core of the precursor star. This structure is expected to be preserved throughout the initial expansion of the remnant. Recalling that the X-ray spectrum of Tycho's remnant showed half the energy to be in spectral lines, and that the *Advanced X-ray Astronomical Facility (AXAF)* is expected to have a detector capable of giving excellent spectral and spatial resolution, we estimate that it will be possible with *AXAF* to get pictures of these young remnants in the light of ionized silicon, sulphur, argon, etc. This will give a measure of the morphology of different elements within the remnant and will be a definitive test of the layering hypothesis—perhaps a determination of whether or not the precursor star was a massive one.

The Cygnus Loop (0.8 Kpc distant, ~15,000 years old)

The Cygnus Loop is a remarkable object. It has the appearance of a bright shell at all wavelengths and an angular size of 2.5° so observations of moderate resolution can detect detailed structure. Figure C5 shows the IPC X-ray map of the Cygnus Loop. These X-rays are thermal emission from a 3×10^6K gas of approximately cosmic composition (Ku, et al 1984, Charles, Kahn, and McKee, 1985). X-ray lines from ionized oxygen and neon have been detected by Inoue et al (1980). This gas has been heated by a 400 km s^{-1} shock wave traveling in a medium of density 0.2 cm^{-3}. The shell, except in the south, appears to be almost spherical but somewhat retarded in places. X-ray emission is strong in these retarded regions, which are probably regions of higher density where the shock travels more slowly and, since X-ray luminosity goes as the square of the density, emission is brighter. The optical radiation is from clouds compressed by the hot X-ray emitting gas. This material, at density of $\sim 10^2$ cm^{-3} and a temperature of 10^4K, is different from the hot X-ray emitting material. Although the optical and X-ray emission are globally from the same portions of the remnant, there is no correlation on a scale of a few arcseconds and there is little filamentary structure in the X-ray emission.

Within the Cygnus Loop the contrast between emission from bright spots on the rim and the interior is very high. The shock wave is now apparently lighting dense regions of the interstellar medium and we are seeing the structure of the circumstellar material. This SN probably resulted from the explosion of a massive star which, during its lifetime, produced a bubble of radius $\sim$ 15 pc in the interstellar medium. This bubble formed a dense outer shell which is now being illuminated by the SN shock. The high pressure gas behind the shock has further compressed the material of this shell and formed the optical filaments. In the north of the remnant there is a region of strong X-ray emission that is not bright optically, where the shock is apparently propagating in a cloud-free medium.

On a scale of arcsec to arcmin, the optical morphology is complicated. The filaments have been studied with narrow band filters by Raymond et al (1987) and by Hester and Cox (1986). They have recorded the filamentary structure in the light of Hα, [OIII], and [SII]. These bright regions are not thread-like filaments but are more like sheets. The depth of material along the line of sight is comparable to the length of the observed filaments rather than the thickness. Measurements are interpreted as 100 km s^{-1} shocks driven into dense regions of cool material imbedded in the X-ray emitting gas. There is approximate pressure equilibrium between the two regions.

Blair and Panagia (1987) have reviewed *IUE* observations of the Cygnus Loop which is close enough so reddening is low and there is not much attenuation of the ultraviolet. Strong lines are seen from CIII, CIV, OIII, SiIV, and HeII. Intensities of these lines are compared with radiative shock calculations and use to restrict the range of parameters allowed. These lines appear to come from shocks of $\sim$ 130 km/s driven into clouds with density 1-5 cm^{-3}. An interesting connection between the intensity of C lines and dust depletion might provide a link between ultraviolet spectra and infrared emission from the dust heated by the hot post-shock gas.

In the northeast there is a faint optical filament associated with the leading edge of the X-ray emission. This filament shows almost pure H emission, an indication of neutral hydrogen swept up in the high velocity shock.

The radio emission is very closely correlated to the optical (Dickel and Willis 1980). The optical and ultraviolet radiation from clouds compressed by the surrounding hot gas can be strong enough to cool the radiating region, thus lowering the pressure and causing further compression to higher density, which in turn increases the rate of radiation of energy. Thus, clouds are compressed and imbedded magnetic field is compressed and, since synchrotron luminosity goes as B^2, these regions appear bright at radio as well as at optical wavelengths. This correlation was first pointed out by Duin and van der Laan (1975) in IC 443 and has recently been furthered pursued by Mufson et al (1987).

Figure C6 shows *IRAS* observations of the Cygnus Loop. The east limb is clear but the western side of the remnant is obscured by emission from a larger region. Braun and Strom (1986) have derived the intensity of two spectral components, warm dust and cool dust. Figure C7 shows the morphology of the warm dust, and the entire structure of the Cygnus Loop stands out clearly. The infrared emission follows the X-ray emission quite closely, and is bright in the north where the remnant is bright in X-rays but not at radio or optical wavelengths. The infrared emission is largely from dust heated by the X-ray emitting gas and Dwek et al (1987) have again pointed out that the infrared luminosity of the Cygnus Loop is five times the X-ray luminosity, showing that infrared cooling is important for adolescent SNR as well as for the young ones.

Puppis A (2 kpc distant, 3700 years old)

This SNR is one of the brightest soft X-ray sources in the sky and the morphology has been well determined at radio (Green 1971) and at X-ray (Petre et al 1982) wavelengths. It is also bright in the IR band but these data are, at present, not analyzed. A visual inspection indicates that the X-ray and IR appearances are almost the same. This section will use recent optical observations to illustrate the value of optical and X-ray spectra in determining the nature of the remnant. Pup A is too hot to have bright optical filaments. It is interesting that so much information can be obtained from a few faint wisps in that part of the spectrum where the remnant is least visible.

Although the brighter filaments show a high abundance of nitrogen (Dopita, Matthewson, and Ford 1977) and may indicate material with an overabundance of nitrogen, some faint filaments have been recently observed to emit only lines from oxygen and neon. Thus it appears that Pup A can be classified as an "oxygen-rich" SNR as have been Cas A, MSH 11-54, SNR 0102-72.3 and SNR 0540-69.3.

Winkler et al (1987) have made the reasonable assumption that these filaments are ejecta from the inner layers of the exploding star, and have determined their radial and proper motion over a baseline of 6 years. They find velocities of 400-1600 km/sec and have traced the motion back to a common center of expansion. Since they are moving within a low density medium, previously swept by the higher velocity blast wave, these shreds of material have been able to survive and have probably not been decelerated by the accumulation of swept-up mass. Thus Winkler et al have derived an age of 3700 $\pm$ 300 years for the remnant.

The *Einstein Observatory* Focal Plane Crystal Spectrometer (FPCS) obtained extensive high-resolution X-ray spectra of Pup A. Lines from O, Ne and Fe were clearly observed and the relative intensities of both O and Ne to Fe were 3-5 higher than solar. The hot X-ray emitting gas is apparently enriched by the

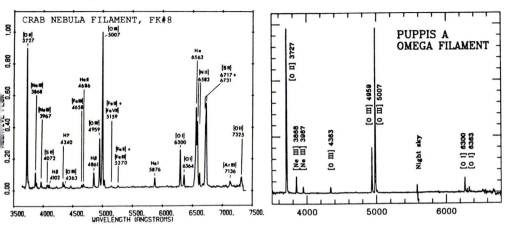

Figure 10 Optical spectra of filaments in the Crab Nebula (Fesen and Kirshner 1982) and in Pup A (Winkler et al 1987). The Crab filament spectrum shows lines from a gas mixture with normal elemental abundances. The gas in Pup A is overabundant in O and Ne.

same material observed optically. Canizares and Winkler (1981) calculate that the explosion produced 3 $M_\odot$ of O and Ne which are now mixed with 100 $M_\odot$ of interstellar material in the remnant. A progenitor star of mass $\sim$ 25 $M_\odot$ is implied.

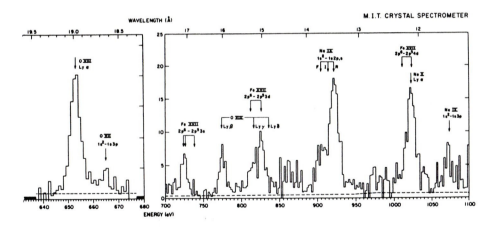

Figure 11 X-ray spectra of the brightest part of Pup A. Line emission from O, Ne and Fe is clear and the relative intensities indicate an over abundance of O and Ne in the hot gas. Data from Winkler et al (1981).

The Crab Nebula (2 Kpc distant, 933 years old)

Now let us consider something completely different. The Crab is a young remnant with low expansion velocity. The filamentary shell we observe is expanding at

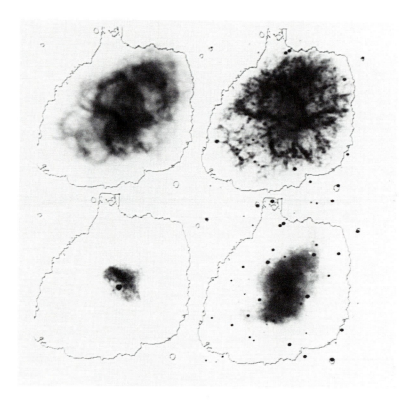

Figure 12 The Crab Nebula at 4 wavelengths. The upper left shows 20 cm radio emission; upper right, Hα; lower right, optical continuum; and lower left, X-ray.

1,500 km s^{-1} and the temperature of shocked material is too low to emit X-rays. Figure 12 shows the Crab seen at radio, optical, and X-ray wavelengths. At upper right is a picture taken by Robert Kirshner through a filter passing only the Hα line showing the optical filaments which are predominantly neutral hydrogen. At upper left is a 21 cm radio continuum map from Andrew Wilson which shows that the radio emission is also filamentary and has structure following closely that of the optical filaments. At lower right is blue continuum light coming from synchrotron radiation associated with energy deposited in the central region by the pulsar. At lower left is the *Einstein* HRI X-ray picture. The bright point source is pulsed radiation from the pulsar itself. This is partially surrounded by a bright diffuse X-ray nebula, which incidently has the highest surface brightness of any diffuse X-ray source yet observed. Note that the high energy radiation is indeed from the center of the nebula and well within the surrounding filamentary structure.

In the Crab Nebula, the pulsar is clearly a high energy phenomenon.

Pulsars

Neutron stars are, of course, remnants of the cores of gravitationally collapsed massive stars. Those which are formed spinning and with high magnetic fields are generally observable as pulsars (although the relative orientation of spin axis and observer probably affects the strength of the observable signal). Pulsars were discovered in the radio band and modern surveys have located over 400 isolated radio pulsars with periods from .0015 to 4 sec. Both period and period derivative are usually measured allowing calculation of pulsar spin-down energy loss and age. Ages of radio pulsars range from 900 to 10^9 years.

Young energetic pulsars emit high energy radiation and several pulsars within supernova remnants were discovered through X-ray emission including one in the LMC that has not yet been detected in the radio band. If the neutron star has an age less than $\sim 10^4$ years, the associated remnant of the shell should, of course, also be detectable. Several old radio pulsars have also been detected as X-ray sources (Seward and Wang 1988).

Only three pulsars have been detected in the optical band: 0531+21, 0833-45, and 0540-69. All are young and are located in the central part of the supernova remnants; the Crab Nebula, Vela XYZ, and 0540-69.3 in the LMC.

The mechanism by which pulsed high-energy radiation is emitted is not well understood. Some processes generally considered are: 1) hot spots on the surface around the magnetic poles due to heat conducted from the interior, 2) hot spots due to gravitational energy of accreting material and 3) synchrotron radiation generated by high energy electrons in the magnetosphere of the rotating neutron star. The detection of optical pulsation from the above mentioned three pulsars identifies the mechanism as synchrotron radiation since the spectra are power laws over many decades of energy.

The high X-ray luminosity of central objects in the SNRs Kes 73 and CTB 109 indicates that these are probably accretion-powered binaries.

The Vela and Crab Pulsars have also been detected in γ-rays by the COS B satellite. Indeed the Vela pulsar radiates most of its pulsed emission at γ-ray energies. It is curious that the pulse shape of the Vela pulsar changes significantly with waveband from the radio to γ energies whereas that of the Crab Pulsar is almost invariant with energy (Bignami and Hermsen 1983).

Future more sensitive observations should add more high-energy pulsar detections and waveforms to the handful now available. With more data, chances of developing a model for the pulse emission mechanism will be greatly improved.

Summary

There is now an abundance of morphological and spectral information from SNR. Most physical parameters, however, have been derived from single wavebands. The X-ray and optical emission come from different physical regions and observations in one band have so far only been used indirectly to determine characteristics of the other (e.g. by requiring that the pressure be the same throughout the remnant).

Extensive models are required to incorporate detailed observations over several wavebands and this work has only started recently. The existing data have not yet been fully incorporated into calculations. When this has been achieved, we will have a greatly improved understanding of the structure of SNR, but more observations will be necessary before a "full" understanding is achieved.

The principle improvements desired in observations are high spatial-resolution spectra in X-ray and infrared wavebands. Present spectra of young SNR are averaged over many morphological features and sometimes over the entire remnant. The observatories of the 1990's: *HST*, *AXAF*, and *SIRTF* will perform these observations. These spectra will reveal the composition and temperature of the gas on an arcsecond scale throughout the remnant. Thus the physics of the interface between regions at different temperatures can be detailed. The material in the gas, determined by optical, ultraviolet, and X-ray spectroscopy, can also be related to the distribution of dust, determined by infrared emission.

The radio band also holds great promise for tying all regions of the remnant together. Radio emission seems to originate throughout most remnants indicating that magnetic fields and charged particles are pervasive. The fields are generated by compression of previous fields and by turbulent processes. The particles are also accelerated by these processes and within shocks. Thus the interface between regions is expected to be bright in radio waves, as is the primary shock in Tycho's SNR.

We look forward to finally being able to "see" and identify shocked interstellar clouds and clumps of ejecta at all temperatures throughout the remnants. We look forward to being able to identify remnants as originating in Type I and Type II explosions. We expect that detailed observations in all wavebands will be necessary to accomplish this.

This work was supported by NASA contract NAS8-30751 (FDS) and NASA grant NAGW-764 (RAC).

References

Arnett, W. D. 1987, *Ap. J.*, **319**, 136.
Axelrod, T. S. 1980 in *Type I Supernovae*, ed. J. C. Wheeler (Austin: Univ. of Texas), p. 80.
Bandiera, R., Pacini, F., and Salvati, M. 1984, *Ap. J.*, **285**, 134.

Bartel, N. 1986, *Highlights of Astr.*, **7**, 655.

Bartel, N., Rogers, A. E. E., Shapiro, I. I., Gorenstein, M. V., Gwinn, C. R., Marcaide, J. M., and Weiler, R. W. 1985, *Nature*, **318**, 25.

Becker R. H., Holt S. S., Smith, B. W., White N. E., Boldt, E. A., Mushotzky, R. F., and Serlemitsos, P.J., 1980, *Ap.J.*, **235**, L5.

Bignami, G., and Hermsen, W. 1983 *Ann. Rev. Astron. Ap.*, **21**, 67.

Bionta, R. M. *et al.* 1987, **Phys. Rev. Lett.**, **58**, 1494.

Blair, W.P., and Panagia, N., 1987, *Exploring the Universe with the IUE Satellite*, Y. Kondo ed., D. Reidel, 549.

Branch, D., Falk, S. W., McCall, M. L., Rybski, P., Uomoto, A. K., and Wills, B. J. 1981, *Ap. J.*, **244**, 780.

Branch, D. 1984, *Ann. N. Y. Acad. Sci.*, **422**, 186.

Branch, D., Doggett, J. B., Nomoto, K., and Thielemann, F.-K 1985, *Ap. J.*, **294**, 619.

Branch, D. and Nomoto, K. 1986, *Astr. Ap.*, **164**, L13.

Braun, R., 1985, The interaction of Supernovae with the Interstellar Medium, Thesis, University of Leiden.

Braun, R., and Strom, R.G., 1986, *Ast. Ap.*, **164**, 208.

Canizares, C. R., Kriss, G. A. and Feigelson, E. D. 1982, *Ap. J. (Letters)*, **253**, L17.

Cassatella, A. 1987, in ESO Workshop on SN 1987A, ed. I. J. Danziger (Garching: ESO), p. 101.

Cassatella, A., Fransson, C., van Santvoort, J., Gry, C., Talavera, A., Wamsteker, W., and Panagia, N. 1987, *Astr. Ap.*, **177**, L29.

Charles, P.A., Kahn, S.M., and McKee, C.F., 1985, *Ap.J.*, **295**, 456.

Chevalier, R. A. 1981, *Fund. Cosmic. Phys.*, **7**, 1.

Chevalier, R. A. 1982a, *Ap. J.*, **258**, 790.

Chevalier, R. A. 1982b, *Ap. J.*, **259**, 302.

Chevalier, R. A., 1982c, *Ap.J. Letters*, **259**, L85.

Chevalier, R. A. 1984, *Ap. J. (Letters)*, **285**, L63.

Chevalier, R. A. 1986, *Ap. J.*, **308**, 225.

Chevalier, R. A. 1987a, in ESO Workshop on SN 1987A, ed. I. J. Danziger (Garching: ESO), p. 481.

Chevalier, R. A. 1987b, *Nature*, **329**, 611.

Chevalier, R. A. 1988, *Nature*, submitted.

Chevalier, R. A. and Fransson, C. 1987, *Nature*, **328**, 44.

Cohen, *et al.* 1987, IAU Circ. No. 4500.

Colgate, S. A. 1975, *Ap. J.*, **198**, 439.

Colgate, S. A. and McKee, C. 1969, *Ap. J.*, **157**, 623.

Danziger, I. J., Bouchet, P., Fosbury, R. A. E., Gouiffes, C., Lucy, L. B., Moorwood, A. F. M., Oliva, E., and Rufener, F. 1988, in George Mason Workshop on Supernova 1987A, ed. M. Kafatos (Cambridge: C.U.P.), in press.

Dickel, J.R., and Jones, E.M., 1985, *Ap.J.*, **288**, 707.

Dickel, J.R., and Willis, A.G., *Ast. Ap.*, **85**, 85.

Dopita, M.A., Mathewson, D.S., and Ford, V.L., 1977, *Ap.J.*, **214**, 179.

Dotani, T. *et al.* 1987, *Nature*, **330**, 230.

Duin, R.M., and van der Laan, H., 1975, *Ast. Ap.*, **39**, 33.

Duin, R.M., and Strom, R.G., 1975, *Ast. Ap.* **40**, 111.

Dwek, E., Szymkowak, A., Petre, R., and Rice, W.L., 1987, *Ap.J. Letters*, **320**, L27.

Dwek, E. 1983, *Ap. J.*, **274**, 175.

Dwek, E. 1985, *Ap. J.*, **297**, 719.

Elias, J. H., Matthews, K., Neugebauer, G., and Persson, S. E. 1985, *Ap. J.*, **296**, 379.

Feast 1988, in George Mason Workshop on SN 1987A, ed. M. Kafatos (Cambridge: C.U.P.), in press.

Fesen, R.A., and Kirshner, R.P., 1982, *Ap.J.* **258**,1.

Fransson, C. 1984, *Astr. Ap.*, **133**, 264.

Fransson, C., Benvenuti, P., Gordon, C., Hempe, K., Palumbo, G. G. C., Panagia, N., Reimers, D., Wamsteker, W. 1984, *Astr. Ap.*, **132**, 1.

Fransson, C. and Chevalier, R. A. 1988, *Ap. J.*, submitted.

Fransson, C., Grewing, M., Cassatella, A., Panagia, N., and Wamsteker, W. 1987, *Astr. Ap.*, **177**, L33.

Gaskell, C. M., Capellaro, E., Dinerstein, D. R., Harkness, R. P., and Wheeler, J. C. 1987, *Ap. J. (Letters)*, **306**, L77.

Gilmozzi, R., Cassatella, A., Clavel, J., Fransson, C., Gonzalez, R., Gry, C., Panagia, N., Talavera, A., and Wamsteker, W. 1987, *Nature*, **328**, 318.

Graham, J. R., Meikle, W. P. S., Allen, D. A., Longmore, A. J., and Williams, P. M. 1986, *M.N.R.A.S.*, **218**, 93.

Green, D.A., and Gull, S.F., 1983, Supernova Remnants and their X-ray Emission, ed. J. Danziger and P. Gorenstein, D. Reidel, 329.

Hamilton, A.J.S., Sarazin, C.L., Szymkowiak, A.E., and Vartanian, M.H., 1985 *Ap.J. Letters*, **297**, L5.

Hamuy, M., Suntzeff, N. B., Gonzalez, R., and Martin, G. 1988, *A. J.*, **95**, 63.

Hanuschik, R. W. and Dachs, J. 1987, in ESO Workshop on SN 1987A, ed. I. J. Danziger (Garching: ESO), p. 153.

Harkness, R. P., Wheeler, J. C., Morgan, B., Downes, R. A., Kirshner, R. P., Uomoto, A., Barker, E. S., Cochran, A. L., Dinerstein, H. L., Garnett, D. R., and Levreault, R. M. 1986, *Ap. J.*, **317**, 355.

Harvey, P., Lester, D., and Joy, M. 1987, IAU Circ. No. 4518.

Hester, J.J., and Cox, D., 1986, *Ap.J.*, **300**, 675.

Hirata, K. *et al.* 1987, *Phys. Rev. Lett.*, **58**, 1490.

Inoue, H., Koyama, K., Matsuoka, M., Ohashi, T., Tanaka, Y., and Tsunemi, H., 1980, *Ap.J.*, **238**, 886.

Itoh, M., Kumagai, S., Shigeyama, T., Nomoto, K., and Nishimura, J. 1987, *Nature*, **330**, 233.

Jones, E. M., Smith B. W, and Straka, W. C. 1981, *Ap. J.*, **249**, 185.

Kirshner, R. P. and Oke, J. B. 1975, *Ap. J.*, **200**, 574.

Kirshner, R. P., Oke, J. B., Penston, M. V., and Searle, L. 1973, *Ap. J.*, **185**, 303.

Kirshner, R. P., Sonneborn, G., Cassatella, A., Gilmozzi, R., Wamsteker, W., and Panagia, N. 1987, IAU Circ. No. 4435.

Kirshner, R.P., Winkler, P.F., and Chevalier, R.A., 1987, *Ap.J.*, to be published.

Klein, R. I. and Chevalier, R. A. 1978, *Ap. J. (Letters)*, **223**, L109.

Ku, W.H.-M., Kahn, S.M., Pizarski, R., and Long, K.S., 1984, *Ap.J.*, **278**, 615.

Lundqvist, P. and Fransson, C. 1987, in ESO Workshop on SN 1987A, ed. I. J. Danziger (Garching: ESO), p. 495.

Lundqvist, P. and Fransson, C. 1988, *Astr. Ap.*, in press.

Maeder, A. 1987, in ESO Workshop in SN 1987A, ed. I. J. Danziger (Garching: ESO), p. 251.

Makino, F. 1987, IAU Circ. No. 4336.

Matsui, Y., Long, K.S., Dickel, J.R., and Greisen, E.W., 1984, *Ap.J.*, **287**, 167.

Matz, S. M.,Share, G. H., Leising, M. D., Chupp, E. L., and Vestrand, W. T. 1987, IAU Circ. No. 4510.

McCray, R., Shull, J. M., and Sutherland, P. 1987, *Ap. J. (Letters)*, **317**, L73.

Meikle, W. P. S., Matcher, S. J., and Morgan, B. L. 1987, *Nature*, **329**, 608.

Merrill, K. M. 1980, IAU Circ. No. 3444.

Meyerott, R. E. 1980, *Ap. J.*, **239**, 257.

Moseley, H., Glaccum, W., Loewenstein, R., Silverberg, R., Dwek, E., and Graham, J. 1987, IAU Circ. No. 4500.

Mufson, S.L., McCullough, M.L. Dickel, J.R., Dwek, E., Petre, R., White, R., and Chevalier, R., 1987, *Ap.J.*, to be published.

Nisenson, P., Papaliolios, C., Karovska, M., and Noyes, R. 1987, *Ap. J. (Letters)*, **320**, L15.

Oliva, E. 1987, *Ap. J. (Letters)*, **321**, L45.

Pacini, F. and Salvati, M. 1981, *Ap. J. (Letters)*, **245**, L107.

Palumbo, G. G. C., Maccacaro, T., Panagia, N., Vettolani, G., and Zamorani, G. 1981, *Ap. J.*, **247**, 484.

Panagia, N. 1985 in *Supernovae as Distance Indicators*, ed. N. Bartel (Berlin: Springer-Verlag), p. 14.

Panagia, N. *et al.* 1980, *M.N.R.A.S.*, **192**, 861.

Panagia, N., Gilmozzi, R., Cassatella, A., Wamsteker, W., Kirshner, R. P., and Sonneborn, G. 1987, IAU Circ. No. 4514.

Pinto, P. A. and Woosley, S. E. 1988, *Ap. J.*, in press.

Porter, A. C. and Filippenko, A. V. 1987, *A. J.*, **93**, 1372.

Raymond, J.C., Hester, J.J., Cox, D., Blair, W.P., Fesen, R.A., and Gull, T.R., 1988, *Ap.J.*, **324**, 869.

Rosenberg, I., and Scheuer, P.A.G., 1973 *M.N.R.A.S.*, **161**, 27.

Rupen, M. P., van Gorkom, J. H., Knapp, G. R., Gunn, J. E., and Schneider, D. P. 1987, *A. J.*, **94**, 61.

Sandie, W., Nakano, G., Chase, L., Fishman, G., Meegan, C., Wilson, R., Paciesas, W., and Lasche, G. 1988, IAU Circ. No. 4526.

Seward, F.D., Gorenstein, P., and Tucker, W., 1983, *Ap.J.* **266**, 287.

Seward, F.D., and Wang, Z.R., 1988 *Ap.J.*, submitted.

Shapiro, I. *et al.* 1987 in IAU Symp. No. 129, The Impact of VLBI on Astrophysics and Geophysics, ed. M. J. Reid and J. M. Moran (Dordrecht: Reidel), in press.

Sonneborn, G., Altner, B., and Kirshner, R. P. 1987, *Ap. J. (Letters)*, **323**, L35.

Sramek, R. A., Panagia, N., and Weiler, K. W. 1984, *Ap. J. (Letters)*, **285**, L59.

Sunyaev, R. *et al.*, 1987, *Nature*, **330**, 227.

Strom, R.C., Goss, W.M., and Shaver, P.A., 1982, *M.N.R.A.S.*, **200**, 473.

Turtle, A. J., Campbell-Wilson, D., Bunton, J. D., Jauncey, D. L., Kesteven, M. J., Manchester, R. N., Norris, R. P., Storey, M. C., and Reynolds, J. E. 1987, *Nature*, **327**, 38.

Walborn, N. R., Lasker, B. M., Laidler, V. G., and Chu, Y-H. 1987, *Ap. J. (Letters)*, **321**, L41.

Wamsteker, W., Gilmozzi, R., and Cassatella, A. 1987, IAU Circ. No. 4410.

Wamsteker, W., *et al.* 1987, *Astr. Ap.*, **177**, L21.

West, R. M., Lamberts, A., Jorgensen, H. E., and Schuster, H. E. 1987, *Astr. Ap.*, **177**, L1.

Weiler, K. W., Sramek, R. A., Panagia, N., van der Hulst, J. M., and Salvati, M. 1986, *Ap. J.*, **301**, 790.

Williams, R. E. 1987, *Ap. J. (Letters)*, **320**, L117.

Winkler, P.F., Canizares, C.R., Clark, G.W., Markert, T.H., Kalata, K., and Schnopper, H.W., 1981, *Ap.J.*, **246**, L27.

Winkler, P.F., Tuttle, J.H., Kirshner, R.P., and Irwin, M.J., 1987, proceedings of IAU Colloquium 101, Penticton, Canada, June 1987.

Woosley, S. E. 1988, *Ap. J.*, in press.

Woosley, S. E. and Weaver, T. A. 1986, *Ann. Rev. Astr. Ap.*, **24**, 205.

Interstellar Matter Within Elliptical Galaxies

Michael Jura

Department of Astronomy

University of California, Los Angeles

1. Introduction

The traditional view of elliptical galaxies is that they are essentially devoid of interstellar matter. Unlike spirals which conspicuously display dust patches and lanes, optical pictures of elliptical galaxies generally exhibit smooth isophotes (Sandage 1961, Sandage and Tammann 1981). Also, ellipticals emit relatively little blue and ultraviolet light (O'Connell 1976, Sandage and Visvanathan 1978, Bertola, Capaccioli and Oke 1982). Although somewhat uncertain in view of recent analysis of the ultraviolet fluxes from these galaxies (O'Connell and McNamara 1987, Rocca-Volmerange and Guiderdoni 1987, Kjaergaard 1987), the traditional interpretation of their colors is that ellipticals contain very few hot, young stars. Consequently, a self-consistent view of ellipticals is that they consist of only a population of old stars with no interstellar matter to form new stars. In this model, it is realistic to imagine that since there is no recent star formation, elliptical galaxies are sufficiently simple that they are not undergoing dramatic and unpredictable changes in their light output. Therefore, it is hoped to use elliptical galaxies as standard candles to map the geometry of space-time and the large scale structure of the universe (Gunn, Stryker and Tinsley 1981).

As technology has advanced, it has become increasingly well-established that a significant fraction ($> 50\%$) of all ellipticals contain detectable amounts of interstellar matter (see Schweizer 1987). Although, in the past, it has been proposed that interstellar matter within an elliptical implies that the object has simply been misidentified, here we take the classification of an elliptical galaxy to be reliable on the basis of high-quality optical photographs (see, for example, Sandage and Tammann 1981). If such objects are subsequently discovered to contain interstellar matter, this does not imply that they were misidentified, but rather that the physics of these systems is not well understood in the traditional simple model.

More than ten years ago, as it was recognized that the simple picture of elliptical galaxies as being essentially without interstellar matter was, at the very least, oversimplified. As stars evolve, they lose mass back into their surroundings. Planetary nebulae have been observed in elliptical galaxies (see the review by Ford 1982) providing direct evidence for a source of interstellar matter. Recent infrared observations also indicate the presence of mass-losing stars within ellipticals (Impey, Wynn-Williams and Becklin 1986, Soifer *et al.* 1986, Jura *et al.* 1987). The appropriate question to ask is not whether ellipticals contain interstellar matter, but how much and in what forms. We need to develop an understanding of the sources and sinks of the interstellar matter.

One possibility to explain the lack of conspicuous interstellar matter within ellipticals, proposed by Johnson and Axford (1971) and Mathews and Baker (1971), is that supernovae explosions heat the gas, and then the material flows from the elliptical galaxy as a wind. There is now strong observational evidence against this point of view. In particular, X-Ray emission from ellipticals (Forman, Jones and Tucker 1985) indicates that often there is a substantial amount of hot gas within these galaxies. Therefore, in ellipticals, gas cannot be rapidly flowing out, because both the density and the resulting X-ray emission would be very low. Theoretically, it is now thought that there is considerable amount of dark (non-luminous or very low luminosity) matter within ellipticals. This result implies that the gravitational potentials of these objects are much deeper than thought 10 years ago (Faber and Gallagher 1979), and it is much more difficult to invoke supernovae explosions as an explanation for driving a wind (Mathews and Loewenstein 1986). If there is no wind or other removal mechanism, it is not surprising that the matter lost from stars still resides within the galaxy. In some active galaxies, there are outflows associated with jets and nuclear activity. However, the mass flow rates in these outflows are usually much less than would be required to expel from the galaxy all the matter that resides in the interstellar medium.

Although most of the gas that is lost from stars is apparently not flowing out of ellipticals, the ultimate fate of matter that is injected into the interstellar media of ellipticals is still not known. Possibilities that have been proposed include (i) the gas simply exists as a hot, extended interstellar medium (Forman, Jones and Tucker 1985); (ii) star formation proceeds and consumes the interstellar matter although perhaps with a different luminosity function than in the Milky Way (Jura 1977, Fabian, Nulsen and Canizares 1982, Sarazin and O'Connell 1983, White and Sarazin 1987a,b,c, O'Connell and McNamara 1987) and (iii) matter falls into the center, perhaps into a black hole (see Rees 1984). Below, we discuss the observations which pertain to these possibilities.

Mass loss from stars is not the only possible source of interstellar matter within ellipticals because these galaxies are preferentially found in groups and clusters (see, for example, Abell 1975). Interstellar matter within ellipticals may also result from collisions with gas-rich companions (Knapp, Turner and Cunniffe 1985) and by cooling flows (Fabian, Nulsen and Canizares 1984, Sarazin 1986), the accretion of gas from an intercluster medium.

Regardless of its origin, as in the Milky Way, the interstellar media within elliptical galaxies may exist with a very wide variety of temperatures and densities (see, for example Spitzer 1978, Jura 1987). In the Milky Way, we can identify at least three phases of interstellar gas, and, as described in more detail below, observations show that there are at least three phases of interstellar material within ellipticals. There is hot, ionized gas with temperatures around 10^7 K, warm gas with temperatures near 10^4 K, and very cold gas that may be either atomic hydrogen or even molecular. To study the material in these different phases requires observations from radio to X-ray wavelengths. In Section II, we discuss observations of the relatively cold gas. In Section III, we discuss the warm gas and in Section IV we discuss the hot gas. In Section V, we discuss models for the dynamics, and the origin and fate of the matter in these galaxies, and in Section VI, we present our conclusions.

2. Cold (T < 100 K) Interstellar Matter in Ellipticals

There is a variety of techniques which can be used to study cold interstellar matter within elliptical galaxies. Here we discuss observations of both the gas and dust.

Atomic and Molecular Gas

The 21 cm line of neutral hydrogen can be used to measure the amount of atomic gas within an elliptical. There have been a number of extensive surveys of these objects (Dressel, Bania and O'Connell 1982, Lake and Schommer 1984, Knapp, Turner and Cunniffe 1986); only about 10% or 15% of all ellipticals are sources. In contrast, S0's are much easier to detect (Wardle and Knapp 1986), while essentially all nearby spirals are currently detectable (Haynes, Giovanelli and Chincarini 1984). The elliptical galaxies which are 21 cm sources typically contain 10^7 or 10^8 $M_\odot$ of gas (Bottinelli and Gouguenheim 1977, Gallagher *et al.* 1977, Whiteoak and Gardner 1977, Knapp, Gallagher and Faber 1978, Fosbury *et al.* 1978).

More recently, besides simply detecting the 21 cm emission from ellipticals, maps have been made to describe the detailed morphology of the gas in individual objects including NGC 1052 (van Gorkom *et al.* 1986), NGC 4278 (Raimond *et al.* 1981), NGC 2974 (Kim *et al.* 1987) and several other objects (Lake, Schommer and van Gorkom 1987). These data can be used to compare the gas dynamics with optical observations of the stellar dynamics. These results often indicate that the gas within ellipticals has an external origin because the rotation axis of the stars and the gas are not parallel. If the gas simply resulted from mass loss from the stars, we would expect such an alignment. Finally, infall motions are derived from 21 cm absorption against the nuclear non-thermal background for a few early-type galaxies (van der Hulst, Galisch and Haschick 1983, Shostak *et al.* 1983).

Except for NGC 5128, molecules have not been detected in elliptical galaxies (Morris and Rickard 1982) despite a number of searches (Verter 1985, Jaffe 1987). Although there is not enough data to reach a definite conclusion, at the moment, there is no reason to think that there is much more H_2 than H within ellipticals. In particular, extrapolations from the amount of dust inferred from the far infrared emission (see below) typically do not indicate much more interstellar material than inferred from observations at 21 cm that determine the amount of atomic hydrogen.

Dust

Interstellar dust has been inferred to be present in ellipticals both from optical photometry which shows dust patches and infrared emission which results from thermal emission from the grains. Other means to search for interstellar dust grains such as searching for optical polarization (see Jura 1978) have not been exploited.

There is now an extensive literature reporting optical observations of dust patches in ellipticals and other early-type galaxies (Bender and Mollenhoff 1987, Bertola and Galletta 1978, Ebneter and Balick 1985, Gallagher 1986, Hansen, Norgaard-Nielsen and Jorgensen 1985, Kotanyi and Ekers 1979, Mollenhoff and Bender 1987, Sadler and Gerhard 1985, Sparks *et al.* 1985). The dust is recognized either by dark patches, or red patches or both. With CCD's it is now much easier to detect relatively inconspicuous patches of dust than was possible with photographic photometry; as many as 50% of all ellipticals contain detectable amounts of dust (Sadler and Gerhard 1985, Sparks *et al.* 1985). It should be recognized that, in principle, the presence of an optically-thin dust patch may lead to a bright reflection nebula rather than a dark patch of obscuration. However, no such bright reflection nebula associated with dust has yet been identified in external galaxies.

The advantages of studying the dust optically is that one has a good angular scale so one can hope to measure the morphology of the dust distribution. Often, for example, the dust is aligned along the minor axis of the galaxy. However, a major difficulty with analysis of the optical dust patches is that it has proven to be difficult to derive quantitative information about the mass of dust and cold interstellar matter. We generally have essentially no information about where along the line of sight the dust happens to lie, and it is therefore difficult to determine with any accuracy the amount of obscuration of background starlight and the amount of scattering of the integrated starlight of the galaxy.

An alternative way to study the amount of dust within an elliptical is to measure the far-infrared flux. As with grains in the Milky Way (Spitzer 1978), dust in the interstellar medium in an elliptical should be very cold, typically between 10 and 30 K (Jura 1982). With the data from the *IRAS* satellite, it became possible to detect this far infrared emission (Jura 1986, Tytler 1988, Wrobel, Neugebauer and Miley 1987). It is now established that over half of all ellipticals are far infared sources and the most likely explanation for this emission is that they contain appreciable amounts of cold intestellar matter (Jura *et al.* 1987). In particular, with the simplifying assumptions that (i) the dust to gas ratio has a value in ellipticals similar to that in the interstellar medium in the local Milky Way, and (ii) that all the dust lies at a temperature of 20 K, we expect (Jura 1986) that if the mass of interstellar matter is denoted M then:

$$M = 1.6 \times 10^5 \ D^2 \ F_\nu(100 \ \mu m) \ M_\odot$$

In this equation, D is the distance to the galaxy in Mpc and $F_\nu(100 \ \mu m)$ is the observed flux at 100 μm in Jy. Typically, the inferred mass of cold interstellar matter lies between 10^7 and 10^8 $M_\odot$.

3. Warm (T $\sim$ 10,000 K) Interstellar Matter

The most direct probe of warm gas in elliptical galaxies is their optical emission-line spectra. Humason, Mayall and Sandage (1956) reported that about 15% of all ellipticals exhibited optical emision lines in their nuclear regions. With more sensitive observing techniques, the detection rate of nuclear emission lines is now greater than 50% (Caldwell 1984b, Phillips *et al.* 1986). Also, emission lines can now be detected outside the immediate nucleus (Demoulin-Ulrich, Butcher and Boksenberg 1984). As reviewed by Schweizer (1987), the ionized gas is these galaxies typically has T $\sim$ 10,000 K and n_e $\sim$ 10^3 cm^{-3}. The mass of ionized gas is then inferred to be between 10^3 and 10^6 $M_\odot$. Therefore, it seems that there is considerably more cold interstellar matter than warm gas, a result that also obtains in the Milky Way (Spitzer 1978).

One great advantage of optical emission lines is that they can be studied with high spatial and spectral resolution. Consequently, it is possible to perform detailed studies of the distribution and kinematics of the gas (see, for example, Caldwell 1984a, Ford and Butcher 1979; Caldwell, Kirshner and Richstone 1986).

4. Hot (T $>$ 10^6 K) Interstellar Matter

It has recently been established, primarily through analysis of the Einstein data archives, that elliptical galaxies typically emit X-rays (Nulsen, Stewart and Fabian

1984, Dressel and Wilson 1985, Forman, Jones and Tucker 1984, Trinchieri and Fabbiano 1985, Trinchieri, Fabbiano and Canizares 1986, Canizares, Fabbiano and Trinchieri 1987). Because the X-ray luminosity increases roughly as the square of the blue-light luminosity, it seems that at least for the more luminous X-ray emitting galaxies, there is an additional source of emission besides the superposition of binary X-ray sources (such as those that occur in globular clusters) that have been proposed as the sources for the total X-ray emission from some galaxies (Katz 1975). Because the X-rays are extended and because the spectra are consistent with thermal emission from a hot gas, it is thought that the X-ray emission arises from a hot interstellar medium.

From the crude spectral data that are available, the temperature of the hot gas is estimated to be approximately 10^7 K which implies, for the observed surface brightnesses, that typical densities in the inner ~ 1 kpc of these galaxies of hot gas are n_e in the range $\sim 10^{-2}$ to 10^{-1} cm^{-3} (Forman, Jones and Tucker 1985). These two quantities indicate a central gas pressure in these ellipticals of nT $= 10^5$ to 10^6 cm^{-3} K, very substantially larger than the thermal pressure in the local interstellar medium of the Milky Way which is variable and mostly ranges between 10^3 and 10^4 cm^{-3}K (Jenkins, Jura and Loewenstein 1983). This difference in the thermal pressure of the hot gas between ellipticals and spirals may be quite important in the dynamic evolution of the gas in the two systems.

The derived masses of hot gas in the interstellar media of ellipticals range from 10^8 to 10^{10} $M_\odot$ (Forman, Jones and Tucker 1985). However, these derived masses are uncertain by at least an order of magnitude. A major source of uncertainty is that the inferred density varies as a slow function of galactic radius. Therefore, the bulk of the mass of the hot gas is inferred to be at great distances (~ 50 kpc) from the center of the galaxy. However, these outer regions have very low X-ray surface brightnesses and therefore are very difficult to study with high precision. For example, Forman, Jones and Tucker (1985) estimate an order of magnitude more mass within NGC 1365 than do Nulsen, Stewart and Fabian (1984). Nevertheless, despite these uncertainties, it is clear that there is a substantial amount of hot interstellar gas within elliptical galaxies, and, in particular, the mass of hot gas appears to be significantly larger than the mass of cold gas.

5. Theoretical Models

There is no definitive global model for the interstellar medium of the Milky Way, and therefore, it should not be surprising that the interstellar media of ellipticals are not well understood either. Although there is considerable uncertainty (Cox and McCammon 1986), currently, the most promising models to describe the overall structure of the Milky Way require three phases (cold, warm and hot gas) that are regulated by supernova explosions (Cox and Smith 1974, McKee and Ostriker 1977). The applicability of these models to ellipticals is not clear, at least in part because the supernova rate in ellipticals is much lower than in spirals (Tammann 1982). An important additional source of gas heating in ellipticals is that mass lost from stars in ellipticals collides with the ambient interstellar medium at $\sim$300 km s^{-1} instead of $\sim$30 km s^{-1} as occurs in the disk of the Milky Way. That is, because we are not considering a relatively cold disk of stars and gas, the injected matter from mass-losing stars behaves very differently in an elliptical.

Here, we first note that the interstellar material in ellipticals can be used to probe the overall structure of these objects. Second, we sketch some ideas on the origin and evolution of interstellar matter within ellipticals.

Interstellar Matter as a Probe of Galactic Structure

The internal dynamics of elliptical galaxies is, of course, a subject that has received an enormous amount of attention. Here we just note that interstellar matter can also serve as a probe of the dynamics of these objects.

One proposal is that since the X-ray emission results from hot, extended (r $\sim$50 kpc) gas, there must be a substantial amount of mass to explain this gravitationally-bound material and therefore elliptical galaxies have a heavy halo (Forman, Jones and Tucker 1985). While this particular model of a hot gas in hydrostatic equilibrium may be oversimplified because it does not include thermal instabilities, the observed extended hot gas may still require a heavy halo of some sort (Sarazin and White 1987).

Another way to use the interstellar gas in ellipticals is to compare the gas dynamics with the stellar dynamics (Caldwell, Kirshner and Richstone 1986). For example, in the elliptical galaxy NGC 2974, Kim *et al.* (1987) have measured a circular rotational velocity of the 21 cm emitting gas of 350 km s^{-1}. The stellar velocity dispersion in the galactic center is 220 km s^{-1} (Davies *et al.* 1987). Putting these two results together can provide insight into the nature of the distribution of both the stars and the gas. For example, according to equation (16a) of Richstone and Tremaine (1984), we expect for a spherical galaxy with an isotropic stellar velocity dispersion, σ_r, that:

$$\rho^{-1} d/dr(\rho\sigma_r^2) = -[GM/r^2]$$

In this equation, ρ is the density of stars and M is the total mass of the galaxy within radius r. We define the gas circular velocity, v_c, such that:

$$v_c^2/r = GM/r^2$$

Now suppose that the density of stars, ρ, varies as a power law such that:

$$\rho = \rho_0 \, r^{-a}.$$

We make the further simplifying assumption that σ_r is independent of r as is characteristic of an isothermal sphere; this implies that the total mass density varies as r^{-2}. (The gravitational mass need not follow the light distribution.) We then expect from the above equations that:

$$\sigma_r = v_c(a)^{-1/2}$$

The observational result that $\sigma_r/v_c = 0.58$ for NGC 2974 implies that $a = 2.9$, a very plausible value for the density distribution of stars necessary to explain the typical isophotes of ellipticals. Obviously, a more sophisticated analysis of the gas and stars is possible; nevertheless, the data are consistent with the view that the total density varies as r^{-2} while the density of luminous stars varies approximately as r^{-3}.

Origin and Evolution of Interstellar Matter

As discussed above, it seems quite unlikely that material currently flows out of ellipticals as a wind (Mathews and Loewenstein 1986). Consequently, except in some galaxies that are in rich clusters where stripping of the interstellar matter is important, it seems that most ellipticals retain most of their interstellar material.

There are at least three ways by which ellipticals can acquire interstellar gas. First, there is mass loss from evolving stars which typically amounts to $\sim 0.3\ M_\odot$ yr^{-1} (Faber and Gallagher 1976, Jura *et al.* 1987). Also, there are cooling flows where the rate of accumulation can be very large, but a rate of 0.1 to 1 $M_\odot$ yr^{-1} is not unlikely (see Sarazin 1986). Finally, the incorporation of stars and gas from companions probably occurs several times in the history of many ellipticals (Quinn 1984, Knapp, Turner and Cunniffe 1985); it is difficult to estimate a mass inflow rate for this process.

With accumulation rates of at least 0.3 $M_\odot$ yr^{-1}, it is necessary to consider the ultimate fate of this matter. During a Hubble time, we might anticipate that an elliptical galaxy accumulates $\sim 10^{10}\ M_\odot$ of interstellar matter. As discussed above, it is possible that there is this much hot gas within the very extended halos of ellipticals; however, observationally this is not well established. Also, since the hot gas is thermally unstable (see, for example, Sarazin and White 1987), it may not simply reside in steady, hydrostatic equilibrium as has been proposed. Evidence that the gas is thermally unstable is in fact provided by the observed presence of warm and cold material in the interstellar media of these galaxies.

If the hot gas does not flow out of the galaxy and does not simply accumulate, then it must go somewhere. At least some of this gas may flow into the center to produce nuclear activity. However, an accumulation rate of 0.3 $M_\odot$ yr^{-1} would produce much greater activity than is normal within most ellipticals (Rees

1984). Also, the central masses of ellipticals are generally nowhere near 10^{10} $M_\odot$ (Richstone and Tremaine 1985).

Another possiblity is that there is star formation occuring within ellipticals (Jura 1977, Fabian, Nulsen and Canizares 1982, White and Sarazin 1987a,b,c). While there is little evidence for O and B stars in most of these galaxies and little direct evidence for current star formation (O'Connell and McNamara 1987), it is quite possible that stars form with a different initial mass function within ellipticals than within spirals such as the Milky Way. The X-ray data within ellipticals do indicate that the thermal pressures in much of their interstellar media are much greater than within the Milky Way. It therefore may be possible to compress low mass interstellar clouds into stars and therefore, on the average, form much lower mass stars. Also, because of these higher pressures, star formation may be much more efficient within ellipticals than within spirals. Sandage (1986) has argued that the main difference between ellipticals and spirals is that ellipticals have undergone a more thorough consumption of their gas through star formation. This picture might explain why there is more hot gas than cold gas in ellipticals in contrast to the situation in the Milky Way where there is much more cold gas than hot gas. In this conception, the bulk of the interstellar medium is hot both because of heating by supernovae and the heating that results from the high initial kinetic energy for all the newly injected interstellar matter. The observed cold interstellar matter is material that is in the process of collapsing into stars, while the warm matter is in transition between the hot and the cold phases. Further observations of ellipticals may provide insights both into the nature of the interstellar media of these galaxies and the general process of star formation.

6. Summary

In order to characterize fully the interstellar matter within elliptical galaxies, it is necessary to study the cold, warm and hot gas. This requires observations at radio, infrared, optical, ultraviolet and X-ray wavelengths. Multi-spectral studies have proven essential even to have reached our currently incomplete understanding elliptical galaxies.

The accumulation of a large amount of data for ellipticals shows (i) that mass from evolved stars is occuring, (ii) accretion from companions is significant and (iii) cooling inflows often occur. There is no evidence for large mass outflow rates from most of these objects, and, in fact, the observations show that the substantial amount of interstellar matter which does reside within ellipticals exists in at least three physical phases. However, the relationships among these three phases and the ultimate fate of this interstellar matter is still an unsolved problem.

This work has been partly supported by NASA. I thank C. Canizares, D. W. Kim, J. Knapp, D.

Richstone and J. Turner for their comments.

References

Abell, G. O. 1975, in *Galaxies and the Universe*, ed. A. Sandage, M. Sandage, and J. Kristian (Chicago: University of Chicago), p. 601.

Bender, F., and Mollenhoff, C. 1987, *Astr. Ap.*, **177**, 71.

Bertola, F., Capaccioli, M., and Oke, J. B. 1982, *Ap. J.*, **254**, 494.

Bertola, F., and Galletta, G. 1978, *Ap. J. (Letters)*, **226**, L115.

Bottinelli, L., and Gouguenheim, L. 1977, *Astr. Ap.*, **54**, 641.

Caldwell, N. 1984a, *Ap. J.*, **278**, 96.

Caldwell, N. 1984b, *P. A. S. P.*, **96**, 287.

Caldwell, N., Kirshner, R. P., and Richstone, D. O. 1986, *Ap. J.*, **305**, 136.

Canizares, C. R., Fabbiano, G., and Trinchieri, G. 1987, *Ap. J.*, **312**, 503.

Cox, D. P., and McCammon, D. 1986, *Ap. J.*, **304**, 657.

Cox, D. P., and Smith, B. W. 1974, *Ap. J. (Letters)*, **189**, L105.

Davies, R. L., Burstein, D., Dressler, A., Faber, S. M., Lynden-Bell, D., Terlevich, R. J., Wegner, G. 1987, *Ap. J. Suppl.*, **64**, 581.

Demoulin-Ulrich, M. H., Butcher, H. R., and Boksenberg, A. 1984, *Ap. J.*, **285**, 527.

Dressel, L. L., Bania, T. M., and O'Connell, R. W. 1982, *Ap. J.*, **259**, 55.

Dressel, L. L., and Wilson, A. S. 1985, *Ap. J.*, **291** , 668.

Ebneter, K., and Balick, B. 1985, *Astr. J.*, **90**, 183.

Faber, S. M., and Gallagher, J. S. 1976, *Ap. J.*, **204**, 365.

Faber, S. M., and Gallagher, J. S. 1979, *Ann. Rev. Astr. Ap.*, **17**, 135.

Fabian, A. C., Nulsen, P. E., and Canizares, C. R. 1982, *M.N.R.A.S.*, **201**, 933.

Fabian, A. C., Nulsen, P. E., and Canizares, C. R. 1984, *Nature*, **310**, 773.

Ford, H. C. 1982, in *IAU Symposium 103, Planetary Nebulae*, ed. D. R. Flower (Dordrecht: Reidel), p. 443.

Ford, H. C., and Butcher, H. 1979, *Ap. J. Suppl.*, **41**, 147.

Forman, W., Jones, C., and Tucker, W. H. 1985, *Ap. J.*, **293**, 102.

Fosbury, R. A. E., Mebold, U., Goss, W. M., and Dopita, M. A. 1978, *M.N.R.A.S.*, **183**, 549.

Gallagher, J. S. 1986, *P.A.S.P.*, **98**, 81.

Gallagher, J. S., Knapp, G. R., Faber, S. M., and Balick, B. 1977, *Ap. J.*, **215**, 643.

van Gorkom, J., Knapp, G. R., Raimond, E., Faber, S. M., and Gallagher, J. S. 1986, *Astr. J.*, **91**, 791.

Gunn, J. E., Stryker, L. L., and Tinsley, B. M. 1981, *Ap. J.*, **249**, 48.

Hansen, L., Norgaard-Nielsen, H. V., and Jorgensen, H. E. 1985, *Astr. Ap.*, **149**, 442.

Haynes, M. P., Giovanelli, R., and Chincarini, G. L. 1984, *Ann. Rev. Astr. Ap.*, **22**, 445.

van der Hulst, J. M., Golisch, W. F., and Haschick, A. D. 1983, *Astr. Ap.*, **264**, L37.

Humason, M. L., Mayall, N. U., and Sandage, A. R. 1956, *Astr. J.*, **61**, 97.

Impey, C. D., Wynn-Williams, C. G., and Becklin, E. E. 1986, *Ap. J.*, **309**, 572.

Jaffe, W. 1987, *Astr. Ap.*, **171**, 378.

Jenkins, E. B., Jura, M., and Loewenstein, M. 1983, *Ap. J.*, **270**, 88.

Johnson, H. E., and Axford, W. I. 1971, *Ap. J.*, **165**, 381.

Jura, M. 1977, *Ap. J.*, **212**, 634.

Jura, M. 1978, *Ap. J.*, **223**, 421.

Jura, M. 1982, *Ap. J.*, **254**, 70.

Jura, M. 1986, *Ap. J.*, **306**, 483.

Jura, M. 1987, in *Interstellar Processes*, eds. D. J. Hollenbach and H. A. Thronson, (Dordrecht: Reidel), p. 3.

Jura, M., Kim, D. W., Knapp, G. R., and Guhathakurta, P. 1987, *Ap. J. Letters*, **312**, L11.

Katz, J. I. 1976, *Ap. J.*, **207**, 25.

Kim, D. W., Guhathakurta, R., van Gorkom, J., Jura, M., and Knapp, G. R. 1987, *Ap. J.*, in press.

Kjaergaard, P. 1987, *Astr. Ap.*, **176**, 210.

Knapp, G. R., Turner, E. L., and Cunniffe, P. E. 1985, *Astr. J.*, **90**, 454.

Knapp, G. R., Gallagher, J. S., and Faber, S. M. 1978, *Astr. J.*, **83**, 11.
Kotanyi, C. G., and Ekers, R. D. 1979, *Astr. Ap.*, **73**, L1.
Lake, G., and Schommer, R. A. 1984, *Ap. J.*, **280**, 107.
Lake, G., Schommer, R. A., and van Gorkom, J. H. 1987, *Ap. J.*, **314**, 57.
Mathews, W. G., and Baker, J. C. 1971, *Ap. J.*, **165**, 381.
Mathews, W. G., and Loewenstein, M. 1986, *Ap. J. (Letters)*, **306**, L7.
McKee, C. F., and Ostriker, J. P. 1977, *Ap. J.*, **218**, 148.
Mollenhoff, C., and Bender, R. 1987, *Astr. Ap.*, **174**, 63.
Morris, M., and Rickard, L. J. 1982, *Ann. Rev. Astr. Ap.*, **20**, 517.
Nulsen, P. E., Stewart, G. C., and Fabian, A. C. 1984, *M.N.R.A.S.*, **208**, 185.
O'Connell, R. W. 1976, *Ap. J.*, **206**, 370.
O'Connell, R. W., and McNamara, B. R. 1987, preprint.
Phillips, M. M., Jenkins, C. R., Dopita, M. A., Sadler, E. M., and Binette, L. 1986, *Astr. J.*, **91**, 1062.
Quinn, P. J. 1984, *Ap. J.*, **279**, 596.
Raimond, E., Faber, S. M., Gallagher, J. S., and Knapp, G. R. 1981, *Ap. J.*, **246**, 708.
Richstone, D. O., and Tremaine, S. 1984, *Ap. J.*, **286**, 27.
Richstone, D. O., and Tremaine, S. 1985, *Ap. J.*, **296**, 370.
Rocca-Volmerange, B., and Guiderdoni, B. 1987, *Astr. Ap.*, **175**, 15.
Sadler, E.M., and Gerhard, O. E. 1985, *M.N.R.A.S.*, **214**, 177.
Sandage, A. 1961, *The Hubble Atlas of Galaxies* (Washington D. C.: Carnegie Institution of Washington).
Sandage, A. 1986, *Astr. Ap.*, **161**, 89.
Sandage, A., and Tammann, A. 1981, *A Revised Shapley-Ames Catalog of Bright Galaxies* (Washington D. C.: Carnegie Institution of Washington).
Sandage, A., and Visvanathan, N. 1978, *Ap. J.*, **225**, 742.
Sarazin, C. L. 1986, *Rev. Mod. Phys.*, **58**, 1.
Sarazin, C. L., and O'Connell, R. W. 1983, *Ap. J.*, **268**, 552.
Schweizer, F. 1987, in *IAU Symp. No. 127, Structure and Dynamics of Ellipticals Galaxies*, ed. T. de Zeeuw (Dordrecht: Reidel), p. 109.
Shostak, G. S., van Gorkom, J., Ekers, R. D., Sanders, R. H., Goss, W. M., and Cornwell, T. J. 1983, *Astr. Ap.*, **119**, L3.
Soifer, B. T., Rice, W. L., Mould, J. R., Gillett, F. C., Rowan-Robinson, M., and Habing, H. J. 1986, *Ap. J.*, **304**, 651.
Sparks, W. B., Wall, J. V., Thorne, D. J., Jorden, P. R., van Breda, E. G., Rudd, P. J., and Jorgensen, H. E. 1985, *M.N.R.A.S.*, **217**, 87.
Spitzer, L. 1978, *Physical Processes in the Interstellar Medium*, (New York: J. Wiley).
Trinchieri, G., and Fabbiano, G. 1985, *Ap. J.*, **296**, 447.
Trinchieri, G., Fabbiano, F., and Canizares, C. R. 1986, *Ap. J.*, **310**, 637.
Tytler, D. 1988, *Ap. J.*, in press.
Verter, F. 1985, *Ap. J.Suppl.*, **57**, 261.
Wardle, M., and Knapp, G. R. 1986, *Astr. J.*, **91**, 23.
White, R. E., and Sarazin, C. L. 1987a, *Ap. J.*, **318**, 612.
White, R. E., and Sarazin, C. L. 1987b, *Ap. J.*, **318**, 621.
White, R. E., and Sarazin, C. L. 1987c, *Ap. J.*, **318**, 629.
Whiteoak, J. B., and Gardner, F. F. 1977, *Aust. J. Phys.*, **30**, 187.
Wrobel, J. M., Neugebauer, G., and Miley, G. K. 1986, *Ap. J. (Letters)*, **309**, L11.

Active Galactic Nuclei

C. Megan Urry

Space Telescope Science Institute

1. Introduction

It is hard to imagine an area of astrophysics where multiwavelength studies could be more important than the case of active galactic nuclei — indeed the physics dictates that many wavebands are of comparable importance energetically. It is only relatively recently, however, with the advent of space-borne instrumentation which opens up previously unobservable spectral regions in the infrared, ultraviolet, X-ray, and γ-ray bands that multiwavelength studies have begun to improve our understanding of active galactic nuclei. This chapter explains what we know about active galactic nuclei, in large part because of multiwavelength data, and describes what we can hope to learn with future multiwavelength experiments.

The term active galactic nucleus, or AGN, can signify any extragalactic object with substantial emission in excess of normal stellar processes (starlight, H II regions, supernovae, etc.), including low luminosity LINERs (Low-Ionization Nuclear Emission Regions), starburst systems, narrow-line ($\leq 1000\,\mathrm{km/s}$) and broad-line ($\geq 1000\,\mathrm{km/s}$) Seyfert galaxies, narrow-line and broad-line radio galaxies, radio-loud and radio-quiet quasars, optically violently variable (OVV) quasars, and BL Lacertae objects. These range in bolometric luminosity from a modest $10^8\,L_\odot$ up to $10^{15}\,L_\odot$ for the most luminous quasars and BL Lacs (assuming isotropic radiation). Many, though not all, AGN appear to contain a strong source of nonthermal radiation, presumably a population of relativistic electrons accelerating in a magnetic field. This nonthermal radiation can span several decades in frequency but in some cases is masked by other components or is reprocessed from one energy to another. (For a nice review of the properties of various AGN, and of possible connections among them, see Lawrence 1987.) Here we will concentrate on those AGN that have strong central engines, *i.e.* most of their energy is produced on scales smaller than a few parsecs.

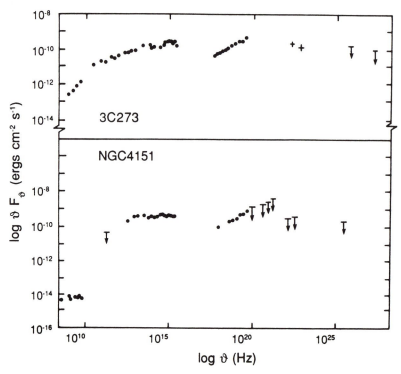

Figure 1 Radio-through-γ-ray spectra of two well-studied AGN, the high-luminosity quasar 3C273 and the low-luminosity Seyfert galaxy NGC4151. Data are from Ramaty and Lingenfelter (1982) and references therein. In both cases, emission over a wide range of wavelengths contributes significantly to the bolometric luminosity.

The large luminosities of AGN are thought to derive ultimately from gravitational potential energy. A deep potential well, as would be found in the vicinity of a black hole, can convert matter to energy with relatively high efficiency. Also, in at least some cases, the primary energy source must occupy a relatively small volume, on the order of the black hole size for Eddington-limited accretion, since AGN can vary on short time scales.

There are observational reasons to prefer cylindrical symmetry to spherical symmetry in models of some AGN (*e.g.* Lawrence and Elvis 1982, Antonucci and Miller 1985), and there are also theoretical considerations. For one thing, spherical accretion is less efficient than other geometries (see review by Rees 1984). For another, the fuel for the gravitational engine, which ultimately comes from the host galaxy and so has angular momentum, may form an accretion disk around the central compact object. Most of the energy radiated from such a disk would come from $3-30\,R_S$, where $R_S \equiv 2GM/c^2$ is the Schwarzschild radius, although the disk itself could have a much larger radius. For a black hole mass $M = 10^8\,M_\odot$, this distance is $\sim 10^{14}\text{-}10^{15}\,\text{cm}$. In contrast, the size of the Broad Line Region (BLR) is estimated to be $\sim 10^{16}\text{-}10^{18}$ cm, and is in turn 10-100 times smaller than the Narrow Line Region (NLR). If the accretion disk is heated by viscous processes, thermal radiation could be observed directly and/or could serve as a seed photon distribution for scattering or reprocessing (absorption and re-emission). For example, the optical-to-ultraviolet emission from an accretion disk or any other mechanism could be absorbed by dust and re-radiated in the infrared.

All these thermal and nonthermal processes produce radiation at wavelengths from a meter down to picometers; or, in terms of energy, from 10^{-6} to 10^6 eV. The spectral energy distribution across much of this range is described very crudely by equal power per logarithmic bandwidth; that is, a flat distribution in $\log\nu F_\nu$ versus $\log\nu$, or a spectral index of $\alpha \sim 1$ ($F_\nu \propto \nu^{-\alpha}$). Two examples are given in Figure 1, which shows the spectral energy distributions from radio to γ-ray wavelengths of the famous quasar 3C273 and the well-studied Seyfert galaxy NGC4151 (Ramaty and Lingenfelter 1982, and references therein).

Multiwavelength studies have greatly contributed to our understanding of AGN. These studies can be conveniently separated into the areas discussed in §§ 2 and 3, respectively: the continuum of blazars (a class of AGN defined below); and the continuum and emission lines of the remaining emission-line objects. The temporal and spectral properties of these two classes, blazars and non-blazars, are significantly different and may indicate fundamental underlying physical distinctions, although some similarities do exist. Consequently, different observing strategies for each class are required, as will become obvious in the following two sections. In § 4 we summarize the information gained from multiwavelength studies of AGN spectra and variability, and discuss some of the questions to be addressed by future experiments.

2. Blazars

Blazar Characteristics and Variability

Blazars comprise both emission-line AGN known as Optically Violently Variable quasars (OVVs) and BL Lacertae objects, which have little or only weak line emission (small equivalent widths). They are extreme examples of the "active" phenomenon in that they tend to be highly polarized, they have large amplitude variability in both intensity and polarization, on short time scales[1],

---

[1] For the purposes of this paper, the term "time scale" will the mean doubling time scale, which is given by $t \sim F/(dF/dt)$. For a factor of 2 change in flux, this definition

and they are compact flat-spectrum radio sources that sometimes exhibit super-luminal motion.

The variability time scales of blazars seem to vary with wavelength. In the radio, the time scales range from a few months to a few years, with most sources having the longer time scales; the overall intensity change for most sources is less than ~20% over several years (Medd *et al.* 1972; Dent and Hobbs 1973; Dent, Kapitzky, and Kojoian 1974; Dent and Kapitzky 1976; Altschuler and Wardle 1975; Andrew *et al.* 1978; Aller *et al.* 1981, 1985; Altschuler 1982; Waltman *et al.* 1986). A few sources are clearly exceptions For example, BL Lac itself, which is very well monitored in the radio, increased by ~50% in less than a month (Reich and Steffen 1982; Aller, Aller, and Hughes 1985; Waltman *et al.* 1986).

At shorter wavelengths, the time scales are usually shorter, typically days or weeks in the infrared-through-optical bands. The Rosemary Hill Observatory has 18 years of photometric data for over two hundred extragalactic objects, of which roughly three dozen of the most variable are blazars (McGimsey *et al.* 1975, Scott *et al.* 1976, Pollock *et al.* 1979, Pica *et al.* 1980, Webb *et al.* 1986). Many different types of optical activity have been seen, from short-term flicker-ing to long-term trends. Individual sources can range in brightness over four or five magnitudes. Examples of the shortest time scale events are: in A0235+164, a decay of 2.3 B-magnitudes in 12 days; in AP Lib, a 0.7 mag change in 5 hours; and in NRAO530, a 0.85 mag decay in 1 night. In general, intensity changes of $\pm\frac{1}{2}$ mag night-to-night are not uncommon, and time scales range from days to months in the more active objects (see also Smith *et al.* 1987). The shortest reported time scale in the optical-infrared regime is 50 seconds for a change of ±1 mag at 1.25μm in OJ287 (Wolstencroft, Gilmore, and Williams 1982).

At still shorter wavelengths, time scales appear to decrease further, with the caveat that ultraviolet time scales are not well-measured. In X-rays, variability seems more pronounced although long-term data are more sparse. (For a sum-mary of observations of X-ray-bright BL Lac objects, see Urry, Mushotzky, and Holt 1986 and Singh and Garmire 1985.) While the X-ray intensity of a source over many years spans a range comparable to the optical variations, usually less than a factor of 10, the time scales are generally shorter. Some well-studied

---

gives a time scale that can reasonably be expected to be characteristic of the system. For smaller changes, this definition of a time scale can be misleading, in the sense that the observed change could be a random fluctuation that has little to do with characteris-tic time scales of the system (see Lawrence *et al.* 1987 and McHardy and Czerny 1987).

sources vary on time scales of hours to a day, and most X-ray-observed blazars do vary on time scales shorter than a year (Barr and Mushotzky 1986). Some X-ray variability on time scales of an hour or less have also been seen. At the Taos Workshop on Multiwavelength Astrophysics, Aldo Treves showed an X-ray light curve of the BL Lac object PKS2155-304 in which the intensity doubled in one hour, corresponding to the high value $\Delta L / \Delta t \sim 2 \times 10^{42}$ ergs/s^2 (Morini *et al.* 1986a). The fastest variation ever observed in any extragalactic object is the 30-second decrease by a factor of ≥ 10 in the X-ray intensity of the BL Lac object H0323+022, for a $\Delta L / \Delta t \sim 10^{44}$ ergs/s^2 (Doxsey *et al.* 1983, Feigelson *et al.* 1986). This greatly exceeds the efficiency limit discussed by Cavallo and Rees (1978) and Fabian (1979).

Correlations between continuum variations in different wavebands may indicate whether the continua arise from related processes, and may reflect other properties of the source as well. An intriguing example involving optical and radio variability in the blazar A0235+164 was discussed by Larry Molnar at the Taos Workshop (Molnar 1988). The optical light curve of A0235+164 is double-peaked, the radio light curve is broader and single-peaked, and the cross-correlation function is consistent with zero lag. (Interpolation of the sparsely sampled optical light curve could cause spurious signal in the correlation function, but Molnar carefully determined the statistical significance of the peak by correlating the observed radio light curve, which was more or less continuous, with ≥ 1000 simulated optical light curves.) Following the suggestion that BL Lac objects might be gravitationally mini-lensed OVV quasars (Ostriker and Vietri 1985), Molnar proposed that the observed radio and optical variability could be explained by motion of the nearby mini-lens. The radio source has to be larger than the optical continuum source so that the amplification pattern of the lens causes double peaks in the optical light curve but smears them out in the radio light curve. The required velocity of the mini-lens for the observed variations is uncomfortably large, $v \sim 0.3c$, but the model is an interesting suggestion, and may well be useful in explaining long-term variability in BL Lac objects.

The examples of variability discussed here emphasize the shortest time scales in order to make the point that, at least for the most active objects, multifrequency observations must be simultaneous. Specifically, observations at different wavelengths must occur closer in time than the typical variability time scale for the source being observed. For most blazars, within the hour is best, within the day is probably all right, and within weeks could be misleading, since incommensurate variations at different wavelengths may well have occurred.

Observed Broad-Band Continuum

Early multiwavelength observations were seldom simultaneous (*cf.* Kondo *et al* 1981), and broad-band spectra were constructed from flux densities at a few radio frequencies, visual magnitudes in a few colors, ultraviolet spectra from one or both of the IUE spectrographs, and an occasional X-ray spectrum or X-ray flux density (Ledden *et al.* 1981, Weistrop *et al.* 1981, and Bregman *et al.* 1981, Urry and Mushotzky 1982). These results were interesting enough to cause several different groups, despite the difficulties, to arrange coordinated observations by large teams using many instruments, some of them space-based. Two dozen BL Lacs and OVVs have been observed in these programs, many of them more than once. These are listed in Table 1 (Bregman *et al.* 1981, 1982, 1984, 1986a,b; Kondo *et al.* 1981; Worrall *et al.* 1982, 1984a,b,c, 1986; Glassgold *et al.* 1983; Mufson *et al.* 1984; Maraschi *et al.* 1985; Pollock *et al.* 1985; Feigelson *et al.* 1986; Landau *et al.* 1986; Makino *et al.* 1987). Additional broad-band studies of large samples of blazars have used non-simultaneous data (Madejski and Schwartz 1983; Cruz-Gonzalez and Huchra 1984; Urry 1984; Madejski 1985; Ledden and O'Dell 1985; Maraschi *et al.* 1986; Ghisellini *et al.* 1986; Madau, Ghisellini, and Persic 1987).

In general, the continuum spectra of OVVs and BL Lacs look remarkably similar, with flat or rising spectra near 10^9 Hz (meaning the power-law spectral index is $\alpha \lesssim 0$, where α was defined in § 1), gradually steepening to a spectral index of 1 or greater at frequencies somewhere in the range 10^{11}–10^{14} Hz, and steepening still further in the ultraviolet to $\alpha \gtrsim 1.5$. The continuation of the spectrum to X-ray frequencies is less regular; the ultraviolet spectrum sometimes joins smoothly to the X-ray spectrum, but it may also fall above or below the X-ray flux. Two typical multiwavelength blazar spectra are shown in Figure 2. For the radio-selected OVV quasar 3C345 (panel a), the X-ray flux lies above the extrapolation of the ultraviolet spectrum, while for the optically-selected BL Lac object Mrk421 (panel b), it is more or less on the extrapolation.

In some OVV and HPQ (highly polarized quasar) spectra, including 3C345, weak blue bumps (discussed in § 3) have been detected above the smooth blazar continuum (Malkan and Moore 1986, Bregman *et al.* 1986b; the bump in 3C345 is so weak that it is barely visible in Figure 2.) Some intermediate objects like 3C273 and 3C120 have stronger blue bumps but also have the blazar-like characteristic of superluminal motion. It is possible that OVVs and HPQs are a mixture of pure-blazar and ordinary-quasar continua (see § 3); in this section we concentrate on "pure" blazar continua.

Table 1 Simultaneous Multifrequency Observations of Blazars

IAU Name	Other Name	Reference
0219+428	3C 66A	1
0323+022	HEAO1	2
0422+004	OF 038	3
0537-441	PKS	4
0735+178	PKS	5, 6
0736+178		6
0754+100	OI 090.4	1
0829+046	OJ 049	6
0851+202	OJ 287	6, 7, 8
0906+430	3C 216	6
0912+297	OK 222	3
0954+658		6
1101+384	Mrk 421	9
1133+704	Mrk 180	10
1156+295	4C 29.45, Ton 599	6, 11, 12
1215+303	ON 325	12
1219+285	ON 231, W Comae	3, 6
1253-055	3C 279	6
1308+326	B2	6
1413+135		13
1418+546	OQ 530	6, 12
1514-241	Ap Lib	6
1538+149	4C 14.60	6
1641+399	3C 345	6, 14
1652+398	Mrk 501	10, 15
1727+502	I Zw 187	6, 16
1749+096		6
1807+698	3C 371	6, 17
2155-304	PKS	18
2223-052	3C 446	19

References — 1. Worrall *et al.* 1984c. 2. Feigelson *et al.* 1986. 3. Worrall *et al.* 1986. 4. Maraschi *et al.* 1985. 5. Bregman *et al.* 1984. 6. Landau *et al.* 1986. 7. Worrall *et al.* 1982. 8. Pollock *et al.* 1985. 9. Makino *et al.* 1987. 10. Mufson *et al.* 1984. 11. Glassgold *et al.* 1983. 12. Worrall *et al.* 1984b. 13. Bregman *et al.* 1981. 14. Bregman *et al.* 1986b. 15. Kondo *et al.* 1981. 16. Bregman *et al.* 1982. 17. Worrall *et al.* 1984a. 18. Morini *et al.* 1986a. 19. Bregman *et al.* 1986a.

Often the overall spectral curvature of blazars is described by power laws that meet at so-called "break" frequencies, even though the curvature is really gradual. Landau *et al.* (1986) modeled high-quality radio-through-ultraviolet spectra of more than a dozen blazars with the next most complicated function after a single power law, namely a parabola in $\log F_\nu$ versus $\log \nu$. It fit reasonably well, to within 10-15% at most frequencies, which in gross terms means

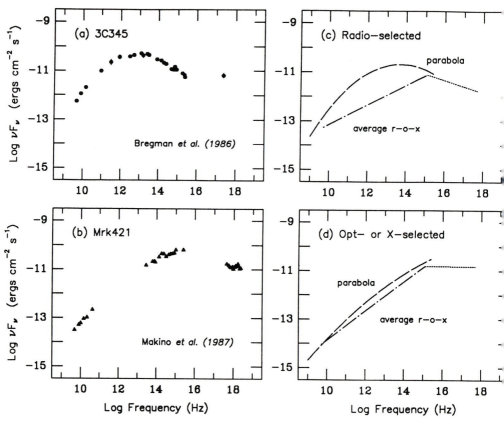

Figure 2 Broad-band spectra of blazars. The ordinate is log νF_ν, so a flat curve in this plot corresponds to a spectral index $\alpha=1$ and represents equal energy output per decade of frequency. (a) The radio-selected OVV quasar 3C345, with peak emission in the infrared (Bregman et al. 1986b). (b) The optically-selected BL Lac object Mrk421, which peaks in the ultraviolet (Makino et al. 1987). The spectral difference is related to the selection criteria rather than to any OVV-BL Lac distinction. (c,d) Best-fit parabolas (in log F_ν versus log ν) for two blazars in the Landau et al. (1986) sample, converted to log νF_ν versus log ν. The straight lines are average values of α_{ro} and α_{ox} for the appropriate sample (Ledden and O'Dell 1985, Maraschi et al. 1986). Again, the curves illustrate the difference between the spectrum of a radio-selected object (dominated by infrared emission) and an optically- or X-ray-selected object (dominated by ultraviolet emission).

that power-law models are good representations of the data over limited frequency ranges, that the power law gradually steepens with increasing frequency, and that the greatest steepening occurs in a narrow wavelength region. Figure 2 shows two best-fit parabolic models from Landau *et al.* (1986), converted to log νF_ν versus log ν. The more curved one (panel c) is typical of a radio-selected blazar like 3C345 and the flatter one (panel d) is appropriate for an optically- or X-ray-selected blazar like Mrk421. Also shown are the average radio-to-optical and optical-to-X-ray ratios for each kind of blazar. (For a discussion of global spectral properties and selection effects, see Ledden and O'Dell 1985 and Maraschi *et al.* 1986.)

The peak energy production is in the infrared-optical for radio-selected blazars, and in the optical-ultraviolet for the optically- or X-ray-selected blazars. Some X-ray-bright BL Lac objects have hard X-ray tails above ~ 5 keV (Urry 1986b), which suggests that hard X-rays could contribute a large fraction of the bolometric luminosity. However, such hard tails are variable, and the time-averaged contribution to the total energy output is probably small (Bezler, Gruber, and Rothschild 1988).

The relative success of power-law models over limited bandwidths, as well as the high degree of polarization observed in these objects at radio and visual frequencies, suggests that the radio-through-optical and perhaps higher-energy emission is produced by relativistic particles, most probably via the synchrotron process. The slow curvature could then be produced by loss processes which preferentially deplete the higher energy electrons (Tucker 1967), or by contributions of stellar light from a luminous host galaxy (Ledden *et al.* 1981, Weistrop *et al.* 1981). More extreme curvature could be due to intrinsic changes in the relativistic particle distribution (Beichman *et al.* 1981, Bregman *et al.* 1981, Zdziarski 1986).

It is likely that the very flat spectra in the radio regime arise not from a flat distribution of electrons but from opacity effects due to an inhomogeneous structure (Cotton *et al.* 1980). VLBI maps of resolved sources with flat radio spectra frequently show that the apparent flatness is due to superposition of components with different self-absorption frequencies (e.g. Readhead *et al.* 1983, Unwin *et al.* 1983). There is no *a priori* reason that flat radio spectra cannot be produced by flat electron energy distributions, particularly since a large fraction of the radio emission from the objects under discussion is usually unresolved, but by analogy with the resolved sources, inhomogeneities are likely. We also expect that the spectra are optically thin above a frequency where they steepen to $\alpha \sim 0.5$, and that further spectral curvature results from the short synchrotron cooling times of high energy electrons.

The smoothness of blazar spectra over such a wide frequency range suggests that observed emission in different bands could arise in a common process. Some coordinated variability events support this picture while counterexamples reveal that the connection is not so clear (see review by Bregman, Maraschi, and Urry 1987; see also discussion Molnar 1988 above), possibly because of details of the energy loss processes. If the increases in intensity in these sources are due to the injection or reacceleration of relativistic particles, then the whole spectrum ought to rise more or less simultaneously. However, when the intensity begins to fall due to cessation of particle replenishment or acceleration, the spectrum will not evolve uniformly since the loss time scale are highly energy dependent. Thus a broad-band spectrum obtained during the decay of a flare, when the source is not in a steady state, may not directly indicate the energy dependence of the radiating particles. This means repeated observations densely spaced in time may be as important as simultaneity in understanding the underlying emission mechanisms.

Simple Synchrotron Self-Compton Models

A number of different emission mechanisms have been proposed to explain the multifrequency spectra listed in Table 1. Because of the likelihood that the radio-through-optical flux is synchrotron radiation, and because of the attractive simplicity of a single component model, it is natural to consider synchrotron models for the entire spectrum, taking into account the possible importance of inverse Compton emission at X-ray frequencies. Support for this picture derive from the two-component structure, steep at low energies and flat at high energies, occasionally seen in the X-ray spectra of BL Lac objects (Urry 1986a,b and references therein). This kind of analysis has been done for most of the sources in Table 1. The simplest synchrotron model is a stationary spherical or slab volume filled with a uniform density of relativistic particles radiating isotropically. If the particle distribution is a power law in energy, (*i.e.*, the particle density is proportional to γ^{-p}), to first order the optically-thin emitted flux will be a power law with index $\alpha = (p-1)/2$, which steepens at short wavelengths due to losses (see dashed curves in Fig. 3). Where the X-ray spectrum is a smooth extrapolation of the ultraviolet spectrum, it can be due to synchrotron radiation; where it lies below the extrapolation, then further steepening, perhaps due to a cutoff in electrons above some energy, must occur independent of the source of the X-rays; and where it lies above the extrapolation, it may result from either inverse Compton scattering of synchrotron photons (the synchrotron self-Compton, or SSC model) or from some other process.

Homogeneous, isotropic SSC models have been calculated by Jones O'Dell, and Stein (1974) and Gould (1979). The magnetic field strength in the source depends on the flux density, F_{sa}, and frequency, ν_{sa}, at which the synchrotron radiation is self-absorbed; the angular size of the emitting region, θ, and the self-absorption frequency; the Doppler factor, δ (see below); and, only very weakly, the optically-thin spectral index, α, and the redshift, z:

$$B \approx k_1(\alpha)\ \theta^4\ \nu_{sa}^5\ F_{sa}^{-2}\ \frac{\delta}{1+z} \ . \tag{1}$$

Using these parameters, plus the so-called break frequency, ν_b, where the greatest steepening of the optically-thin spectrum occurs (Tucker 1967), one can predict the flux density of self-Compton emission,

$$F^{sc} = k_2(\alpha) \ln(\nu_b/\nu_{sa}) \; \theta^{-4\alpha-6} \; \nu_{sa}^{-3\alpha-5} \; F_{sa}^{2\alpha+4} \left[\frac{1+z}{\delta}\right]^{2\alpha+4} , \qquad (2)$$

as well as the energy range spanned by the scattered radiation. The angular size, θ, can be estimated from the distance plus the variability time scale, which introduces additional powers of δ. The efficacy of the SSC model can be evaluated by comparing the prediction to the observed self-Compton flux (or an upper limit to it) in the appropriate energy range. (Equations giving the predicted self-Compton flux were first derived by Marscher *et al.* 1979 and Marscher 1983; related formulae, self-consistently based on the formalisms of Gould 1979 and of Jones, O'Dell, and Stein 1974, were given by Urry 1984, and some of these appear in Feigelson *et al.* 1986.)

A common result of such a calculation is that the predicted self-Compton flux is orders of magnitude larger than the observed flux. This is simply a restatement of the so-called "Compton catastrophe", or brightness-temperature problem: these blazars are so compact and so luminous that according to standard SSC theory, all of the radiation should emerge as Compton-scattered photons rather than synchrotron photons. This contradiction of observations can be resolved if the emission region (or the emission pattern) is moving at relativistic speeds toward the observer. (In the following, we discuss primarily anisotropic relativistic bulk motion, although certainly relativistically outflowing spherical shells can produce the same effect, albeit at a high cost in energy.) The principal effect of such motion is to beam the emitted radiation in the forward direction, so that the observed flux density can be enhanced by orders of magnitude over the rest frame emitted flux density. The enhancement factor is $\delta^{3+\alpha}$, where δ, the kinematic Doppler factor, is defined by $\delta = (1-\beta)^{1/2}/(1-\beta\cos\phi)$. This has its maximum when the angle to the line-of-sight, ϕ, is small, and when the velocity of the motion, βc, is large. The time scale on which variability is observed is also shortened by a factor δ, increasing the likelihood that we could observe [apparently] rapid events.

The ratio of predicted self-Compton flux density (assuming no bulk relativistic motion) to the actual self-Compton flux density is proportional to δ^p, where $p \sim 5$ if the angular size of the synchrotron source is measured directly and $p \gtrsim 10$ if the size is estimated from the variability time scale according to $\theta \sim \delta ct/D$. The observed X-rays are an upper limit to the self-Compton component, so give a lower limit to δ. (Higher Doppler factors correspond to lower intrinsic self-Compton fluxes.)

Using this approach, Urry *et al.* (1981) found using hard X-ray data that $\delta >> 1$, implying on the basis of the spectra alone that relativistic motion is characteristic of the class. However, the assumption of homogeneity, *i.e.* that the flat radio spectra were optically thin and that the self-absorption frequency was very low, led to these large values of δ. Using the larger sample of BL Lac observed with the lower-energy *Einstein Observatory* Imaging Proportional Counter (IPC), Madejski and Schwartz (1983) repeated the calculation and found, in contrast, that for approximately half the objects, $\delta \lesssim 1$. Their calculation was biased toward lower values of δ because the IPC flux was probably dominated by synchrotron emission rather than self-Compton emission (Madejski 1985), and their assumptions about other input parameters were not unique. Clearly this disparity of conclusions means that values of δ derived in this manner are quite sensitive to unconstrained input parameters and cannot be used to demonstrate or rule out the presence of bulk relativistic motion (Urry 1984).

Inhomogeneous Models

The simple SSC models are not realistic in any case. As explained earlier, flat-spectrum synchrotron sources are likely to be inhomogeneous, and perhaps even clumpy. The inhomogeneity of the source can be handled in several ways. In a simple discrete approach, the spectrum is decomposed into a finite number of optically-thin components, each of which represents isotropic synchrotron emission from a simple homogeneous radiating volume. One such spectral decomposition is illustrated in Figure 3. Some of the input parameters, like the optically-thin spectral slope and the self-absorption frequency of each component, must be assumed if the individual components are not actually observed. Of course, using multifrequency VLBI individual components can be resolved, and their spectra measured, at least at radio wavelengths.

Using Equation (1) it is straightforward to show that the components with the highest-frequency turnovers are likely to be the most compact. It is also clear that for a constant magnetic field, these compact, high-frequency components generate the highest self-Compton flux (see Eqn. 2). If the magnetic field varies among components, however, the origin of the dominant self-Compton flux is less clear. In an "onion-skin" geometry the components are concentric. Suppose the strength of the magnetic field falls as a power of a radial coordinate, $B \propto r^{-m}$. Since $\theta \propto r$, one can use Equations (1) and (2) to derive the ratio between self-Compton flux densities from two arbitrary components:

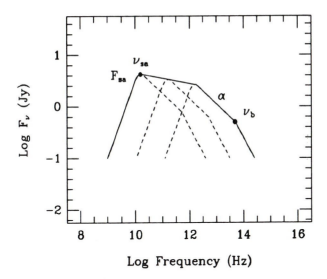

Figure 3 A simple model for spectral decomposition into three homogeneous synchrotron sources. The flatness of the radio spectrum is due to a superposition of components with different turnover frequencies. Above the highest turnover frequency the spectrum is optically thin; below this it is optically thick. The higher frequency components are almost certainly more compact, but the dominant contribution to the self-Compton emission depends on a combination of field strength, electron density, and photon density in each component.

$$\frac{F_1^{sc}}{F_2^{sc}} \approx \left[\frac{r_1}{r_2}\right]^{\frac{m(3\alpha+5)-8\alpha-10}{5}} . \tag{3}$$

(The weak dependence on F_{sa} cancels out because the superposition of components is approximately flat.) We can evaluate this ratio for likely values of the optically-thin spectral index, *i.e.* $\alpha \sim 0.5$. (The answer is not extremely sensitive to this value.) For $m \lesssim 2.2$, the exponent is negative and the ratio decreases with increasing r_1/r_2. In other words, when B is constant or falls off slowly, the compact, high-frequency component dominates, while for a steeper dependence of the magnetic field, $m \gtrsim 2.2$, the larger, low-frequency component produces more self-Compton flux.

Alternatively, one can model a continuous inhomogeneity with spatial dependences of both the magnetic field and the particle density, and this has been done most often for a jet geometry (Blandford and Königl 1979; Marscher 1980; Königl 1981; Reynolds 1982; Ghiselini, Maraschi, and Treves 1985; Hutter and Mufson 1986). Of necessity, these approaches have a generous number of free parameters, although plausible guesses for some can be made on physical grounds. Königl (1981) calculated a conical jet model in which the electron density and magnetic field are described by arbitrary power laws decreasing along the jet: $N(r,\gamma_e)=N_o\gamma_e^{-p}r^{-n}$ and $B(r)=B_o r^{-m}$. Electrons are injected with a fixed energy distribution, and are continuously accelerated along the jet to offset adiabatic losses from the expanding volume, although radiation losses are taken into account. Since the highest energy electrons are maximally depleted in the inner regions of the jet, the maximum electron energy, γ_{max}, increases along the jet. The earlier Blandford and König (1979) model was a special case with $m=1$, $n=2$, and $p=2$. Hutter and Mufson (1986) generated a model similar to Königl's but differing in details of the calculation, particularly with respect to Compton-scattering. Reynolds (1982) generalized the inverse Compton calculations of Jones, O'Dell, and Stein (1974) to ordered geometries, taking into account the non-random angles between interacting photons and electrons in relativistically outflowing winds and jets.

Marscher (1980) considered a geometry suggested by physical considerations of the actual production of a jet (Blandford and Rees 1974), consisting of an inner nozzle from which the jet emanates, a parabolic segment in which the flow is accelerated outward such that the bulk Lorentz factor increases linearly with r, and an outer, free-expanding, constant velocity conical segment, similar to the Königl construct. He considered the case $m=2$, $n=2$ for the field and particle radial power-law dependences. Ghisellini, Maraschi, and Treves (1985) also considered a series of models combining a parabolic inner jet and a conical outer jet, but used general power-law dependences of electron density and magnetic field. In contrast to the Königl picture, in the Ghisellini *et al.* models the speed of the flow increases along the jet and the maximum electron energy decreases along the jet. In effect this model describes a wide inner jet, sometimes called a fan beam, which at a fixed radial coordinate r (and so for a given B and N), corresponds to a smaller emitting volume than the cone to which it is smoothly connected. Thus relative to a cone, a paraboloid contributes less synchrotron emission from the inner component than from the outer for the same m and n.

The broad-band spectra of many blazars have been evaluated according to one or more of the above models. The requirement for relativistic motion is one of the most interesting concerns. A relativistic jet pointing nearly along the line-of-sight is an attractive explanation for blazars for reasons unrelated to the Compton catastrophe (Blandford and Rees 1978): it could explain the rapid variability, the high degree of polarization, the rapid variations in polarization fraction and direction, the high luminosities, the apparent super-Eddington flares, and the small equivalent widths of lines. (The similarity of the continua of OVVs and BL Lacs is a bit puzzling in this picture, since the line emission properties appear quite different. Perhaps small blue bumps in OVVs have an important effect on the lines. However, this problem has not been addressed in detail because the observable quantity is the equivalent width, *i.e.* the ratio of line photons to local continuum photons, while the physically important ratio is the number of line photons to ionizing continuum photons.) Analyzing spectra in the context of the Königl model, Urry and Mushotzky (1982) and Worrall *et al.* (1982, 1984a,b,c, 1986) obtained good fits for Doppler factors $\delta \gtrsim 5$, although these fits were not well constrained. However, Mufson *et al.* (1984), Ghisellini, Maraschi, and Treves (1985), and Bregman *et al.* (1984) found that other jet models allow values of δ closer to, or equal to, 1. Since the spectra are all very similar, the extent to which the fit parameters differ indicates the latitude of jet models in describing the observations.

The actual reasons for the disparity of results can be found in a careful comparison of the different jet models. (The implications of the jet geometry were illuminated by the work of Ghisellini, Maraschi, and Treves 1985.) Imagine the jet as three emission regions along the jet — inner, middle, and outer — and then look at the contribution of each of these regions to the integrated jet spectrum. For a reasonable choice of parameters in the Königl model, synchrotron emission at the lowest radio frequency and in the infrared-optical is dominated by the large outer emitting region. Shorter radio wavelengths and longer infrared wavelengths have emission from progressively smaller jet radii, and the middle of the optically-thick part of the synchrotron spectrum comes from the innermost part of the jet. The conical and parabolic geometries are compared in Figure 2 of Ghisellini, Maraschi, and Treves (1985). As explained above, the relative importance of the inner and outer regions in a fixed geometry depends on m and n, the power-law dependences of magnetic field and electron density. Decreasing either parameter means a slower decrease of B or N with radius and increases the contribution of outer regions to the synchrotron spectrum.

It is interesting to look at the variation of self-absorption turnover frequency with radial coordinate. The optically-thin synchrotron flux is proportional to $NB^{1+\alpha}$ (Blumenthal and Gould 1970). Combining this with Equation (1) gives

$$\nu_{sa} \propto r^{\frac{-m(3+2\alpha)-2n-4}{5}} . \tag{4}$$

Clearly, increasing either m or n makes more lower frequency emission at a given distance along the jet. Conversely, a rapid decline in magnetic field or electron density reduces the low frequency emission. Possibly this could explain the difference between radio-loud and radio-quiet quasars (see § 3 below).

The Promise of Multiwavelength Blazar Spectra

If all these jet models can be adjusted to fit the data, and yet they reflect very different physical conditions, such as different values of the Doppler factor, how can we hope to learn about the physics of jets from multiwavelength spectra? The answer is that these models make very different predictions about variability and about spectral evolution during intensity changes. For example, for the Königl model, the shortest time scale variations should occur in the millimeter regime, with less rapid variability in the radio and optical, whereas in the parabolic model of Ghisellini *et al.*, variability time scales increase with decreasing wavelength. Similarly, observations of coordinated variations in the X-ray and other wavelengths can reveal which part of the synchrotron spectrum corresponds to the observed self-Compton emission, and thus can distinguish between the conical and parabolic models of Ghisellini, Maraschi, and Treves (1985).

Even for very simple models like smooth inhomogeneous jets, the predictions of spectral variations are complicated, but repeated multiwavelength observations with intervals shorter than the variability time scale should provide clear support for one scenario or another. The differences among blazars may be as simple as the gross geometry of the jet. This brings us full circle to the variability discussion at the beginning of this section, and it would be nice to be able to make distinctions based on the available data. However, at present any physically reasonable model is not well-constrained. Indeed, estimates of physical quantities from analyses listed in Table 1 cover a wide range of parameter space: magnetic fields from 10^{-6} to 10^2 Gauss, electron densities in the range 10^3–10^6 cm^{-3}, relativistic Doppler factors that range from 1 to more than 100, and solutions that are near equipartition of field and particle energies, or very, very far from equipartition. If these ranges of parameters actually exist, that is, if the models are correct, then conditions in the jets are very different, and the means by which they propagate and/or are confined by an external medium or an external magnetic field must vary widely. If these jets were resolved, they would presumably look very different. Alternatively, the presently available broad-band data are not of sufficiently high quality to constrain jet models, and the estimates of physical parameters are very poor. It would be useful to relate the properties of unresolved jets, determined from spectral analyses, to the properties of resolved, more-or-less transverse jets, determined from radio and optical mapping.

3. Non-Blazar Emission-Line AGN

Observed Broad-Band Continuum

In contrast to the blazars, the continuum emission from non-blazars appears much more complex. At the same time, broad-band observations are somewhat simpler. Since normal quasars and Seyfert galaxies are less variable, on longer time scales, than blazars, the requirement of simultaneity is less restrictive. (This refers only to the continuum; for discussion of emission-line studies, see below.) Some large, quasi-simultaneous observing campaigns have been arranged for a handful of well-studied AGN like the Seyfert galaxy NGC4151 (Beall *et al.* 1981; Perola *et al.* 1982, 1986; Bassani *et al.* 1986) and 3C273 (Clegg *et al.* 1983; Robson *et al.* 1983, 1986; Camenzind and Courvoisier 1984; Courvoisier *et al.* 1987; Cutri *et al.* 1985; Landau *et al.* 1986; Smith *et al.* 1987).

In a global sense, however, the real advances in our understanding of the continuum of these AGN have come from studies of large samples, for which most of the available data are non-simultaneous (Neugebauer *et al.* 1979, Malkan and Sargent 1982, Malkan 1983, Elvis *et al.* 1986, Edelson and Malkan 1986, Carleton *et al.* 1987). Particularly important was the addition of previously missing data in the infrared with IRAS, in the ultraviolet with IUE, and in the X-ray with various satellites (HEAO1, the *Einstein Observatory*, EXOSAT, *Tenma, Ginga*) — all space-based experiments. Information in these spectral regimes has provided valuable diagnostics for understanding the emission processes in these objects.

From ~ 10 μm to ~ 1 keV, and sometimes over a much broader range, these AGN emit *roughly* equal energy per decade (Edelson and Malkan 1986, Elvis *et al.* 1986). Two examples were shown in Figure 1. However, the broad-band spectra of non-blazars are anything but smooth and describable by a single function, as was the case for blazars (*e.g.*, Perry *et al.* 1987). Instead, they are characterized by bumps and wiggles that are signatures of different thermal and nonthermal emission processes. Three high-quality infrared-through-ultraviolet spectra number of AGN are shown in Figure 4, in order of decreasing luminosity from 3C273 to NGC4151, along with the best multi-component fit to the data (Edelson and Malkan 1986). Approximately six different components have regularly been identified in radio-through-X-ray spectra of emission-line AGN. They are outlined briefly below, and then the two high-energy components are discussed in more detail.

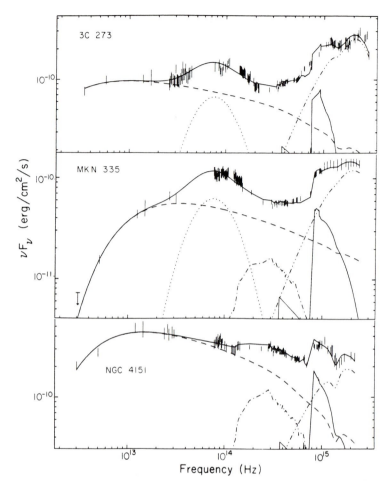

Figure 4 Infrared-through-optical spectra of three AGN that span roughly three orders of magnitude in luminosity, from ~5×10^{43} ergs/s for NGC4151 to ~4×10^{46} ergs/s for 3C273 (Edelson and Malkan 1986). The best multi-component fit is shown as a heavy solid line. The components included in the fit are: (dashed line) an infrared power law; (dotted line) a near-infrared bump; (dot-dash line) starlight; (thin solid line) Balmer and Paschen recombination continuum; (dot-dot-dot-dash line) an ultraviolet blackbody. All components have been corrected for Galactic and internal reddening as described by Edelson and Malkan (1986).

A. Radio-to-infrared power law.

A power law from radio to infrared wavelengths is seen in radio-loud AGN, which constitute ~5-10% of all AGN. As was true for blazars, this power law is flatter in the radio and steeper in the infrared. In radio-quiet quasars, a power law can be important in the infrared, but by definition is absent in the radio, with cutoffs typically occurring around $\lambda\sim$ 30-300 μm (*e.g.* Edelson and Malkan 1986). For quasars as for blazars, this power law may be due to synchrotron radiation from a core, jets, and/or lobes. It is less polarized and less variable than in blazars, and often has a steeper spectrum, and so may reflect a smaller contribution from a compact jet and more from an extended component. It is interesting that there is little difference between radio-loud and radio-quiet spectra at wavelengths shortward of the millimeter region, with the exception of the spectral shape near ~ 1 keV (Wilkes and Elvis 1987). In § 2 we suggested that radio-loud versus radio-quiet quasars could be explained by synchrotron emission in the presence of different radial dependences of the magnetic field and/or electron density. That was in the context of an idealized jet model. Perhaps more likely is that the physical conditions in radio-quiet objects preclude the formation or collimation of a jet yet do not affect the rest of the emission.

B. Dust Emission in the Infrared

Some AGN show strong infrared emission from dust. This component is particularly strong in Seyfert 2 galaxies and LINERs, although a few type 1 Seyferts are dusty as well (Stein and Weedman 1976; Rieke 1978; Rieke and Lebofsky 1981; Malkan and Oke 1983; Neugebauer *et al.* 1985; Willner *et al.* 1985; Ward *et al.* 1987; Carleton *et al.* 1987; Edelson, Malkan, and Rieke 1987). The observed infrared dust emission is more sharply peaked in wavelength than a simple power law with an exponential high frequency cutoff, but is broader than isothermal dust would imply. Probably there is a range of dust temperatures, of order ~ 50-100 K (*e.g.* Neugebauer *et al.* 1985). This means the far-infrared continuum (for $\lambda \gtrsim 10\,\mu m$) is steep. It is also correlated with reddening in the optical and ultraviolet (de Grijp *et al.* 1985, Edelson and Malkan 1986), indicating that the optical and ultraviolet radiation is at least partially responsible for heating the dust, and that the dust is well-mixed with the ionized gas.

C. Near-Infrared Excess

Depending on the spectral decomposition at other wavelengths, and on the choice of units when plotting spectra, there may be excess emission in a large fraction of AGN at $5\,\mu m$ (more apparent in νF_ν plots; Edelson and Malkan 1986, Carleton *et al.* 1987, Perry *et al.* 1987) or ~ $3\,\mu m$ (more apparent in F_ν plots; Stein and Weedman 1976; Cutri *et al.* 1981, 1985; Malkan and Filippenko 1983; Robson *et al.* 1986), or a minimum (an inflection-point) at $1\,\mu m$ (Neugebauer *et al.* 1979, Elvis *et al.* 1986). This near-infrared "component" is not very well defined, since observers have yet to characterize it uniformly. Carleton *et al.* (1987) suggest that this excess is due to hot dust, which presumably could be present in some AGN and not in others. In any case, when present, the near-infrared excess should correlate with other reddening indicators.

Time variability can provide a clue as to whether a spectral region represents one component or a blend of components. With UBVRIJHK photometry Cutri *et al.* (1985) found that for most quasars, the amplitude of variability increased to the blue, and was negligible at 2.2 μm. In these objects, the infrared power law was more or less constant while the blue bump varied. 3C273 was atypical, varying equally at K through U, possibly because of a relatively strong contribution in the infrared from a variable, blazar-like power-law component. Courvoisier *et al.* (1987) concluded that the spectrum of 3C273 has a break at 5 μm rather than a bump since the observed variability characteristics for λ< 5 μm and λ> 5 μm were very different, but again, it may be an exception. The time variability properties of dust emission are likely to be complicated, since they can result from heating or cooling of the dust, or destruction of grains. Presumably the strength of dust emission should vary in phase with the appropriate reddening indicators, thus confirming and constraining dust models, but such observations have not been done. Clearly it would be useful to do variability studies longward of the K-band, but it is important to point out that the appearance of a near-infrared excess (or deficit) depends crucially on the correct identification of overlapping spectral components at shorter and at longer wavelengths.

D. Starlight from the Host Galaxy

Obviously the relative contribution of light from a host galaxy is greatest in low luminosity AGN (excepting starbursts), since the range of galaxy luminosities is much smaller than the range of active-nucleus luminosities. Modeling a galaxy and estimating its contribution to the total flux is straightforward but the methods used in the literature are not yet standardized (*e.g.* see comments by Ward *et al.* 1987). Some use galaxy spectra only; others use imaging data as well. Multicolor imaging to measure the surface brightness profile, in combination with spectroscopic observations of stellar absorption features (*e.g.* Malkan and Filippenko 1983), can minimize uncertainties in the galaxy contribution.

E. "Blue Bumps"

Unlike blazars, which have convex (downwardly curving) spectra in the optical-ultraviolet, non-blazars usually have concave-up spectra (Richstone and Schmidt 1980). Indeed, whatever spectral steepening has been seen in the ultraviolet emission of high-redshift quasars can be attributed to absorption by intervening intergalactic matter (Weyman, Carswell, and Smith 1981; Bechtold *et al.* 1984; Steidel and Sargent 1987). In the one far-ultraviolet observation of a low-redshift quasar, 3C273, which should not be affected by intervening absorbers (*cf.* Maraschi *et al.* 1988), Reichert *et al.* (1988) did see either a change in the power-law slope or Lyman continuum absorption at the location of the quasar, but the implications for other quasars are unclear since 3C273 is hardly typical.

The flat optical-through-ultraviolet spectra are due to two effects. One is the contribution of Balmer continuum emission from ~ 3600-4000 Å (Kwan and Krolik 1981, Grandi 1982, Malkan and Sargent 1982) and blended FeII emission from ~ 2000-4000 Å (Wills, Netzer, and Wills 1985), which is collectively known as the "little blue bump" and which is reasonably well-understood. The other is an additional, higher energy, usually more luminous, continuum component known as the "big blue bump". As can be seen from Figure 4, it probably dominates the radiation from the central engine, at least in high-luminosity AGN, although if dust is present the ultraviolet photons may be absorbed and re-radiated in the infrared. The big blue bump is described in more detail below.

F. X-Ray Power Law

Non-blazar AGN observed above 2 keV, mostly Seyfert 1 galaxies, have remarkably uniform, hard spectra, which are well-fit by a power-law model with mean spectral index $\alpha \sim 0.65 \pm 0.15$ (the latter is the dispersion about the mean; Rothschild *et al.* 1983, Mushotzky 1984). Few quasars are bright enough at these energies to have been observed with existing detectors but their spectra are generally consistent with this canonical value (Pounds 1985, Worrall *et al.* 1980, Halpern 1982), although 3C273 has a slightly flatter spectral index $\alpha \sim 0.4$ (Worrall *et al.* 1979). At lower X-ray energies, the picture becomes complicated, and since X-ray emission has important consequences for the energetics and emission-line properties of AGN, it is discussed in more detail below.

The Big Blue Bump

The big blue bump may represent most of the luminosity from quasars. (For a detailed review of the properties of the blue bump, see O'Dell 1986.) Because emission from AGN spans quite a few decades in wavelength, many models proposed to explain their spectra are nonthermal, such as synchrotron and SSC models (see § 2). However, the big blue bump may be different. Shields (1978) suggested that the flat optical-ultraviolet spectral component in quasars could be thermal radiation from an accretion disk, which if optically thick and geometrically thin should have a flat spectrum with $\alpha \sim 1/3$ from the superposition of blackbody emission at different disk radii. The emergent spectrum depends primarily on the black-hole or compact-object mass, M, and the accretion rate, $\dot{M}$, and to a lesser degree on the disk viscosity and the electron opacity (Lynden-Bell 1969, Pringle and Rees 1972, Shakura and Sunyaev 1973, Shields 1978, Pringle 1981, Wandel and Petrosian 1988). Shields applied his accretion disk hypothesis to the optical-through-ultraviolet spectrum of 3C273. Malkan and Sargent (1982) and Malkan (1983) expanded on this idea, fitting spectra of 15 quasars and Seyfert galaxies with blackbody and/or accretion disk spectra. (Malkan's accretion disk model included general relativistic effects discussed by Novikov and Thorne 1973 and Page and Thorne 1974.) Single-temperature blackbody fits generally have temperatures in the narrow range $2\text{-}3\times 10^4\,\mathrm{K}$, while the maximum disk temperatures for composite disk spectra can be much hotter, $T \gtrsim 10^5\,\mathrm{K}$. The relative contribution of the thermal emission is greater in high-luminosity ($L \gtrsim 10^{30}\,\mathrm{ergs/s/Hz}$) than in low-luminosity ($L \lesssim 10^{30}\,\mathrm{ergs/s/Hz}$) AGN, although no trend with luminosity is apparent on either side of this break-point (Kriss 1988). This means, crudely, that the big blue bump is more important in quasars than in Seyfert galaxies, but that its strength does not increase systematically relative to the nonthermal component (see Edelson and Malkan 1986). This is apparent in Figure 4.

There are problems with simple accretion disk models, one of which is that the implied accretion rates are sometimes near the Eddington limit or super-Eddington,[2] which is internally inconsistent with the thin disk approximation. The temperature of a disk at its innermost radius scales as $T \propto \dot{M}^{1/4}M^{-1/2}$ (*e.g.* Shakura and Sunyaev 1973), and the wavelength of the Wien peak depends linearly on temperature. The emitted luminosity is proportional to $\dot{M}$ and the Eddington limit is proportional to M, so for the simplest disks, a harder spectrum means a higher temperature, which means a higher ratio $\dot{M}/M$, which means a higher L/L_{Edd}. The sharp Wien turnover is not seen in ultraviolet spectra, even for high-redshift quasars (Bechtold *et al.* 1984, O'Brien, Gondhalekar, and Wilson 1987). Perhaps the entire ultraviolet-through-soft-X-ray regime is a single accretion-disk component (see Elvis 1988 for a thorough review), but for simple disks such hard spectra grossly violate the Eddington limit. However, as with all first approaches, these simple models are not very realistic. Several effects that have been ignored are likely to be important, including electron opacity, gas and dust opacity, radiative transfer effects,

---

[2] The Eddington limit applies to spherical accretion only, and so might seem irrelevant to a discussion of accretion disks, which are inherently nonspherical. However, the Eddington limit is a useful fiducial mark, and observations suggest that the luminosities of most AGN are sub-Eddington (*e.g.*, Wandel and Yahil 1985, Wandel and Mushotzky 1986).

thermal instabilities, disk thickening in high luminosity systems, and general relativistic effects in the vicinity of a Kerr (*i.e.* rotating) black hole (Kolykhalov and Sunyaev, 1984, 1988; Czerny and Elvis 1987; Sun and Malkan 1987; Wandel and Petrosian 1988).

Czerny and Elvis (1987) pointed out that electron opacity is significant in any realistic accretion disk. They investigated the effects of electron scattering and found that it hardened the disk spectrum without increasing the maximum disk temperature, thus alleviating the Eddington-limit problem. It also made the predicted turnover in the ultraviolet less precipitous, commensurate with strong extreme-ultraviolet emission (and perhaps soft X-ray excesses; Pravdo *et al.* 1981, Bechtold *et al.* 1987). The turnover energy was also relatively insensitive to the maximum disk temperature, explaining why all blackbody fits could give similar temperatures.

Czerny and Elvis also considered inclination effects, since disks radiating near the Eddington limit are probably puffed up at their innermost edge. The direction of the effect is that the disk spectrum softens as the inclination angle increases, (as the disk goes from face-on to edge-on), because the hot inner disk is occulted. Finally, Czerny and Elvis suggested that by analogy to galactic sources, AGN accretion disks may have hot coronae that redistribute photons to higher energies via inverse-Compton scattering, hardening the disk spectrum further.

The central compact object in the Czerny and Elvis model was a non-rotating black hole. Rotating black holes can have nearly an order of magnitude higher efficiency (Rees 1984), and moreover are likely since the accreting matter has enough angular momentum to spin-up a non-rotating black hole in a relatively short time. Both general relativistic effects near a Kerr black hole and opacity were considered by Kolykhalov and Sunyaev (1984, 1988) and Sun and Malkan (1987), who summed stellar atmospheres as well as blackbody models to produce disk spectra. Strong general relativistic effects cause a significant inclination dependence in a sense opposite to that found by Czerny and Elvis (1987). The radiation near a Kerr black hole is both Doppler boosted and gravitationally focussed along the disk plane, so that edge-on disks actually contribute *more* emission from the hot inner region, and the spectrum *hardens* with increasing inclination angle. These competing inclination effects, the Doppler boosting and gravitational bending of light, and the occultation of the inner region by a puffed-up disk, will both affect the relative frequency with which soft X-ray excesses are detected in AGN (see below) if the soft X-rays come from a hot accretion disk.

Wandel and Petrosian (1988) applied accretion disk models similar to those of Czerny and Elvis (1987) to the optical-ultraviolet spectra of ~ 70 quasars and Seyfert galaxies. The best fit models had higher black hole masses for quasars than for Seyfert galaxies, 10^8-$10^{9.5}$ $M_\odot$ compared to $10^{7.5}$-$10^{8.5}$ $M_\odot$, as if quasars had been accreting at a higher rate over their lifetimes. The tendency for higher luminosity AGN to have flatter optical-through-ultraviolet spectra (*e.g.* Cheng, Kinney, and Fang 1987, but see O'Brien, Gondhalekar, and Wilson 1987 who argue that the effect correlates more strongly with redshift than with luminosity) may be explained by these models since increasing $\dot{M}$ for a given M hardens the spectrum (Fig. 1 of Wandel and Petrosian 1988, *cf.* Kriss 1988).

Despite the appeal of the accretion disk picture, the big blue bump is not necessarily thermal. O'Dell, Scott, and Stein (1987) argue that the blue bump can be explained equally well by synchrotron radiation from a particular broken-power-law distribution of secondary electrons that arises from a sufficiently high density of relativistic protons. (They did not try fitting their model to any observations, however.)

The particular scenario for electron production has been described by a number of authors (Eichler 1979, Protheroe and Kazanas 1983, Kazanas and Ellison 1986, Zdziarski 1986). In the vicinity of the black hole, protons are accelerated to relativistic energies by a first-order Fermi process. The resulting proton energy distribution is assumed to resemble the observed cosmic ray spectrum in our Galaxy, a power law $N_p(\gamma_p) \propto Q_p(\gamma_p) \propto \gamma_p^{-r}$), with index $r \sim 2.4$. (Since proton energy losses are proportional to energy, the steady-state distribution, N_p, has the same form as the injected spectrum, Q_p.) The kinetic energy of the protons is much greater than their gravitational potential energy, and the scattering mean free path is much larger than the black hole, so the protons escape and undergo inelastic collisions, producing pions ($p+p \to p+p+\pi^\pm+\pi^0$), which decay into γ-rays, muons, electrons, positrons, and neutrinos ($\pi^0 \to \gamma\gamma$, $\pi^+ \to \mu^+ + \nu_\mu$ or $\pi^+ \to e^+ + \nu_e + \bar{\nu}_\mu$, and $\pi^- \to \mu^- + \bar{\nu}_\mu$), and the muons decay into electrons or positrons ($\mu^- \to e^- + \bar{\nu}_e + \nu_\mu$). Approximately 50% of the emergent energy is in the form of neutrinos.

The broken power-law arises as follows. Electrons are produced with the same power-law index as the protons, $r \sim 2.4$. In steady-state the production rate is balanced by synchrotron and Compton losses, which are proportional to the square of the electron energy:

$$N_e(\gamma_e) \propto \frac{\int_{\gamma_e}^{\infty} Q_e(\gamma_e)\, d\gamma_e}{\gamma_e^2}. \qquad (5)$$

(This is just the integral of the continuity equation.) Below the pion rest mass energy, which corresponds to an electron Lorentz factor $\gamma_{br} \sim 100$, electrons are produced by cooling only. (Zdziarski 1986 argues that the actual break may be slightly higher, $\gamma_{br} \sim 300$, because of details of the proton spectrum and the energy dependence of the cross-section for the muon decay.) Thus for $\gamma_e < \gamma_{br}$, the lower limit on the integral in Equation (5) is γ_{br}, and the integrand is a constant. Above γ_{br}, the integrand is proportional to γ_e^{-r+1}. Thus the steady-state electron distribution is a broken power law:

$$
N_e(\gamma_e) \propto
\begin{cases}
\gamma_e^{-2}, & \gamma_e < \gamma_{br} \\[2mm]
\gamma_e^{-r-1}, & \gamma_e > \gamma_{br}
\end{cases}
\tag{6}
$$

O'Dell, Scott, and Stein (1987) use $r=2$ rather than $r=2.4$ (as for cosmic rays), so their electron distribution has indices 2 and 3.

The relevance of this process for AGN spectra is that, to first order, the synchrotron spectral shape reflects the electron spectral shape via $\alpha = (p-1)/2$, where p is the index of the electron distribution. The O'Dell *et al.* synchrotron model therefore has a broken power-law form with $\alpha = 0.5$ and $\alpha = 1$. The mapping of the electron break energy to the synchrotron break energy depends on the ambient magnetic field. For $B \sim 4000\,\text{Gauss}$, the peak frequency of the big blue bump can be identified with the pion-rest-mass break in the electron spectrum. Synchrotron losses in such a strong magnetic field are severe, and electrons must be continuously reaccelerated or re-injected in order to maintain the spectral distribution. The optical-through-ultraviolet radiation would not be polarized because of Faraday depolarization due to strong fields and high particle densities; similarly, the self-absorption frequency is high and no radio emission would be produced (O'Dell, Scott, and Stein 1987). Other consequences of this nonthermal model are that the spectrum is featureless, unlike the spectra of accretion disks from which absorption features may be expected (*e.g.* Sun and Malkan 1987); also, rapid variability is possible, whereas in accretion disks some of the optical and ultraviolet radiation comes from large radii and so should be relatively slow to react to changes in the accretion rate. Detailed fits of this nonthermal model to observed blue bumps will demonstrate its viability and if successful, will allow estimates of the magnetic field strengths and electron distributions in different AGN.

Broad-Band X-ray Emission

As described earlier, the hard X-ray spectra of AGN, at least those measured, are fairly flat. In fact, the energy output per band increases with decreasing wavelength. However, at lower X-ray energies, 0.2-4 keV IPC spectra of Seyfert galaxies (Urry *et al.* 1986, Kruper *et al.* 1988) and radio-quiet quasars (Wilkes and Elvis 1987) are steeper, with $\alpha \sim 0.9$. The radio-loud quasars have a distinctly flatter index than the radio-quiet quasars, $\alpha \sim 0.5$ (Wilkes and Elvis 1987).

Not only is the lower-energy X-ray spectrum softer, there is evidence in the IPC spectral sample for a still softer excess over the $\alpha \sim 0.9$ power law (Wilkes and Elvis 1987, Kruper *et al.* 1988). Also, HEAO1-A2 and EXOSAT observations of AGN have revealed quite a few soft excesses at energies below 1 keV, in some cases below 0.1 keV: E1615+061 (Pravdo *et al.* 1981); E1821+643 (Pravdo and Marshall 1984); Mrk 841 (Arnaud *et al.* 1985); Mrk 509 (Singh, Garmire, and Nousek 1985); NGC4051 (Lawrence *et al.* 1985); MCG6-30-15 (Pounds, Turner, and Warwick 1986b); NGC4151 (Pounds *et al.* 1986c, Perola *et al.* 1986); NGC7469 (Barr 1986); Fairall 9 (Morini *et al.* 1986b); PG1211+143 (Bechtold *et al.* 1987); NGC 5548 (Branduardi-Raymont 1986); and Mrk335 (Pounds *et al.* 1986a, Turner and Pounds 1987). Some of these soft components are variable.

Soft excesses may be common in the X-ray spectra of AGN. A soft X-ray survey in the direction of the Coma cluster, along which the column density of X-ray-absorbing gas in the interstellar medium of our galaxy is especially low, discovered an anomalously high number of AGN, implying that when Galactic absorption does not prevent us from observing very soft X-rays, AGN are copious producers of them (Branduardi-Raymont 1986).

Reconciling the "canonical" and the "soft excess" pictures can be done in two ways. Wilkes and Elvis (1987) suggested that the X-ray spectrum of an AGN consists of three components: (i) a continuation of the infrared power law with $\alpha \sim 1$ that dominates the spectrum of radio-quiet AGN in the IPC band; (ii) a harder power law with $\alpha \sim 0.5$ that appears at $E \geq 2$ keV for radio-quiet and at $E \geq 0.5$ keV for radio-loud AGN; and (iii) a soft excess below ~ 0.3 keV, with spectral index $\alpha \sim 2\text{-}3$, although the form of this component need not be a power law. A combination of (i) and (ii) could produce the $\alpha \sim 0.65$ power law seen in the 2-20 keV data.

One of the attractions of the Wilkes and Elvis three-component explanation is that in some samples the infrared and X-ray fluxes appear to be linearly correlated (Malkan 1984, Fabbiano *et al.* 1986, Carleton *et al.* 1987), implying that a single power law with $\alpha=1$ can explain both the infrared and the X-ray emission. (This proposed underlying power law is energetically unimportant in the ultraviolet compared to the big blue bump.) However, recent studies of radio-quiet quasars have shown that in fact the infrared-X-ray correlation is in fact not linear (Worrall 1987, Kriss 1988). Instead, it is the infrared and optical luminosities that are linearly correlated, while X-ray luminosity grows more slowly than linearly with either. That is, $L_x \propto L_{ir}^\varepsilon$ (Worrall 1987, Kriss 1988) or $L_x \propto L_{opt}^\varepsilon$ (Zamorani *et al.* 1981, Reichert *et al.* 1982, Kriss and Canizares 1985, Avni and Tananbaum 1986, Tananbaum *et al.* 1986), where $\varepsilon \sim 0.7$. (The apparent linear correlation of L_x and L_{ir} occurred because upper limits were omitted, and because the samples included radio-loud quasars, which because of a separate and strong L_x-L_r correlation [see discussion of FeII below] were biased toward $\varepsilon \sim 1$.) Thus, since α_{ir-x} is not 1, and is in fact a function of luminosity, there is no reason to expect an extrapolation from the infrared to the X-ray with $\alpha \sim 1$ to predict the X-ray flux, much less to represent the underlying X-ray power law. Also, a spectral break from $\alpha = 1.0$ to $\alpha = 0.5$ has not been seen in high quality 2-20 keV spectra of AGN (R. Mushotzky, private communication). The smoothness of the observed X-ray power law in the energy range 2 to ≥ 20 keV is remarkable (Rothschild *et al.* 1983). Moreover, one might expect a large dispersion of spectral indices for 2-20 keV fits if the observed power law were a mix of two components of varying relative normalizations, while the observed distribution is quite narrow (Mushotzky 1984).

An alternative explanation would be two components, (1) a sometimes variable, steep, low-energy part with $\alpha \sim 1$-3 below ~ 1 keV and (2) the canonical high energy part with $\alpha \sim 0.65$ above ~ 1 keV. In at least some cases the soft- and hard-X-ray components do vary separately, which could explain the wide range in spectral indices seen in the IPC quasar survey. The relatively low-resolution IPC experiment would generally see a blend of these two components, except in the case of radio-loud objects, where the soft component is for some reason systematically weaker relative to the hard component. (There may be a few exceptional AGN that don't have 2-20 keV spectra with $\alpha \sim 0.65$, but so far most do.)

The form of the X-ray continuum has important implications for the energy budget and also for the ionization structure of the broad line region (see discussion of emission lines below). If the soft excesses are high frequency manifestations of the big blue bump, then the unobserved extreme-ultraviolet (EUV) emission probably dominates the bolometric luminosity in most cases. However, the X-ray power laws are fairly flat, and represent rising energy output per decade. Thus, the relative contributions to the bolometric luminosity of the EUV and hard X-ray components depends on the high energy cutoff of the

X-ray power law.

Collimated proportional counters such as HEAO1 and EXOSAT have sufficient independent resolution elements per bandwidth to identify multiple spectral components in the energy range 2-20 keV (and for EXOSAT with added information at ~0.1 keV), but only for the ~40 brightest AGN. The IPC was at the focus of an imaging X-ray telescope, and so could look at many more, fainter AGN, but its spectral resolution was too poor to detect multiple components. To decide between the two scenarios outlined here, we need an X-ray telescope sensitive enough to look at ≥100 AGN, over the bandwidth where these multiple spectral components can be found (~0.1-10 keV or more), with sufficient energy resolution to see the breaks between them. Some of the future experiments discussed by Kriss (this volume), such as BBXRT, XMM, and ASTRO-D, are exactly what is needed.

Emission Lines

Both observational and theoretical considerations suggest that X-ray emission has important consequences for the optical and ultraviolet emission lines in AGN. For example, many AGN have optical and ultraviolet FeII lines (usually blended in low-resolution spectra), with radio-quiet objects having stronger optical FeII than radio-loud objects (Osterbrock 1977; Phillips 1977; Peterson, Foltz, and Byard 1981). Since X-ray luminosity is strongly correlated with radio luminosity (Ku, Helfand, and Lucy 1980; Zamorani *et al.* 1981; Owen, Helfand, and Spangler; Kembhavi, Feigelson, and Singh 1985; Worrall *et al.* 1987), the connection between radio and optical FeII properties probably involves X-rays interacting with the emission-line gas. Indeed AGN with steep X-ray spectra tend to have strong optical FeII lines relative to AGN with flat X-ray spectra (Remillard and Schwartz 1987; Elvis, Wilkes, and McHardy 1987). Although current theory (see next paragraph) draws an attractive connection between X-ray spectrum and emission-line gas, the predicted effect goes in the wrong direction; that is, it fails to explain why soft X-ray spectra correspond to stronger optical FeII, and why ultraviolet lines are not similarly affected.

The theoretical basis for understanding the ionization structure of the emission-line gas in terms of the X-ray spectrum was for many years quite satisfactory. Krolik, McKee, and Tarter (1981) showed that for steady-state conditions, a hard X-ray spectrum can produce a two-temperature structure, with cool, high-density gas (the BLR clouds) in pressure equilibrium with hot, low-density gas (the intercloud medium). The range of ionization seen in emission-line AGN can be interpreted as the result of a combination of photoionization and Compton heating and cooling in the clouds. (There are exceptions: for example, LINERs with diffuse emission-line gas may be ionized by shocks.) Low ionization lines form in the partially-ionized, interior zones of the BLR clouds, and since only X-rays have sufficient energy to penetrate deep into the clouds, their greatest effect is on low-ionization lines like FeII. The two-phase picture has successfully, perhaps too successfully, explained many of the characteristics of the observed line emission. (Not the optical FeII though — it predicts that harder X-rays should make more FeII, whereas the reverse is observed; Wilkes, Elvis, and McHardy 1987.)

However, we have just said that the X-ray spectra of AGN are not uniformly hard at least below ~ 1-2 keV. The softer X-ray spectra recently observed at lower energies, down to an effective energy of $\lesssim 100\,eV$ with the EXOSAT Low Energy detectors, can have a dramatic effect on calculations of the physical state of the gas subjected to such a radiation field. The "extra" soft photons tend to cool the hot gas and destroy its two-phase character (Fabian *et al.* 1986; see also Guilbert, Fabian, and McCray 1983). This and the FeII dilemma suggest that a re-evaluation of the whole simple equilibrium scenario is in order. In fact, careful consideration of the time scale for heating and cooling processes already suggested that the two-phase picture was not a perfect explanation since the equilibrium state was unlikely to exist (Krolik, McKee, and Tarter 1981). The onset of the instability produced by the softer X-ray spectrum may be exactly what is needed to cause condensations of the broad-line clouds in the first place (Krolik 1988a).

Multiwavelength studies are important for understanding the emission-line region. Both its geometry and ionization structure affect the interplay between line emission in the optical and ultraviolet, and ionizing continuum in the ultraviolet and soft-X-ray. Time-resolved variations could therefore be quite illuminating. Most studies to date relate line emission to adjacent, or "local", continuum, and assume that the ionizing continuum at shorter wavelengths behaves similarly. Direct observation of the ionizing continuum would obviously be preferable, particularly as variability time scales in most AGN are a function of wavelength. In either case, the question of simultaneity is complicated. The BLR is large relative to the continuum source, so there should be a lag between continuum and emission-line flares or dips. Then sensitivity to a range of lag times requires that the data-sampling rate be quite a bit higher than the expected lag time. Moreover, if continuum changes are quite rapid, and are not isolated flares, the picture is complicated still further because it is the integrated history of the ionizing continuum that is important. Simply sampling data on times short compared to the continuum fluctuations is generally insufficient.

Gaskell and Sparke (1986), cross-correlating optical and ultraviolet line and continuum fluxes in the active galactic nuclei NGC4151, Akn120, and Mrk509, found lag times ranging from 5 to 20 days, which they interpreted as indicative of the size of the line-emitting regions (see also Ulrich *et al.* 1984, Bromage *et al.* 1985, Gaskell and Peterson 1986, Clavel *et al.* 1987). However, these data were irregularly spaced, and not always simultaneous. Krolik (1988b) has argued that the cross-correlation function in the presence of unevenly spaced data and an unknown underlying character of the continuum fluctuations can give spurious, or at best, misleading, results. High-quality X-ray light curves of two low-luminosity Seyfert galaxies, NGC4051 (Lawrence *et al.* 1987) and NGC5506 (McHardy and Czerny 1987), suggest that the power density spectrum of the intensity fluctuations is $1/f$-like, *i.e.* the amount of variability at various time scales is inversely proportional to frequency. If this is generally true of the ionizing continuum in AGN, then Krolik's simulations show that without regularly spaced data points, at frequencies higher than the inverse of the light-crossing time of the line-emitting region, the lag time from the correlation function does not measure the characteristic time scales, and those minimum time scales cannot be measured. Indeed, the random fluctuations in the continuum emission do not even reflect the characteristic size of the central engine (Lawrence *et al.* 1987, McHardy and Czerny 1987). Furthermore, independent of the character of the underlying fluctuations, the correlated errors of line and local continuum measurements can cause spurious signals in the correlation function at lag times comparable to the mean sampling times (Gaskell and Peterson 1987, Edelson and Krolik 1988).

Whether the photoionizing continuum in most AGN varies as $1/f$ is not known, but clearly cross-correlations in emission-line studies must be done carefully. The situation can be greatly improved if the continuum and line data in the optical and ultraviolet are obtained simultaneously and at regularly spaced intervals, and if the line-continuum correlations are performed on independent data. Also, although there are connections among optical, ultraviolet, and X-ray continua, the correlations are not linear (see above). The physically interesting correlation is between emission-line strength and the ionizing continuum in the ultraviolet-through-X-ray, not just the local continuum. Coordinating such observations may be prohibitively difficult if ground-based observations are involved, but is easy with space-based experiments like *SpEx* that have well-matched, co-aligned optical, ultraviolet, and X-ray telescopes (see Kriss, this volume).

4. Conclusions

Multiwavelength studies have made major contributions to our understanding of AGN. Broad-band spectra have one or more identifiable components, produced by a variety of physical processes, some of which we are beginning to understand. A given spectral component may span one or more decades in wavelength, and separating it from adjacent components can be difficult without substantial wavelength coverage. Also, many AGN vary, and the variability time scales over a range of wavelengths can differ. Therefore, comprehensive, complete wavelength coverage and simultaneity are very important. Of course, simultaneously covering a wide wavelength range is not easy, as the Taos Conference on Multiwavelength Astrophysics made abundantly clear. The collected data, however, are invaluable for elucidating the structures of AGN.

The broad-band spectra of blazars appear nonthermal — at least they are smooth over a very broad wavelength range. SSC-jet models are an attractive explanation for the blazar phenomenon, both on spectral grounds and for other reasons, but detailed comparisons of theory and observations have been hampered by insufficient data. Complete radio-through-X-ray spectra are achieved only with Herculean effort, and so such observations are done for only a handful of AGN; and then they are seldom repeated; and if repeated, only on time scales long compared to the fastest variability time scales.

Repeated multiwavelength spectra of a blazar during a flare and decay would be particularly informative, since the evolution of the spectral shape is a signature of the underlying emission process. The stochastic nature of such events is antithetical to the scheduling of 6-10 telescopes, both ground-based and space-based, unless procedural changes are adopted. Yet this is the information we need to distinguish among, for example, jet models that require relativistic beaming and those that do not.

Broad-band continua of non-blazar AGN consist of many components, understood at various levels of sophistication, between which energy can be exchanged. For example, infrared dust emission, ultraviolet reddening, and X-ray absorbing properties of Seyfert 2 galaxies are on the whole consistent with an obscured-Seyfert 1 picture (*e.g.* Lawrence and Elvis 1982). Here is a case where combining data from more than 5 decades in wavelength suggests a simple relation between Seyfert 1 and Seyfert 2 galaxies. But what are the characteristics of the dust? What are its temperature and spatial distributions? Answers to these questions require more high-quality spectra, and would be easier were there fortuitous variability as well.

Similarly, the big blue bump is energetically important in many AGN, and has a wide range in luminosity, but what is it? Is it thermal or nonthermal? If it is produced by an accretion disk, it would be the first direct sign that accretion is powering the central engines of AGN. It is extremely important, then to look for signatures of an accretion disk, such as the absorption edges predicted by Sun and Malkan (1987) and Kolykhalov and Sunyaev (1984, 1988), since featureless big blue bumps can in principle be produced in other ways.

The X-ray emission from AGN, which could constitute a large fraction of the bolometric luminosity, is less well understood. What is its spectral shape in the 0.1-100 keV band? The answer has important implications for ionization models of the optical and ultraviolet emission-line regions. And how does the spectral shape vary as a function of AGN type, of luminosity, of redshift, and of properties in other wavebands? Few AGN are bright enough to have been observed over a very broad X-ray band with good spectral resolution, and with the current data, there is understandably some confusion.

As for emission lines, it is popular to say that since they represent only a tiny fraction of the total AGN luminosity, using them as clues to the underlying structure and emission mechanisms is like analyzing car exhaust to learn what is under the hood. Even so, emission-line diagnostics provide a wealth of detailed information about ionization, velocity, dust, and so on. Since we understand at least a little about photoionization and X-ray heating of emission-line gas, correlations between lines and continuum are an excellent probe of the emission-line region, even if the line-emitting gas constitutes only a small part of the mass. But better observations are clearly required to eliminate the effects of sparse data, irregular sampling, and correlated errors. This is an inherently multiwavelength project, parts of which (ultraviolet and X-ray) can only be done from space, and other parts of which (optical and infrared) could be done better from space, since the "weather" is more predictable there.

In all types of AGN, variability studies, with multiwavelength coverage, are extremely important for identifying which spectral components are related. And the wavelength coverage has to be good, or the components cannot be separated reliably. Over and over the present discussion of active galactic nuclei comes down to the same three points: more wavelength coverage, of more objects, repeated more often. The kinds of future multiwavelength experiments discussed in this book and at the Taos meeting are very promising, and should provide at least some answers to questions raised in this chapter.

am grateful to several people for critical reading of the manuscript, especially Alison Campbell, Rick Edelson, Jerry Kriss, Matt Malkan, Richard Mushotzky, Colin Norman, and Andrzej 'dziarski, and I thank France Córdova for her endless patience. The Taos Workshop on Mul‌iwavelength Astrophysics in August 1987 was particularly useful for clarifying the issues in ‌road-band studies of AGN, and in suggesting future avenues for progress. This work was sup‌.orted in part by NASA grant NAG 5-495.

References

ʌller, H.D., Aller, M.F., and Hodge, P.E. 1981, *A. J.*, **86**, 325.

ʌller, H.D., Aller, M.F., and Hughes, P.A. 1985, *Ap. J.*, **298**, 301.

ʌller, H.D., Aller, M.F., Latimer, G.E., and Hodge, P.E. 1985, *Ap. J. Suppl.*, **59**, 513.

ʌltschuler, D.R. 1982, *A. J.*, **87**, 387.

ʌltschuler, D.R., and Wardle, J.F.C. 1975, *Nature*, **255**, 306.

ʌndrew, B.H., MacLeod, J.M., Harvey, G.A., and Medd, W.J. 1978, *A. J.*, **83**, 863.

ʌntonucci, R.R.J., and Miller, J.S. 1985, *Ap. J.*, **297**, 621.

ʌrnaud, K.A., Branduardi-Raymont, G., Culhane, J.L., Fabian, A.C., Hazard, C., McGlynn, T.A., Shafer, R.A., Tennant, A.F., and Ward, M.J. 1985, *M.N.R.A.S.*, **217**, 105.

ʌvni, Y., and Tananbaum, H. 1986, *Ap. J.*, **305**, 83.

Ɓarr, P. 1986, *M.N.R.A.S.*, **223**, 29P.

Ɓarr, P., and Mushotzky, R.F. 1986, *Nature*, **320**, 421.

Ɓassani, L., Butler, R.C., di Cocco, G., Della Ventura, A., Perotti, F., Villa, G., Baker, R.E., Dean, A.J., and Lee, T.J. 1986, *Ap. J.*, **311**, 623.

Ɓeall, J.H., Rose, W.K., Dennis, B.R., Crannell, C.J., Dolan, J.F., Frost, K.J., and Orwig, L.E. 1981, *Ap. J.*, **247**, 458.

Ɓechtold, J., Czerny, B., Elvis, M., Fabbiano, G., and Green, R.F. 1987, *Ap. J.*, **314**, 699.

Ɓechtold, J., Green, R.F., Weymann, R.J., Schmidt, M., Estabrook, F.B., Sherman, R.D., Wahlquist, H.D., and Heckman, T.M. 1984, *Ap. J.*, **281**, 76.

Ɓeichman, C.A., Neugebauer, G., Soifer, B.T., Wootten, H.A., Roellig, T., and Harvey, P.M. 1981, *Nature*, **293**, 711.

Ɓezler, M., Gruber, D.E., and Rothschild, R.E. 1988, *Ap. J.*, submitted.

Ɓezler, M., Kendziorra, E., Staubert, R., Hasinger, G., Pietsch, W., Reppin, C., Trümper, J., and Voges, W. 1984, *Astr. Ap.*, **136**, 351.

Ɓlandford, R.D., and Königl, A. 1979, *Ap. J.*, **232**, 34.

Ɓlandford, R.D., and Rees, M.J. 1974, *M.N.R.A.S.*, **169**, 395.

Ɓlandford, R.D., and Rees, M.J. 1978, in *Pittsburgh Conference on BL Lac Objects*, ed. A.M. Wolfe, (Pittsburgh: University of Pittsburgh), p. 328.

Ɓlumenthal, G.R., and Gould, R.J. 1970, *Rev. Mod. Phys.*, **42**, 237.

Ɓranduardi-Raymont, G. 1986, in *The Physics of Accretion onto Compact Objects*, eds. K.O. Mason, M.G. Watson, and N.E. White, (Heidelberg: Springer-Verlag), p. 407.

Ɓregman, J.N., Glassgold, A.E., Huggins, P.J., Aller, H.D., Aller, M.F., Hodge, P.E., Rieke, G.H., Lebofsky, M.J., Pollock, J.T., Pica, A.J., Leacock, R.J., Smith, A.G., Webb, J., Balonek, T.J., Dent, W.A., O'Dea, C.P., Ku, W.H.-M., Schwartz, D.A., Miller, J.S., Rudy, R.J., and LeVan, P.D. 1984, *Ap. J.*, **276**, 454.

Ɓregman, J.N., Glassgold, A.E., Huggins, P.J., and Kinney, A.L. 1986a, *Ap. J.*, **301**, 698.

Ɓregman, J.N., Glassgold, A.E., Huggins, P.J., Neugebauer, G., Soifer, B.T., Matthews, K., Elias, J., Webb, J., Pollock, J.T., Pica, A.J., Leacock, R.J., Smith, A.G., Aller, H.D., Aller, M.F., Hodge, P.E., Dent, W.A., Balonek, T.J., Barvainis, R.E., Roellig, T.P.L., Wisniewski, W.Z., Rieke, G.H., Lebofsky, M.J., Wills, B.J., Wills, D., Ku, W.H.-M., Bregman, J.D., Witteborn, F.C., Lester, D.F., Impey, C.D., and Hackwell, J.A. 1986b, *Ap. J.*, **301**, 708.

Bregman, J.N., Glassgold, A.E., Huggins, P.J., Pollock, J.T., Pica, A.J., Smith, A.G., Webb J.R., Ku, W.H.-M., Rudy, R.J., LeVan, P.D., Williams, P.M., Brand, P.W.J.L., Neugebauer, G., Balonek, T.J., Dent, W.A., Aller, H.D., Aller, M.F., and Hodge, P.E. 1982, *Ap. J.*, **253**, 19.

Bregman, J.N., Lebofsky, M.J., Aller, M.F., Rieke, G.H., Aller, H.D., Hodge, P.E. Glassgold, A.E., and Huggins, P.J. 1981, *Nature*, **293**, 714.

Bregman, J.N., Maraschi, L., and Urry, C.M. 1987, in *Exploring the Universe with the IUE Satellite*, ed. Y. Kondo, (Dordrecht: D. Reidel), p. 685.

Bromage, G.E., Boksenberg, A., Clavel, J., Elvius, A., Penston, M.V., Perola, G.C., Pettini M., Snijders, M.A.J., Tanzi, E.G., and Ulrich, M.H. 1985, *M.N.R.A.S.*, **215**, 1.

Camenzind, M., and Courvoisier, T.J.-L. 1984, *Astr. Ap.*, **140**, 341.

Carleton, N.P., Elvis, M., Fabbiano, G., Willner, S.P., Lawrence, A., and Ward, M. 1987, *Ap. J.*, **318**, 595.

Cavallo, G., and Rees, M.J. 1978, *M.N.R.A.S.*, **183**, 359.

Cheng, F.H., Kinney, A.L., and Fang, L.Z. 1987, in *New Insights in Astrophysics, Proceedings Joint NASA/ESA/SERC Conference*, (ESA SP-263), p. 649.

Clavel, J., Altamore, A., Boksenberg, A., Bromage, G.E., Elvius, A., Pelat, D., Penston M.V., Perola, G.C., Snijders, M.A.J., and Ulrich, M.H. 1987, *Ap. J.*, **321**, 251.

Clegg, P.E., Gear, W.K., Ade, P.A.R., Robson, E.I., Smith, M.G., Nolt, I.G., Radostitz J.V., Glaccum, W., Harper, D.A., and Low, F.J. 1983, *Ap. J.*, **273**, 58.

Cotton, W.D., Wittels, J.J., Shapiro, I.I., Marcaide, J., Owen, F.N., Spangler, S.R., Ruis, A. Angulo, C., Clark, T.A., and Knight, C.A. 1980, *Ap. J. (Letters)*, **238**, L123.

Courvoisier, T.J.-L., Turner, M.J.L., Robson, E.I., Gear, W.K., Staubert, R., Blecha, A. Bouchet, P., Falomo, R., Valtonen, M., and Terašranta, H. 1987, *Astr. Ap.*, **176**, 197.

Courvoisier, T.J.-L., and Ulrich, M.H. 1985, *Nature*, **316**, 524.

Cruz-Gonzalez, I., and Huchra, J.P. 1984, *A. J.*, **89**, 441.

Cutri, R.M., Aitken, D.K., Jones, B., Merrill, K.M., Puetter, R.C., Roche, P.F., Rudy, R.J. Russell, R.W., Soifer, B.T., and Willner, S.P. 1981, *Ap. J.*, **245**, 818.

Cutri, R.M., Wisniewski, W.Z., Rieke, G.H., and Lebofsky, M.J. 1985, *Ap. J.*, **296**, 423.

Czerny, B., and Elvis, M. 1987, *Ap. J.*, **321**, 243.

de Grijp, M.H.K., Miley, G.K., Lub, J., and de Jong, T. 1985, *Nature*, **314**, 240.

Dent, W.A., and Hobbs, R.W. 1973, *A. J.*, **78**, 163.

Dent, W.A., and Kapitzky, J.E. 1976, *A. J.*, **81**, 1053.

Dent, W.A., Kapitzky, J.E., and Kojoian, G. 1974, *A. J.*, **79**, 1232.

Doxsey, R., Bradt, H., McClintock, J., Petro, L., Remillard, R., Ricker, G., Schwartz, D., and Wood, K. 1983, *Ap. J. (Letters)*, **264**, L43.

Edelson, R.A., and Krolik, J.H. 1988, *Ap. J.*, submitted.

Edelson, R., and Malkan, M.A. 1986, *Ap. J.*, **308**, 59.

Edelson, R., Malkan, M.A., and Rieke, G.H. 1987, *Ap. J.*, **321**, 233.

Eichler, D. 1979, *Ap. J.*, **232**, 106.

Elvis, M. 1988, in *Proceedings of the Rutherford-Appleton Laboratory Workshop on Emission Lines in Quasars*, eds. P. Gondhalekar and H. Netzer, in press.

Elvis, M., Green, R.F., Bechtold, J., Schmidt, M., Neugebauer, G., Soifer, B.T., Matthews K., and Fabbiano, G. 1986, *Ap. J.*, **310**, 291.

Elvis, M., Wilkes, B.J., and McHardy, I.M. 1987, *Ap. J. (Letters)*, **321**, L23.

Fabbiano, G., Elvis, M., Carleton, N., Willner, S.P., Lawrence, A., and Ward, M.J. 1986, in *Quasars*, eds. G. Swarup and V.K. Kapahi, (Dordrecht: D. Reidel), p. 85.

Fabian, A.C. 1979, *Proc. Roy. Soc.*, **366**, 449.

Fabian, A.C., Guilbert, P.W., Arnaud, K.A., Shafer, R.A., Tennant, A.F., and Ward, M.J. 1986, *M.N.R.A.S.*, **218**, 457.

Feigelson, E.D., Bradt, H., McClintock, J., Remillard, R., Urry, C.M., Tapia, S., Geldzahler B., Johnston, K., Romanishin, W., Wehinger, P.A., Wyckoff, S., Madejski, G., Schwartz

D. A., Thorstensen, J., and Schaefer, B. E. 1986, *Ap. J.*, **302**, 337.

Gaskell, C. M., and Peterson, B. M. 1986, *A. J.*, **92**, 552.

Gaskell, C. M., and Peterson, B. M. 1987, *Ap. J. Suppl.*, **65**, 1.

Gaskell, C. M., and Sparke, L. S. 1986, *Ap. J.*, **305**, 175.

Ghisellini, G., Maraschi, L., Tanzi, E. G., and Treves, A. 1986, *Ap. J.*, **310**, 317.

Ghisellini, G., Maraschi, L., and Treves, A. 1985, *Astr. Ap.*, **146**, 204.

Glassgold, A. E., Bregman, J. N., Huggins, P. J., Kinney, A. L., Pica, A. J., Pollock, J. T., Leacock, R. J., Smith, A. G., Webb, J. R., Wisniewski, W. Z., Jeske, N., Spinrad, H., Henry, R. B. C., Miller, J. S., Impey, C., Neugebauer, G., Aller, M. F., Aller, H. D., Hodge, P. E., Balonek, T. J., Dent, W. A., and O'Dea, C. P. 1983, *Ap. J.*, **274**, 101.

Gould, R. J. 1979, *Astr. Ap.*, **76**, 306.

Grandi, S. A. 1982, *Ap. J.*, **255**, 25.

Guilbert, P. W., Fabian, A. C., and McCray, R. 1983, *Ap. J.*, **266**, 466.

Halpern, J. 1982, Ph.D. Thesis, Harvard University.

Hutter, D. J., and Mufson, S. L. 1986, *Ap. J.*, **301**, 50.

Jones, T. W., O'Dell, S. L., and Stein, W. A. 1974, *Ap. J.*, **188**, 353.

Kazanas, D., and Ellison, D. C. 1986, *Ap. J.*, **304**, 178.

Kembhavi, A., Feigelson, E. D., and Singh, K. P. 1985, *M.N.R.A.S.*, **220**, 51.

Kolykhalov, P. I., and Sunyaev, R. A. 1984, *Adv. Space Res.*, **3**, 249.

Kolykhalov, P. I., and Sunyaev, R. A. 1988, preprint.

Kondo, Y., Worrall, D. M., Mushotzky, R. F., Hackney, R. L., Hackney, K. R. H., Oke, J. B., Yee, H., Feldman, P. A., and Brown, R. L. 1981, *Ap. J.*, **243**, 690.

Königl, A. 1981, *Ap. J.*, **243**, 700.

Kriss, G. A. 1988, *Ap. J.*, **324**, 809.

Kriss, G. A., and Canizares, C. R. 1985, *Ap. J.*, **297**, 177.

Krolik, J. H. 1988a, *Ap. J.*, **325**, 148.

Krolik, J. H. 1988b, in *Proceedings of the Atlanta Conference on Variability in Active Galactic Nuclei*, eds. H. R. Miller and P. Wiita, in press.

Krolik, J. H., McKee, C. F., and Tarter, C. B. 1981, *Ap. J.*, **249**, 422.

Kruper, J. S., *et al.* 1988, in preparation.

Ku, W. H.-M., Helfand, D. J., and Lucy, L. B. 1980, *Nature*, **288**, 323.

Kwan, J., and Krolik, J. H. 1981, *Ap. J.*, **250**, 478.

Landau, R., Golisch, B., Jones, T. J., Jones, T. W., Pedelty, J., Rudnick, L., Sitko, M. L., Kenney, J., Roellig, T., Salonen, E., Urpo, S., Schmidt, G., Neugebauer, G., Matthews, K., Elias, J. H., Impey, C., Clegg, P., and Harris, S. 1986, *Ap. J.*, **308**, 78.

Lawrence, A. 1987, *Pub. A.S.P.*, **99**, 309.

Lawrence, A., and Elvis, M. 1982, *Ap. J.*, **256**, 410.

Lawrence, A., Watson, M. G., Pounds, K. A., and Elvis, M. 1985, *M.N.R.A.S.*, **217**, 685.

Lawrence, A., Watson, M. G., Pounds, K. A., and Elvis, M. 1987, *Nature*, **325**, 694.

Ledden, J. E., and O'Dell, S. L. 1985, *Ap. J.*, **298**, 630.

Ledden, J. E., O'Dell, S. L., Stein, W. A., and Wisniewski, W. Z. 1981, *Ap. J.*, **243**, 47.

Lynden-Bell, D. 1969, *Nature*, **223**, 690.

Madau, P., Ghisellini, G., and Persic, M. 1987, preprint.

Madejski, G. M. 1985, Ph.D. Thesis, Harvard University.

Madejski, G. M., and Schwartz, D. A. 1983, *Ap. J.*, **275**, 467.

Makino, F., Tanaka, Y., Matsuoka, M., Koyama, K., Inoue, H., Makishima, K., Hoshi, R., Hayakawa, S., Kondo, Y., Urry, C. M., Mufson, S. L., Hackney, K. R., Hackney, R. L., Kikuchi, S., Mikami, Y., Wisniewski, W. Z., Hiromoto, N., Nishida, M., Burnell, J., Brand, P., Williams, P. M., Smith, M. G., Takahara, F., Inoue, M., Tsuboi, M., Tabara, H., Kato, T., Aller, M. F., and Aller, H. D. 1987, *Ap. J.*, **313**, 662.

Malkan, M. A. 1983, *Ap. J.*, **268**, 582.

Malkan, M.A. 1984, in *X-Ray and UV Emission from Active Galactic Nuclei*, eds. W. Brinkman and J. Trümper, (Munich: Max-Planck Report 184), p. 121.

Malkan, M.A., and Filippenko, A.V. 1983, *Ap. J.*, **275**, 477.

Malkan, M.A., and Moore, R.L. 1986, *Ap. J.*, **300**, 216.

Malkan, M.A., and Oke, J.B. 1983, *Ap. J.*, **265**, 92.

Malkan, M.A., and Sargent, W.L.W. 1982, *Ap. J.*, **254**, 22.

Maraschi, L., Blades, J.C., Calanchi, C., Tanzi, E.G., and Treves, A. 1988, *Ap. J.*, submitted.

Maraschi, L., Ghisellini, G., Tanzi, E.G., and Treves, A. 1986, *Ap. J.*, **310**, 325.

Maraschi, L., Schwartz, D., Tanzi, E.G., and Treves, A. 1985, *Ap. J.*, **294**, 615.

Marscher, A.P. 1980, *Ap. J.*, **235**, 386.

Marscher, A.P. 1983, *Ap. J.*, **264**, 296.

Marscher, A.P., Marshall, F.E., Mushotzky, R.F., Dent, W.A., Balonek, T.J., and Hartman M.F. 1979, *Ap. J.*, **233**, 498.

McGimsey, B.Q., Smith, A.G., Scott, R.L., Leacock, R.J., Edwards, P.L., Hackney, R.L. and Hackney, K.R. 1975, *A. J.*, **80**, 895.

McHardy, I., and Czerny, B. 1987, *Nature*, **325**, 696.

Medd, W.J., Andrew, B.H., Harvey, G.A., and Locke, J.L. 1972, *Mem. R.A.S.*, **77**, 109.

Molnar, L. 1988, in preparation, (presented at the Taos Workshop on Multiwavelength Astro physics, August 1987).

Morini, M., Chiappetti, L., Maccagni, D., Maraschi, L., Molteni, D., Tanzi, E.G., Treves, A. and Wolter, A. 1986a, *Ap. J. (Letters)*, **306**, L71.

Morini, M., Molteni, D., Scarsi, L., Salvati, M., Perola, G.C., Piro, L., Simari, G., Boksen berg, A., Bromage, G.E., Clavel, J., Elvius, A., Penston, M.V., Snijders, M.A.J., and Ulrich, M.-H. 1986b, *Ap. J.*, **307**, 486.

Mufson, S.L., Hutter, D.J., Hackney, K.R., Hackney, R.L., Urry, C.M., Mushotzky, R.F. Kondo, Y., Wisniewski, W.Z., Aller, H.D., Aller, M.F., and Hodge, P.E. 1984, *Ap. J.* **285**, 571.

Mushotzky, R.F. 1984, *Adv. Space Res.*, **3**, 157.

Neugebauer, G., Oke, J.B., Becklin, E.E., and Matthews, K. 1979, *Ap. J.*, **230**, 79.

Neugebauer, G., Soifer, B.T., and Miley, G.K. 1985, *Ap. J. (Letters)*, **295**, L27.

Novikov, T.D., and Thorne, K.S. 1973, in *Black Holes*, eds. C. Dewitt and B. Dewitt, (New York: Gordon and Breach), p. 343.

O'Brien, P.T., Gondhalekar, P.M., and Wilson, R. 1987, in *New Insights in Astrophysics Proceedings Joint NASA/ESA/SERC Conference*, (ESA SP-263), p. 601.

O'Dell, S.L. 1986, *Pub. A.S.P.*, **98**, 140.

O'Dell, S.L., Scott, H.A., and Stein, W.A. 1987, *Ap. J.*, **313**, 164.

Owen, F.N., Helfand, D.J., and Spangler, S.R. 1981, *Ap. J. (Letter)*, **250**, L55.

Osterbrock, D.E. 1977, *Ap. J.*, **215**, 733.

Ostriker, J., and Vietri, M. 1985, *Nature*, **318**, 446.

Page, D.N., and Thorne, K.S. 1974, *Ap. J.*, **191**, 499.

Perola, G.C., Boksenberg, A., Bromage, G.E., Clavel, J., Elvis, M., Elvius, A., Gondhalekar, P.M., Lind, J., Lloyd, C., Penston, M.V., Pettini, M., Snijders, M.A.J., Tanzi, E.G. Tarenghi, M., Ulrich, M.H., and Warwick, R.S. 1982, *M.N.R.A.S.*, **200**, 293.

Perola, G.C., Piro, L., Altamore, A., Fiore, F., Boksenberg, A., Penston, M.V., Snijders, M.A.J., Bromage, G.E., Clavel, J., Elvius, A., and Ulrich, M.H. 1986, *Ap. J.*, **306**, 508.

Perry, J.J., Ward, M.J., and Jones, M. 1987, *M.N.R.A.S.*, **228**, 623.

Peterson, B.M., Foltz, C.B., and Byard, P.L. 1981, *Ap. J.*, **251**, 4.

Phillips, M.M. 1977, *Ap. J.*, **215**, 746.

Pica, A.J., Pollock, J.T., Smith, A.G., Leacock, R.J., Edwards, P.L., and Scott, R.L. 1980, *A.J.*, **85**, 1442.

Pollock, A.M.T., Brand, P.W.J.L., Bregman, J.N., and Robson, E.I. 1985, *Space Sci. Rev.* **40**, 607.

Pollock, J.T., Pica, A.J., Smith, A.G., Leacock, R.J., Edwards, P.L., and Scott, R.L. 1979, *A. J.*, **84**, 1658.

Pounds, K.A. 1985, in *Galactic and Extragalactic Compact X-Ray Sources*, eds. Y. Tanaka and W.H.G. Lewin, (Tokyo: Institute of Space and Astronautical Science), p. 261.

Pounds, K.A., Stanger, V.J., Turner, T.J., King, A.R., and Czerny, B. 1986a, *M.N.R.A.S.*, **224**, 443.

Pounds, K.A., Turner, T.J., and Warwick, R.S. 1986b, *M.N.R.A.S.*, **221**, 7p.

Pounds, K.A., Warwick, R.S., Culhane, J.L., and de Korte, P.A.J. 1986c, *M.N.R.A.S.*, **218**, 685.

Pravdo, S.H., and Marshall, F.E. 1984, *Ap. J.*, **281**, 570.

Pravdo, S.H., Nugent, J.J., Nousek, J.A., Jensen, K., Wilson, A.S., and Becker, R.H. 1981, *Ap. J.*, **251**, 501.

Pringle, J.E. 1981, *Ann. Rev. Astr. Ap.*, **19**, 137.

Pringle, J.E., and Rees, M.J. 1972, *Astr. Ap.*, **21**, 1.

Protheroe, R.J., and Kazanas, D. 1983, *Ap. J.*, **265**, 620.

Ramaty, R., and Lingenfelter, R.E. 1982, *Ann. Rev. Nucl. Part. Sci.*, **32**, 235.

Readhead, A.C.S., Hough, D.H., Ewing, M.S., Walker, R.C., and Romney, J.D. 1983, *Ap. J.*, **265**, 107.

Rees, M.J. 1984, *Ann. Rev. Astr. Ap.*, **22**, 471.

Reich, W., and Steffen, P. 1982, *Astr. Ap.*, **113**, 348.

Reichert, G.A., Mason, K.O., Thorstensen, J.R., and Bowyer, S. 1982, *Ap. J.*, **260**, 437.

Reichert, G.A., Polidan, R.S., Wu, C.-C., and Carone, T.E. 1988, *Ap. J.*, **325**, 671.

Remillard, R., and Schwartz, D. 1987, *Bull. AAS*, **18**, 915.

Reynolds, S.P. 1982, *Ap. J.*, **256**, 38.

Richstone, D.O., and Schmidt, M. 1980, *Ap. J.*, **235**, 361.

Rieke, G.H. 1978, *Ap. J.*, **226**, 550.

Rieke, G.H., and Lebofsky, M.J. 1981, *Ap. J.*, **250**, 87.

Robson, E.I., Gear, W.K., Brown, L.M.J., Courvoisier, T.J.-L., Smith, M.G., Griffin, M.J., and Blecha, A. 1986, *Nature*, **323**, 134.

Robson, E.I., Gear, W.K., Clegg, P.E., Ade, P.A.R., Smith, M.G., Griffin, M.J., Nolt, I.G., Radostitz, J.V., and Howard, R.J. 1983, *Nature*, **305**, 194.

Rothschild, R.E., Mushotzky, R.F., Baity, W.A., Gruber, D.E., Matteson, J.L., and Peterson, L.E. 1983, *Ap. J.*, **269**, 423.

Scott, R.L., Leacock, R.J., McGimsey, B.Q., Smith, A.G., Edwards, P.L., Hackney, K.R., and Hackney, R.L. 1976, *A. J.*, **81**, 7.

Shakura, N.I., and Sunyaev, R.A. 1973, *Astr. Ap.*, **24**, 337.

Shields, G.A. 1978, *Nature*, **272**, 706.

Singh, K.P., and Garmire, G.P. 1985, *Ap. J.*, **297**, 199.

Singh, K.P., Garmire, G.P., and Nousek, J. 1985, *Ap. J.*, **297**, 633.

Smith, P.S., Balonek, T.J., Elston, R., and Heckert, P.A. 1987, *Ap. J. Suppl.*, **64**, 459.

Steidel, C.C., and Sargent, W.L.W. 1987, *Ap. J.*, **313**, 171.

Stein. W.A., and Weedman, D.W. 1976, *Ap. J.*, **205**, 44.

Sun, W.-H., and Malkan, M.A. 1987, in *New Insights in Astrophysics, Proceedings Joint NASA/ESA/SERC Conference*, (ESA SP-263), p. 641.

Tananbaum, H., Avni, Y., Green, R.F., Schmidt, M., and Zamorani, G. 1986, *Ap. J.*, **305**, 57.

Tucker, W. 1967, *Ap. J.*, **148**, 745.

Turner, T.E., and Pounds, K.A. 1987, *M.N.R.A.S.*, **224**, 443.

Ulrich, M.H., Boksenberg, A., Bromage, G.E., Elvius, A., Penston, M.V., Perola, G.C., Pettini, M., Snijders, M.A.J., Tanzi, E.G., and Tarenghi, M. 1984, *M.N.R.A.S.*, **206**, 221.

Unwin, S.C., Cohen, M.H., Pearson, T.J., Seielstad, G.A., Simon, R.S., Linfield, R.P., and Walker, R.C. 1983, *Ap. J.*, **271**, 536.

Urry, C. M. 1984, Ph.D. Thesis, The Johns Hopkins University.

Urry, C. M. 1986a, in *Proceedings of the NOAO Workshop on Continuum Emission of Active Galaxies*, (Tucson, Arizona, January 1986), ed. M. L. Sitko, p. 91.

Urry, C. M. 1986b, in *The Physics of Accretion onto Compact Objects*, eds. K. O. Mason, M. G Watson, and N. E. White, (Heidelberg: Springer-Verlag), p. 357.

Urry, C. M., Kruper, J. S., Canizares, C. R., Rohan, M. L., and Oberhardt, M. R. 1986, in *Proceedings of the Conference on Variability in Galactic and Extragalactic X-ray Sources*, ed. A Treves, (Milan: Association for the Advancement of Astronomy), p. 15.

Urry, C. M., and Mushotzky, R. F. 1982, *Ap. J.*, **253**, 38.

Urry, C. M., Mushotzky, R. F., and Holt, S. S. 1986, *Ap. J.*, **305**, 369.

Urry, C. M., Mushotzky, R. F., Kondo, Y., Hackney, K. R. H., and Hackney, R. L. 1981, *Bull APS*, **26**, 583.

Waltman, E. B., Geldzahler, B. J., Johnston, K. J., Spencer, J. H., Angerhofer, P. E., Flor kowski, D. R., Josties, F. J., McCarthy, D. D., and Matsakis, D. N. 1986, *A. J.*, **91**, 231.

Wandel, A. 1987, *Ap. J. (Letters)*, **316**, L55.

Wandel, A., and Mushotzky, R. F. 1986, *Ap. J. (Letters)*, **306**, L61.

Wandel, A., and Petrosian, V. 1988, *Ap. J. (Letters)*, submitted.

Wandel, A., and Yahil, A. 1985, *Ap. J. (Letters)*, **295**, 1.

Ward, M., Elvis, M., Fabbiano, G., Carleton, N. P., Willner, S. P., and Lawrence, A. 1987, *Ap. J.*, **315**, 74.

Webb, J. R., Smith, A. G., Leacock, R. J., Fitzgibbons, G. L., Gombola, P. P., and Shepherd, D. W. 1987, preprint.

Weistrop, D., Shaffer, D. B., Mushotzky, R. F., Reitsema, H. J., and Smith, B. A. 1981, *Ap. J.*, **249**, 3.

Weymann, R. J., Carswell, R. F., and Smith, M. G. 1981, *Ann. Rev. Astr. Ap.*, **19**, 41.

Wilkes, B., and Elvis, M. 1987, *Ap. J.*, **323**, 243.

Wilkes, B. J., Elvis, M., and McHardy, I. M. 1987, *Ap. J. (Letters)*, **321**, L23.

Willner, S. P., Elvis, M., Fabbiano, G., Lawrence, A., and Ward, M. J. 1985, *Ap. J.*, **299**, 443.

Wills, B. J., Netzer, H., and Wills, D. 1985, *Ap. J.*, **288**, 94.

Wolstencroft, R. D., Gilmore, G., and Williams, P. M. 1982, *M.N.R.A.S.*, **201**, 479.

Worrall, D. M. 1987, *Ap. J.*, **318**, 188.

Worrall, D. M., Boldt, E. A., Holt, S. S., and Serlemitsos, P. J. 1980, *Ap. J.*, **240**, 421.

Worrall, D. M., Giommi, P., Tananbaum, H., and Zamorani, G. 1987, *Ap. J.*, **313**, 596.

Worrall, D. M., Mushotzky, R. F., Boldt, E. A., Holt, S. S., and Serlemitsos, P. J. 1979, *Ap. J.*, **232**, 683.

Worrall, D. M., Puschell, J. J., Bruhweiler, F. C., Miller, H. R., Rudy, R. J., Ku, W. H.-M., Aller, M. F., Aller, H. D., Hodge, P. E., Matthews, K., Neugebauer, G., Soifer, B. T., Webb, J. R., Pica, A. J., Pollock, J. T., Smith, A. G., and Leacock, R. J. 1984a, *Ap. J.*, **278**, 521.

Worrall, D. M., Puschell, J. J., Bruhweiler, F. C., Sitko, M. L., Stein, W. A., Aller, M. F., Aller, H. D., Hodge, P. E., Rudy, R. J., Miller, H. R., Wisniewski, W. Z., Cordova, F. A., and Mason, K. O. 1984b, *Ap. J.*, **284**, 512.

Worrall, D. M., Puschell, J. J., Jones, B. Bruhweiler, F. C., Aller, M. F., Aller, H. D., Hodge, P. E., Sitko, M. L., Stein, W. A., Zhang, Y. X., and Ku, W. H.-M. 1982, *Ap. J.*, **261**, 403.

Worrall, D. M., Puschell, J. J., Rodriguez-Espinosa, J. M., Bruhweiler, F. C., Miller, H. R., Aller, M. F., and Aller, H. D. 1984c, *Ap. J.*, **286**, 711.

Worrall, D. M., Rodriguez-Espinosa, J. M., Wisniewski, W. Z., Miller, H. R., Bruhweiler, F. C., Aller, M. F., and Aller, H. D. 1986, *Ap. J.*, **303**, 589.

Zamorani, G., Henry, J. P., Maccacaro, T., Tananbaum, H., Soltan, A., Avni, Y., Liebert, J., Stocke, J., Strittmatter, P. A., Weymann, R. J., Smith, M. G., and Condon, J. J. 1981, *Ap. J.*, **245**, 357.

Zdziarski, A. 1986, *Ap. J.*, **305**, 45.

Part 2: Multiwavelength Tools

Data Bases and Analysis Centers

Thomas J. Chester

Infrared Processing and Analysis Center

California Institute of Technology

1. Introduction

The Space Age has brought astronomy many benefits. The ability to observe astrophysical objects at wavelengths inaccessible from the ground has opened up new windows to astronomers through which to view the Universe. As astronomers became more sophisticated in using those windows, the volume of data produced by space missions has grown enormously. The Space Age has also led to a large increase in the number of astronomers and an explosion in the amount of ground-based data.

Astronomers were initially unprepared for the resulting flood of data, and therefore we did not reap the maximum benefit from that data. Even now, an astronomer who desires to access multiwavelength data has significant difficulties. There is no general directory listing the availability and characteristics of public data. The lack of such a directory forces every astronomer to individually track down what data are available and how to obtain them. Large quantities of data are virtually inaccessible either because they have not been placed in computer-readable form or because the money has not been spent to put the data in a form that can be easily used. Many other examples abound of such difficulties.

Fortunately, many people and agencies are now actively seeking solutions to these problems. As always, the recognition that there is a problem had to come first. This first step is now several years behind us. Solutions to most of these problems require significant advances in computers and networks. These advances either have been made or they are happening now. Finally, major institutions are now taking the lead in solving these problems. Thus within a decade we will almost certainly have in place a data analysis system that will allow astronomers to fully exploit our data bases.

In this chapter, I discuss currently available data bases and analysis centers that can be exploited for multiwavelength work. The intent is to provide a complete list of major data bases and analysis centers. This should serve as a reference for observers wishing to pursue specific objects across the electromagnetic spectrum. In addition, I discuss current plans toward facilities of the future to provide a vision of how astronomers may soon be able to more efficiently work with the growing body of astrophysical data.

2. Data Bases

Tables 1 and 2 list the major astronomical data bases, organized by wavelength, and gives the general characteristics of each one. (A glossary of the abbreviations used in this chapter is given in Table 4 in the Appendix.) I purposely omit the many valuable catalogs organized by source type as well as the many surveys of limited areas of the sky. Here I concentrate on those catalogs and data bases that result from surveys or that contain large collections of data. A directory of many of the data sets not listed in Tables 1 and 2 can be obtained from the Astronomical Data Center (ADC) in Greenbelt, Maryland, U.S.A. or from the Strasbourg Astronomical Data Centre (CDS) in Strasbourg, France. Addresses for these two institutions are given in the Appendix. (The ADC currently has a 47 page listing of available data sets.)

Table 1 lists catalogs and is sorted into four parts. Table 1a gives available machine-readable catalogs. Table 1b lists catalogs available only in print. Table 1c gives available catalogs of target lists for different instruments. Table 1d lists catalogs that are in preparation. Table 2 gives non-catalog data sets such as images or raw data scans, and is presented in a similar fashion, but excluding the non-relevant category of target lists.

Because this compilation is intended to be complete, I appeal to the reader to inform me of any of my mistakes or omissions. I will endeavor to keep this list up to date and make it available on an on-going basis. Note that the major source for the machine-readable data sets was the directory of data sets placed in the ADC and CDS. Thus I may have omitted some data sets that have not been placed with these agencies. I apologize to those authors whose data sets are missing from the Tables due to my ignorance.

As an aid to understanding some general characteristics of the available data, Figures 1-3 illustrate how several important parameters change with wavelength. (Figure 1 is reproduced from Neugebauer 1986 and Figures 2 and 3 from Chester 1988 with corrections.) The relative sensitivities of instruments at different wavelengths are given in Figure 1, the relative resolution in Figure 2, and the number of cataloged sources in Figure 3. These figures reveal that:

1. Only a handful of sources can presently be studied at all wavelengths.
2. Currently, the sensitivity and resolution of gamma-ray and infrared detectors significantly lag other wavelengths.

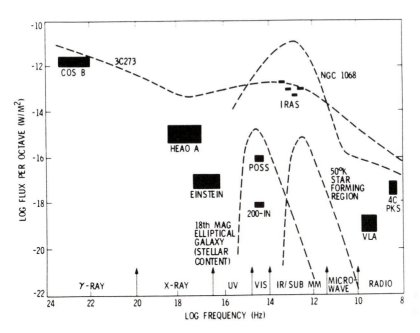

Figure 1 A comparison of the limiting sensitivities for several surveys and non-survey observations over the range of the full electromagnetic spectrum is shown. (Figure reproduced from Neugebauer 1986 with permission.)

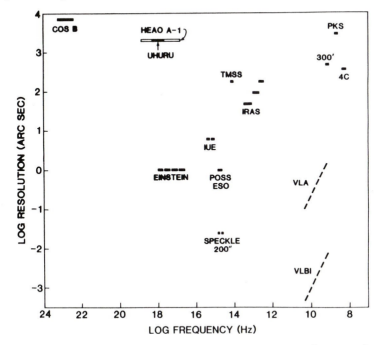

Figure 2 A comparison of the resolution for several published surveys (solid lines) and non-survey (dashed lines) observations over the range of the full electromagnetic spectrum is shown.

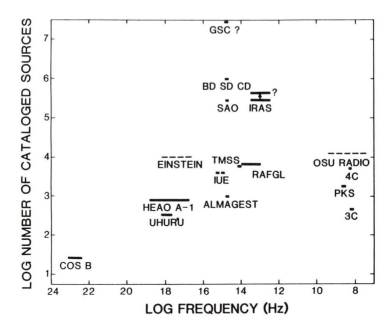

Figure 3 A comparison of the number of sources cataloged by several surveys (solid lines) versus frequency. Some non-survey catalogs (dashed lines) are included for comparison.

3. We are just on the verge of being able to study some classes of sources (such as quasars) at all wavelengths.

4. Some classes of sources (such as 50 K star forming regions whose emission drops very steeply outside a narrow wavelength band) will probably never be directly observed at all wavelengths. However, multiwavelength observations may still prove quite useful in understanding the environment of such sources.

Catalogs of sources observed at different wavelengths are presently a rich resource for astronomers. At least for surveys, the data are readily available. Almost without exception, every survey at every wavelength has produced a catalog of sources that has been made available to the community. Usually, analysis involving catalogs is simple in the sense that data volumes are generally small, the data are usually reduced to conventional units, and only simple software is needed to access the data. Thus astronomers have been able to use these data to produce many good scientific results.

Several tremendously encouraging developments deserve to be singled out:

1. The Set of Identifications, Measurements and Bibliography for Astronomical Data (SIMBAD) data base at the CDS contains published data since 1950 for stars and since 1983 for non-stellar objects. A query system operates on this data so that astronomers now can easily and rapidly access any published data on almost any known astronomical object. This data base can be accessed

Table 1 Astronomical Catalogs Resulting From Surveys or Extensive Compilations
a) Machine Readable Catalogs

Wavelength		Name	Number of Sources	Survey	Sky Coverage	Ref.
X-ray						
.25-25	keV	*HEAO* 1 A-1	842	Yes	100%	1
2-6	keV	*UHURU*	339	Yes	100%	2
2-10	keV	*Ariel V*	251	Yes	100%	3
Ultraviolet						
1330-4250	Å	*OAO 2*	531	No	–	4
1550-3300	Å	*ANS*	3,573	No	–	5
Optical						
5500	Å	SAO	258,997	Yes	100%	6
5500	Å	Durchmust.	~1,000,000	Yes	100%	7
4400	Å	ESO/Uppsala	18,438	Yes	$-90° < \delta < -17.5°$	8
—		OSU	~ 185,000	No	–	9
5500	Å	SKYMAP	248,516	Yes	100%	10
~ 7000	Å	Dearborn	44,076	Yes	$-4.5° < \delta < +90°$	11
Infrared						
2.2	μm	TMSS	5,612	Yes	$-33° < \delta < +81°$	12
4.2-27.4	μm	RAFGL	6,146	Yes	92%	13
8-120	μm	*IRAS* PSC	245,889	Yes	96%	14
8-120	μm	*IRAS* SSS	16,740	Yes	96%	15
8-120	μm	*IRAS* SSC	43,866	No	3%	16
1-~ 20	μm	CIO	~ 13,000	No	–	17
Radio						
178	MHz	3CR	326	Yes	$-5° < \delta < +90°$	18
178	MHz	4C	4,843	Yes	$-7° < \delta < +80°$	19
408	MHz	Parkes	1,736	Yes	$-90° < \delta < +20°$	20
26-2700	MHz	Ohio RM	~ 12,000	No	–	21
5	GHz	5 GHz	518	Yes	$|b| > 10°$	22
Multiwavelength						
Stellar		SIMBAD	~ 700,000	No	–	23

remotely. A similar data base for galaxies is being created at the Infrared Processing and Analysis Center (IPAC) at Caltech with the help of the CDS and will be available to the astronomical community by 1990.

2. The Catalog of Infrared Observations (Gezari, Schmitz, and Mead 1987) contains most published infrared observations from 1965 to 1986 in its current release, and their data base is continuously updated.

3. Both the CDS and the ADC have a continuing program of putting older catalogs into machine-readable form. Thus the *Durchmusterung* catalogs containing over 1,000,000 sources are now available to astronomers in a form in which

Table 1 Astronomical Catalogs Resulting From Surveys or Extensive Compilations
b) Catalogs Available only as Hardcopy

Wavelength		Name	Number of Sources	Survey	Sky Coverage	Ref.
Gamma-ray						
50-5000	MeV	COS-B	25	Yes	~ 50%	24
30-150	MeV	SAS-2	11	Yes	~ 50%	25
X-ray						
.18-2.8	keV	*HEAO 1 A-2*	114	Yes	95%	26
1-40	keV	MIT/*OSO-7*	185	Yes	100%	27
2-60	keV	*HEAO 1 A-2*	60	Yes	100%	28
13-180	keV	*HEAO 1 A-4*	72	Yes	100%	29
Radio						
81.5	MHz	2C	1,936	Yes	$-38^\circ < \delta < +83^\circ$	30
159	MHz	3C	471	Yes	$-22^\circ < \delta < +71^\circ$	31

Table 1 Astronomical Catalogs Resulting From Surveys or Extensive Compilations
c) Target lists

Wavelength		Name	Number of Targets	Ref.
X-ray				
.15-4	keV	*Einstein*	5,611	32
.1-20,000	keV	*EXOSAT*	>2,000	33
Ultraviolet				
1150-3300	Å	*IUE*	~ 4,500	34
Infrared				
8-120	μm	*IRAS* AOs	13,853	35
Radio				
1.3-20	cm	VLA	??	36

they can be used!

4. The soon to be released Space Telescope Guide Star Catalog produced by the Space Telescope Science Institute is a major step toward making available the vast volumes of data currently stored only on photographic plates. The Space Telescope Guide Star Catalog will contain ~20,000,000 stars down to 16th magnitude and is compiled from scans of the ESO plates and the Palomar "Quick Survey".

5. The entire SRC Southern Sky Survey is being scanned by the Royal Observatory Edinburgh. Their catalog will contain ~200,000,000 stars and galaxies

Table 1 Astronomical Catalogs Resulting From Surveys or Extensive Compilations
d) Catalogs In Preparation

Wavelength		Name	Number of Sources	Survey	Sky Coverage	Ref.
X-ray						
	.15-4 keV	*Einstein* SS	~ 10,000?	No	8%	37
Optical						
	4500 Å	COSMOS	~200,000,000?	Yes	$\delta < 0^o$	38
	5500 Å	ST Guide star	22,000,000?	Yes	100%	39
Infrared						
	8-120 μm	*IRAS* FSC	250,000?	Yes	80% ?	40
Multiwavelength						
Extragalactic		IPAC NED	~ 100,000?	No	–	41

down to 22nd magnitude! The POSS is being scanned by the University of Minnesota.

6. There is even a pilot project to create an archive for a ground-based telescope and put that archive into a data base! (See Ochsenbein 1988.)

Access to catalog data could be improved somewhat if catalogs were all written in Flexible Image Transport System (FITS) format (Wells, Greisen, and Harten 1981, Greisen and Harten 1981, Harten, Grosbol, Greisen and Wells 1988a and 1988b). The FITS standards were originally established to allow astronomers to easily share image data between any types of computers. Once one program was written on each computer to read a FITS standard tape, all image data from any computer were immediately transportable. The FITS standards have now been extended to catalog data, and every catalog writer should adhere to this standard. The NSSDC is currently preparing many of the catalogs in their archive in FITS format.

The status of non-catalog data is not nearly as good as the status of catalog data. Several different reasons account for this situation. First, non-catalog data are harder to process and release in an instrument independent form. Second, the keepers of the data worry that the data may be misused by those who do not understand important characteristics of the data. Third, in the past it has not always been considered important to release any non-catalog or non-published data.

Thus computer-readable all-sky images are available only in soft X-rays, the infrared and the radio. Comparable data for hard X-ray and gamma-ray surveys exist but have not been placed in an equivalent readily useable form. Optical images exist only as plates.

However, as with catalogs, several developments are underway that will significantly improve the current status. Some examples:

1. The Space Telescope Guide Star Catalog plate scan data base will be made available.
2. The COSMOS scans of the SRC Southern Sky Survey will be released, as will the Minnesota scans of the POSS.
3. The international *IRAS* collaboration and the *COS-B* and *IUE* satellites have shown how it is possible to release much of their entire data base in a form that can easily be accessed and used by astronomers. *EXOSAT* is also following these examples.

As mentioned above, the introduction and acceptance of standards in writing image data on tape has been another significant improvement. The infrared and radio images are available in FITS format.

Spectroscopy is a special case of non-catalog data. The many optical spectroscopic catalogs are not listed, since that information has been collected in SIM-BAD for stars and will be collected for galaxies at IPAC. Radial velocities derived from spectroscopic data are not listed for the same reason. Thus the only large spectroscopic data base given in Table 2 is that of the *IUE* satellite.

3. Analysis Centers

"The best analysis center is the one connected to the terminal in your office."

The concept of an analysis center is relatively recent. The Space Age for astronomy began with the concept of Principal Investigators designing, building, and operating instruments for themselves, with other astronomers having access only to the publications emerging from the project. When the mission was over, major support for data analysis ended soon thereafter. In those exciting initial days of space astronomy, the evolution of hardware was so rapid that it made more sense to devote resources to the next mission rather than continue to analyze existing data sets. Thus the data from many of the initial planetary missions, for example, were simply sent to an archive and the expertise associated with the data was lost. However, as the returned data sets grew larger and better, and as missions became less frequent, more attention and resources were devoted toward facilitating analysis of existing data.

In addition, the launch of *IUE* and of *Einstein* in 1978 changed astronomer's perceptions of the role of space data. Whereas previously space data were thought of as Principal Investigator data of interest only to specialists in X-ray astron-

Table 2 Astronomical Data Bases Other Than Catalogs
Resulting From Surveys or Extensive Compilations
a) Machine Readable

Wavelength	Name	Type of Product	Survey	Sky Coverage	Ref.
Gamma-ray					
50-5,000 MeV	*COS-B*	150,000 PDB	Yes	~ 50%	42
X-ray					
.13-6.3 keV	Wisconsin	image data	Yes	100%	43
.15-4 keV	*Einstein*	PDB,im. data	No	8%	32
Ultraviolet					
1200-3600 Å	*OAO 2*	spectra	No	–	44
1150-3300 Å	*IUE*	spectra	No	–	45
Infrared					
8-120 μm	*IRAS* imag.	image data	Yes	98%	46
8-120 μm	*IRAS* zodi.	scan data	Yes	98%	46
Radio					
1428 MHz	Heiles	HI line	Yes	$|b| > 10^o, \delta > -30^o$	47
1428 MHz	Colomb	HI line	Yes	$|b| > 10^o, \delta < -25^o$	48

Table 2 Astronomical Data Bases Other Than Catalogs
Resulting From Surveys or Extensive Compilations
b) Hardcopy

Wavelength	Name	Type of Product	Survey	Sky Coverage	Ref.
X-ray					
13-180 keV	*HEAO 1* A-4	plates	Yes	100%	29
Optical					
4500,6500 Å	POSS	plates	Yes	$-36^o < \delta < +90^o$	49
4500,6500 Å	ESO/SRC SS	plates	Yes	$-90^o < \delta < 0^o$	50
–	Harvard Plt.	plates	No	–	–
–	Sonnenberg	plates	No	–	–
–	Bamberg	plates	No	–	–
Infrared					
8-120 μm	*IRAS* imag.	image data	Yes	98%	46
Radio					
1400 MHz	Condon	contour maps	Yes	$-5^o < \delta < +82^o$	51

omy, the general astronomical community became involved in obtaining satellite
data that gave us new information about old familiar astronomical objects. As-

Table 2 Astronomical Data Bases Other Than Catalogs
Resulting From Surveys or Extensive Compilations
c) In Preparation

Wavelength	Name	Type of Product	Survey	Sky Coverage	Ref.
Gamma-ray					
30-150 MeV	*SAS-2*	8,000 PDB	Yes	~ 50%	52
X-ray					
.1-20,000 keV	*EXOSAT*	PDB,im. data	No	5%	53
Optical					
4500 Å	COSMOS	dig. plates	Yes	$\delta < 0^o$	38
5500 Å	ST Guide star	dig. plates	Yes	100%	39
4500,6500 Å	Minnesota	dig. plates	Yes	$\delta > -33^o$	
Radio					
	VLA	image data	No	–	36
4800 MHz	Condon2	image data	Yes	$-1^o < \delta < +73^o$	54

References for Tables 1 and 2: 1. Wood et al 1984. 2. Forman et al 1978. 3. McHardy et al 1981, Warwick et al 1981. 4. Code, Holm, and Bottemiller 1980. 5. Wesselius et al 1982. 6. SAO Star Catalog 1966, revised by ADC 1984. 7. Cape Photographic Durchmusterung - Gill and Kapteyn 1985-1900; Cordoba Durchmusterung - Thome 1892-1932; Southern Durchmusterung - Schoenfeld 1886; Bonner Durchmusterung - Argelander 1859-1862, Kustner 1903, Schmidt 1968. 8. Lauberts 1982. 9. Dixon and Sonneborn 1980. 10. Version 3.3 1986 see Gottlieb 1978. 11. Lee et al 1943, 1944, 1947. 12. Neugebauer and Leighton 1969. 13. Price and Murdock 1983. 14. *IRAS Point Source Catalog* 1986. 15. *IRAS Small Scale Structure Catalog* 1986. 16. Kleinmann et al 1986. 17. Gezari et al 1987. 18. Bennett 1961. 19. Pilkington and Scott 1965, Gower, Scott and Wills 1967. 20. Bolton et al 1964, Price and Milne 1965, Day et al 1966, Shimmins et al 1966, Shimmins and Day 1968. 21. Dixon 1970. 22. Kuhr et al 1981. 23. Egret and Wenger 1988. 24. Swanenberg et al 1981. 25. Houston and Wolfendale 1983. 26. Nugent et al 1983. 27. Markert et al 1979. 28. Marshall et al 1979. 29. Levine et al 1984. 30. Shakeshaft et al 1955 31. Edge et al 1959 32. Seward, F.D. and MacDonald 1986. 33. EXOSAT Observation Log. 34. Catalog of *IUE* Observations. 35. Young et al 1985. 36. Contact Carl Bignell at NRAO. 37. Contact D. Harris at Center for Astrophysics, SAO. 38. MacGillivray et al 1988. 39. Jenkner et al 1988. 40. *IRAS Faint Source Catalog* 1988, in progress. 41. Helou and Madore 1988. 42. Mayer-Hasselwander 1985. 43. McCammon et al 1983. 44. Code and Meade 1978, Meade and Code 1980. 45. Heap 1988. 46. *IRAS* Sky Brightness Images, in *IRAS Explanatory Supplement* 1988. 47. Heiles 1975. 48. Colomb et al 1980. 49. Lund and Dixon 1973. 50. West 1974. 51. Condon and Broderick 1986 52. Murthy and Wolfendale 1986. 53. White 1988. 54. Contact J.J. Condon at NRAO.

tronomers journeyed to the satellite operations centers to make observations and to reduce data. These were the first rudimentary analysis centers.

An analysis center should have the following characteristics:
1. Its overall purpose is to advance science.
2. It has resident experts in the use of the center facilities and resources, **who actively use the data,** to provide assistance to users.

3. It offers resources to assist analysis that are not readily available elsewhere. While this nearly always involves unique data sets, it can also involve special expertise in, for example, comparing data sets.
4. It is generally available to the astronomical community.
5. The data resource of the center is kept alive through on-going scientific work at the center.
6. The center should be accessible by remote users through dial-up or network facilities.

The issue of expertise is perhaps the most fundamental one. An analysis center is not fulfilling its role if it merely archives data. Experts at the centers can assist astronomers in obtaining the maximum science return from the data and allow efficient analysis as part of the package.

Table 3 Data Analysis Centers

Name	Location	Expertise	Remote Access
ESO	Garching, FRG	Optical	No?
ESTEC/ESOC	Nortwijk, NETH	*EXOSAT* X-ray	Yes*
Galaxy data base+	IPAC,Pasadena,CA	Galaxy data	Yes*
Univ. of Groningen	Groningen,NETH	*IRAS*	No?
Infrared data base	Goddard,MD	Infrared literature	No?
IPAC	Caltech,Pasadena,CA	*IRAS*, Extragalactic	Yes*
IPMAF	R.A.L.,England	*IRAS*	Yes
IUE	Boulder,CO	*IUE*	Yes
IUE	Goddard,MD	*IUE*	Yes
IUE	Madrid, Spain	*IUE*	No?
Univ. of Leiden	Leiden,NETH	*IRAS*	No?
NRAO	Charlottesville,VA	Radio	No?
NOAO	Tucson,AZ	Optical	No
NSSDC	Goddard,MD	Catalogs, Space data	Yes
Einstein Data Bank	SAO,Cambridge,MA	*Einstein*	Limited
StSci	Baltimore,MD	*Hubble* ST	Yes*
ST Coord. Facility	Garching,FRG	*Hubble* ST	Yes*
SIMBAD	Strasbourg,France	Stellar data	Yes

+ In preparation, available in 1990? * Not yet available but imminent

Stimulated by plans for the "Great Observatories of the Future" (see below), it has become clear that analysis centers must go hand in hand with the Observatories. Each wavelength has its own peculiarities that require experts to understand them. Thus the logical step is to establish permanent analysis centers that are responsible for data from different wavelengths.

This concept seems opposed to the ideal of multiwavelength astrophysics since it fragments astronomy into wavelength arenas rather than unifying different wavelengths. The reality is that this approach is the best way of doing multiwavelength astrophysics. First, one goal of these analysis centers is to produce data sets that can be used by non-experts in any given field and that can thus be made available to all other analysis centers and individual astronomers. Second, it will always require experts in each wavelength arena in order to help others to achieve better science with those data. Unifying the data sets may superficially seem better, but without an enormous increase in funding it will not be possible to have sufficient expertise at any one location.

The requirement of off-center access is extremely important to any research, but especially multiwavelength research. Once off-center access is fully developed, a user will be immediately able to access all available astronomical data for a given object from the terminal in the astronomer's own office. Questions or problems can immediately be referred to experts in the use of those data at the appropriate analysis center.

Table 3 lists the major existing analysis centers that help astronomers to access and to understand some of the data bases listed above.

4. Plans for the Future

NASA has ambitious plans for establishing a series of Great Observatories that will truly bring about the era of multiwavelength astronomy. These include the Hubble Space Telescope, the Gamma Ray Observatory, the Advanced X-ray Astrophysics Facility, and the Space Infrared Telescope Facility. NASA is also wisely planning how to manage the data base that these missions will produce and how to improve the capability of astronomers to analyze these data.

NASA is currently planning an Astrophysics Data System that will make significant improvements in the ability to obtain astrophysics data from space missions. Two workshops have been held in 1987 with the astrophysics community to design the system (Squibb and Cheung 1988). A brief description of the current design follows.

A master directory of all space data obtained in NASA missions as well as some non-space data will be established and maintained at the NSSDC. The directory will also contain an index of all software available to astronomers as well as a bibliography of all publications relevant to such data.

Three Science Archive Coordination Centers will be established for infrared, high energy, and UV and optical data that will be responsible for ensuring the long-term access to the data and that will maintain the crucial body of scientific expertise with their data sets. These centers will be primary nodes in the NASA Science Internet and will be connected to the Science Operations Centers for NASA's series of Great Observatories. The system will also be connected to

U.S. ground-based observatories, an international network and to astronomical users at their home institutions.

Research will be supported in the areas of distributed data bases, catalog access techniques, and algorithm developments for maximizing retrieval of information. Standards will play a major role in the system for such items as data format, data quality, software, and communication protocols.

When this system is in place, the ease of doing multiwavelength astronomy will be vastly improved, at least when using space-based data. The challenge remains for other organizations to provide the same support for ground-based data.

I wish to acknowledge the participants in the Taos Conference on Multiwavelength Astrophysics, and notably E. Schreier, for their comments and insights into these problems during the conference. The participants in the NASA Astrophysics Data System Workshops have made valuable contributions through their recommendations to NASA to deal with these issues. I would particularly like to thank G. Squibb for conveying the gist of these recommendations prior to the publication of the Workshop Reports. I thank W. Rice, T. Soifer, G. Squibb and W. Warren for their comments on the manuscript.

APPENDIX

The two major world data centers are:

Astronomical Data Center
National Space Science Data Center
Code 633
Goddard Space Flight Center
Greenbelt, Maryland 20771
U.S.A.
(301) 286-8310 or -6695
Telex: 89675 NASCOM GBLT
TWX: 7108289716
SPAN: NSSDC::REQUEST
BITNET: TEADC@SCFVM

Centre de Donnees Stellaires
Observatoire Astronomique
11, rue de l'Universite
F-6700 Strasbourg
FRANCE
88.35.82.00
890 506 STAROBS F

EARN: U01117@FRCCSC21

References

Argelander 1859-1862 *Bonner Sternverzeichnis*. Erst bis dritte Sektion, Astron. Beobachtungen aug der Sternw. der Knoiglichen Rhein. Friedrich-Wilhems-Univ. zu Bonn, Bands 3- 5.
Bennett, A.S. 1961 *Mem.R.A.S.* **68**:163.
Bolton, J.G., Gardner, F.F., and Mackey, M.B. 1964, *Austr.J.Phys.* **17**:340 (Parkes Radio Survey -20° to -60°.)
Catalog of IUE Observations IUE NASA/ESA Observatories, updated regularly.
Chester, T. 1988, in *Mapping the Sky: I.A.U. Symposium No. 133*, in press.
Code, A.D., and Meade, M.R. 1978, *Ap.J.Suppl.* **39**:195.
Code, A.D., Holm, A.V., and Bottemiller, R.L. 1980 *Ap.J.Suppl.* **43**:501.
Colomb, F.R, Poppel, W.G.L., and Heiles, C. 1980 *Astr.Ap.Suppl.* **40**:47.
Condon, J.J., and Broderick, J.J. 1986 *A 1400 MHz Sky Atlas Covering -5° < δ < +82°*, National Radio Astronomy Observatory.

Table 4 Glossary of Abbreviations

Short Name	Full Name
ADC	Astronomical Data Center
ANS	Astronomical Satellite of the Netherlands
CIO	Catalog of Infrared Observations
Colomb	HI Southern Survey
Condon	A 1400 MHz Sky Atlas Covering $-5^o < \delta < +82^o$
Condon2	4800 MHz Sky Survey, in progress
Dearborn	Dearborn Observatory Catalogue of Faint Red Stars
Durchmust.	Bonner Durchmusterung
	Cape Photographic Durchmusterung
	Cordoba Durchmusterung
	Southern Durchmusterung
Einstein SS	*Einstein* Serendipitous Survey Catalog
EGDB	Extragalactic Data Base, IPAC
Equatorial	Equatorial Infrared Catalog
ESO/SRC SS	European Southern Observatory / Science Research Council Sky Survey
ST Guide star	Space Telescope Guide Star Catalog
HEAO	*High Energy Astronomy Observatory*
Heiles	HI Northern Survey
IPAC	Infrared Processing and Analysis Center
IPMAF	*IRAS* Post Mission Analysis Facility
IRAS	*Infrared Astronomical Satellite*
AO	Additional Observations
FSC	Faint Source Catalog
imag.	Sky Brightness Images
PSC	Point Source Catalog
SSC	Serendipitous Survey Catalog
SSS	Small-Scale Structure Catalog
zodi	Zodiacal History File
IUE	*International Ultraviolet Explorer*
NED	National Extragalactic Database
NSSDC	National Space Science Data Center
OAO-2	*Orbiting Astronomical Observatory 2 (Copernicus)*
Ohio RM	Ohio Radio Master Catalog
OSU	Ohio State University Master List of Nonstellar Astronomical Objects
Parkes	Parkes Catalogue of Radio Sources
PDB	Photon Data Base (energies, times, directions)
POSS	Palomar Observatory Sky Survey
RAFGL	Revised Air Force Geophysical Laboratory Catalog
RAL	Rutherford Appleton Laboratory
SAO	Smithsonian Astrophysical Observatory Star Catalog
StSci	Space Telescope Science Institute
Strasbourg	Center for Stellar Studies
SIMBAD	Set of Identifications, Measurements and Bibliography for Astronomical Data, Strasbourg

Table 4 Glossary of Abbreviations (continued)

Short Name	Full Name
TMSS	Two Micron Sky Survey
VLA	Very Large Array
Wisconsin	Wisconsin Soft X-ray Sky Survey
2C	Second Cambridge Radio Catalog
3C	Third Cambridge Radio Catalog
3CR	Revised Third Cambridge Radio Catalog
4C	Fourth Cambridge Radio Catalog
5 GHz	Catalog of Extragalactic Radio Sources Having Flux Densities Greater than 1 Jy at 5 GHz

Day, G.A., Shimmins, A.J., Ekers, R.D., and Cole, D.J. 1966 *Austr.J.Phys.* **19**:35. (Parkes Radio Survey 0° to +20°.)

Dixon, R.S. 1970 *Ap.J.Suppl.* **20**:1.

Dixon, R.S., and Sonneborn, G. 1980 *A Master List of Nonstellar Optical Astronomical Objects*, Ohio State University Press.

Edge, D.O., Shakeshaft, J.R., McAdam, W.B., Baldwin, J.E., and Archer, S. 1959 *Mem. R.A.S.* **68**:37.

Egret, D., and Wenger, M. 1988, in *Astronomy From Large Databases*, ed. F. Murtagh and A. Heck (Munich: ESO).

EXOSAT Observation Log, EXOSAT Observatory, ESOC, Darmstadt, FRG.

Forman, W., Jones, C., Cominsky, L., Julien, P., Murray, S., Peters, G., Tananbaum, H., and Giacconni, R. 1978 *Ap.J.Suppl.* **38**:357.

Gezari, D.Y., Schmitz, M. and Mead, J. 1987 *Catalog of Infrared Observations*, NASA Reference Publication 1196, Goddard Space Flight Center, Greenbelt, MD.

Gill, and Kapteyn 1895-1900 *Cape Ann.* **3-5**.

Gottlieb, D.M. 1978 *Ap.J.Suppl.* **38**:287.

Gower, J.F.R., Scott, P.F., and Wills, D. 1967 *Mem. R.A.S.* **71**:49.

Griesen, E.W., and Harten, R.H. 1981 *Astr.Ap.Suppl.* **44**:371.

Harten, R.H., Grosbol, P., Greisen, E.W., and Wells, D.C. 1988a *Astr.Ap.Suppl.*, in press.

Harten, R.H., Grosbol, P., Greisen, E.W., and Wells, D.C. 1988b *Astr.Ap.Suppl.*, in press.

Heap, S.R. 1988, in *Astronomy From Large Databases*, ed. F. Murtagh and A. Heck (Munich: ESO).

Heiles, C. 1975 *Astr.Ap.Suppl.* **20**:37.

Helou, G., and Madore, B. 1988, in *Astronomy From Large Databases*, ed. F. Murtagh and A. Heck (Munich: ESO).

Houston, B.P., and Wolfendale, A.W. 1983 *Astr.Ap.* **126**:22.

IRAS Catalogs and Atlases, Explanatory Supplement 1988, edited by Beichman, C.A., Neugebauer, G., Habing, H.J., Clegg, P.E., and Chester, T.J., (Washington D.C.: U.S. Government Printing Office).

IRAS Small Scale Structure Catalog 1986, prepared by Helou, G., and Walker, D., (Washington D.C.: U.S. Government Printing Office).

IRAS Point Source Catalog 1986, Joint *IRAS* Science Working Group, (Washington D.C.: U.S. Government Printing Office).

Jenkner, H., Lasker, B.M., and McLean, B.J. 1988, in *Astronomy From Large Databases*, ed. F. Murtagh and A. Heck (Munich: ESO).

Kleinmann, S.G., Cutri, R.M., Young, E.T., Low, F.J., and Gillett, F.C. 1986 *Explanatory Supplement to the IRAS Serendipitous Survey Catalog* (Washington D.C.: U.S. Government Printing Office).

Kuhr, H., Witzel, A., Pauliny-Toth, I.I.K., and Nauber, U. 1981 *Astr.Ap.Suppl.* **45**:367.

Kustner 1903 *Bonner Durchmusterung des Nord. Himmels, zweite berichtigte Auflage*, Bonn Univ. Sternw.

Lauberts, A. 1982 *The ESO/Uppsala Survey of the ESO(B) Atlas*, European Southern Observatory.

Lee, O.J., Baldwin, R.J., and Hamlin, D.W. 1943 *Ann. Dearborn Obs.* **5**, Part 1A.

Lee, O.J., and Bartlett, T.J. 1944 *Ann. Dearborn Obs.* **5**, Part 1B.

Lee, O.J., Gore, G.D., and Bartlett, T.J. 1947 *Ann. Dearborn Obs.* **5**, Part 1C.

Levine, A.M., Lang, F.L., Lewin, W.H.G., Primini, F.A., Dobson, C.A., Doty, J.P., Hoffman, J.A., Howe, S.K., Scheepmaker, A., Wheaton, W.A., Matteson, J.L., Baity, W.A., Gruber, D.E., Knight, F.K., Nolan, P.L., Pelling, R.M., Rothschild, R.E., and Peterson, L.E. 1984 *Ap.J.Suppl.* **54**:581.

Lund, J.M., and Dixon, R.S. 1973 *P.A.S.P.* **85**:230.

McCammon, D., Barrows, D.N., Sanders, W.T., and Kraushaar, W.L. 1983 *Ap.J.* **269**:107.

MacGillivray, H.T., Dodd, R.J., and Beard, S.M. 1988, in *Astronomy From Large Databases*, ed. F. Murtagh and A. Heck (Munich: ESO).

McHardy, I.M., Lawrence, A. Pye, J.P., and Pounds, K.A. 1981 *M.N.R.A.S.* **197**:893

Markert, T.H., Winkler, P.F., Laird, F.N., Clark, G.W., Hearn, D.R., Sprott, G.F., Li, F.K., Bradt, H.V., Lewin, W.H.G., and Schnopper, H.W. 1979 *Ap.J.Suppl.* **39**:573.

Marshall, F.E., Boldt, E.A., Holt, S.S., Mushotzky, R.F., Pravdo, S.H., Rothschild, R.E., and Serlemitsos, P.J. 1979 *Ap.J.Suppl.* **40**:657.

Meade, M.R., and Code, A.D. 1980 *Ap.J.Suppl.* **42**:283.

Mayer-Hasselwander, H.A. 1985 *Explanatory Supplement to the COS-B Final Data Base*, obtainable from K. Bennett, Space Science Dept. of ESA, ESTEC, Noordwijk, The Netherlands.

Murthy, P.V.R., and Wolfendale, A.W. 1986 *Gamma-Ray Astronomy*, Cambridge University Press, Cambridge, England.

Neugebauer, G. 1986 *Proc. Am. Phil. Soc.* **130**:155.

Neugebauer, G., and Leighton, R. 1969 NASA SP-3047.

Nugent, J.J., Jensen, K.A., Nousek, J.A., Garmire, G.P., Mason, K.O., Walter, F.M., Bowyer, C.S., Stern, R.A., and Riegler, G.R. 1983 *Ap.J.Suppl.* **51**:1.

Ochsenbein, F. 1988, in *Astronomy From Large Databases*, ed. F. Murtagh and A. Heck (Munich: ESO).

Pilkington, J.D.H., and Scott, P.F. 1965 *Mem. R.A.S.* **69**:183.

Price, R.M., and Milne, D.K. 1965 *Austr.J.Phys.* **18**:329. (Parkes Radio Survey -60° to -90°.)

Price, S.D., and Murdock, T.L. 1983 *The Revised AFGL Infrared Sky Survey Catalog*, Air Force Geophysical Laboratory, AFGL-TR- 83-0161.

Schoenfeld 1886 *Astron. Beob.* **8**, Part IV.

Seward, F.D., and MacDonald, A. 1986 *Einstein Catalog of Observations*, 5th edition.

Schmidt 1968 *BD* vierten Auflage, Ferd. Dummlers Verlag, Bonn.

Shakeshaft, J.R., Ryle, M., Baldwin, J.E., Elsmore, B., and Thomson, J.H. 1955 *Mem. R.A.S.* **67**:186.

Shimmins, A.J., and Day, G.A. 1968 *Austr.J.Phys.* **21**:377. (Parkes Radio Survey +20° to +27°.)

Shimmins, A.J., Day, G.A., Ekers, R.D., and Cole, D.J. 1966 *Austr.J.Phys.* **19**:837. (Parkes Radio Survey 0° to -20°.)

Smithsonian Astrophysical Observatory Star Catalog (4 vols.) 1966 Washington, D.C.: Smithsonian Institution, revised ADC 1984.

Squibb, G.F., and Cheung, C.Y. 1988, in *Astronomy From Large Databases*, ed. F. Murtagh and A. Heck (Munich: ESO).

Swanenberg, B.N., Bennett, K., Bignami, C.F., Bucchers, R., Carareo, P., Hermson, W., Kanbach, G., Lichti, G.G., Masnou, J.L., Mayer-Hasselwander, H.A., Paul, J.A., Sacco, B., Scarsi, L., and Wills, R.D. 1981 *Ap.J.(Letters)* **243**:L69.

Thome 1892-1932 *Resultados del Obs. Nac. Argentino* **16,17,18,21**.

Warwick, R.S., Marshall, N., Fraswer, G.W., Watson, M.G., Lawrence, A., Page, L.G., Pounds, K.A., Ricketts, M.J., Sims, M.R., and Smith, A. 1981 *M.N.R.A.S.* **197**:865.

Wells, D.C., Griesen, E.W., and Harten, R.H. 1981 *Astr.Ap.Suppl.* **44**:363.

Wesselius, P.R., van Duinen, R.J., de Jonge, A.R.W., Aulders, J.W.G., Luinge, W., and Wildeman, K.J. 1982 *Astr.Ap.Suppl.* **49**:427.

West, R.M. 1974 *ESO Bull.* **10**:25.

White, N.E. 1988, in *Astronomy From Large Databases*, ed. F. Murtagh and A. Heck (Munich: ESO).

Wood, K.S., Meekins, J.F., Yentis, D.J., Smather, H.W., McNutt, D.P., Bleach, R.D., Byram, E.T., Chubb, T.A., Friedman, H., and Meidav, M. 1984 *Ap.J.Suppl.* **56**:507.

Young, E.T., Neugebauer, G., Kopan, E.L., Benson, R.D., Conrow, T.P., Rice, W.L., and Gregorich, D.T. 1985 *A User's Guide to IRAS Pointed Observation Products*, IPAC preprint PRE-008N.

Multiwavelength Observing Today

France A. Córdova

Earth and Space Sciences Division

Los Alamos National Laboratory

1. Introduction

Previous chapters demonstrate that coordinated, multiwavelength observations of many types of astrophysical objects are of great scientific value. This chapter deals with the methods of multiwavelength observing. In Section 2, I discuss the capabilities of the facilities that are most commonly used for correlative observations. Some of these are active observatories of the late 1980s; others are inactive but still provide large archival data bases for use in multiwavelength comparative studies (see also the previous chapter by T. Chester). In Section 3, I look at a few examples of multiwavelength observing campaigns: the organization of these campaigns, their rate of success, and the difficulties incurred. In Section 4, I summarize suggestions for improving such observations, based on the experience from coordinated observations. The recommendations include improving the organization of campaigns, facilitating communications, modifying current proposal reviewing and scheduling procedures, increasing the funding for data analysis and the coordination of space-ground efforts, expanding remote and service observing, dedicating telescopes or partial observing time for coordinated observations, building space platforms with multiwavelength capability, and adopting a high Earth orbit for space missions. Many of these recommendations were proposed by participants in the 1987 Workshop on "Multiwavelength Astrophysics" in Taos, New Mexico, and the subsequent 1987 Colloquium on "Coordination of Observing Projects" in Strasbourg, France. Some similar resolutions on the particular aspect of multiwavelength data analysis were adopted by participants in NASA's 1987 Astrophysical Data Systems Workshops. This Section leads naturally into the next chapter by G. Kriss, which reviews proposed future space instrumentation that promise a greater return for coordinated, multiwavelength observations.

2. Capabilities of the Observing Instruments

There are almost no truly multiwavelength facilities operating today, the *Solar Maximum Mission (SMM)* being an outstanding exception. Its capability extends from gamma-rays (100 MeV) to the optical ($\sim$ 6600 Å). Its primary objective is to study solar flares and flare-related phenomena. The National Aeronautics and Space Administration (NASA) has a guest observer program to augment the programs of the *SMM* experiment teams. Other orbiting facilities with some multiwavelength aspects are the *International Ultraviolet Explorer (IUE)* which has an optical sensor, and the Japanese *Ginga* satellite, whose Gamma-Ray Experiment has both X-ray and gamma-ray detectors to sample gamma-ray bursts. While many ground-based telescopes can observe over a broad wavelength range (e.g., infrared/optical, or millimeter/submillimeter), strictly simultaneous observations in different bands usually cannot be made.

The present usefulness of particular wavebands for multiwavelength coordination varies greatly. Here follows a short synopsis of capability by wavelength, as of 1988 January.

Gamma-rays

For ultrahigh-energy observations above 10^{14} eV large air shower arrays are used. Examples are the Akeno, Baksan, Cygnus, and Haverah Park arrays in Japan, the Soviet Union, the U.S., and the United Kingdom, respectively. These detectors continuously survey the sky $\pm 40°$ off the zenith so that prior notice of a campaign is not usually necessary. The sensitivity of such experiments will improve with the construction of much larger arrays, such as the *ANI* project on Mr. Aragats in Armenia and the University of Chicago array around the Fly's Eye in Utah. For very high-energy observations ($10^{11} - 10^{13}$ eV), atmospheric Cherenkov telescopes, which have acceptance angles of 1–2 degrees, are used; these telescopes can operate only in the absence of moonlight and they do require multiwavelength coordination. Examples of such receivers are at Narrabri in Australia, Haleakala on Maui, the Whipple Observatory in Arizona, and the Crimean Astrophysical Observatory. Thus far, only a few X-ray binaries and radio pulsars, as well as the Crab Nebula and Cen A, may have been detected above 10^{11} eV.

In the 1970s, *SAS 2* and *COS B* made surveys of part of the gamma-ray sky below 100 GeV. The latest *COS B* catalog contains 25 sources between 100 MeV and 3 GeV, of which only a few have secure identifications (Swanenburg *et al.* 1981). Some of these sources are isolated neutron stars (e.g., the Crab and Vela pulsars and possibly Geminga, first discovered with *SAS 2*); others are molecular cloud complexes. In the late 1980s there is no gamma-ray observing capability in this wavelength region, a situation that will change with the launch of *GRO*, the *Gamma-ray Observatory*, in about 1990.

X-rays

Since the launch of the *Uhuru* satellite in 1971 December, X-ray observatories in space have provided opportunities for both comparative multiwavelength studies and coordinated observations of a large number of galactic and extragalactic sources. Much of the data from X-ray surveys and pointed observations have been archived, providing a rich data base that can be explored anew when discoveries in other wavebands are made. The problem is that most X-ray missions can be exploited only a small fraction of the time for coordinated observations because of difficulties in making a final schedule much in advance of an observation, restrictions on maneuvering the spacecraft due to consumption of attitude control gas, loss of observing time due to Earth occultations in low-Earth orbits, and sun-angle constraints which often require that simultaneous ground-based observations be done through large air masses. A few coordinated observations were made with the *High Energy Astronomical Observatories* (the *HEAO* series, all launched in the last half of the 1970s). These were principally of active galactic nuclei and variable stars like dwarf novae and flare stars. The chief benefit of *HEAO 2* (the *Einstein Observatory*) to multiwavelength astrophysics was its sensitive surveys of whole classes of objects, like main sequence stars, normal galaxies, and radio pulsars, for example, which could be used to compare with the properties of these classes in other wavelengths. This led to a much better understanding of, among other things, the physics of stellar coronae, cluster gas dynamics, and hot, isolated neutron stars.

It was the European Space Agency's *EXOSAT* Observatory, operating between 1983 and 1986, that was really the X-ray workhorse of coordinated multiwavelength campaigns. A review of its prolific use in coordinated observations is given by White and Barr (1987). *EXOSAT's* capability for simultaneous observations was greatly enhanced by its 90 hour, widely elliptical orbit (apogee 190,000 km, perigee 350 km), which allowed continuous observing for long periods and little particle contamination of the data. *EXOSAT* made time-critical observations of about 50 sources, including novae, X-ray transients, and active galactic nuclei during their active states; some of these were targets of opportunity. Out of a total of 1643 total *EXOSAT* observations, 503 were coordinated with other astronomical facilities, or 31%. Most of the *EXOSAT* coordinated observations were with optical observatories or the *IUE*; a small number were with radio, infrared, and other facilities, including *IRAS*, *Tenma*, *Voyager*, balloon flights, and gravitational-wave detectors. The majority of multiwavelength observations (80%) were coordinated with only one other waveband, almost 20% with two or three other wavebands, and a few observations were coordinated with four other wavebands.

In the late 1980s only the Japanese satellite *Ginga* is available for guest observer use in the X-ray band (see Makino 1987 for details of the instrumentation). In the first 60 days of testing the satellite after its 1987 February 5 launch, 5 of 6 requests for participation in multiwavelength campaigns were honored (Makino 1987, private communication). The targets of observation were well-known, bright X-ray sources: the compact binaries 4U 1916-056 and Cygnus X-3, the RS CVn stars AR Lac and UX Ari, and the active galaxy NGC 4593. Further multiwavelength observations were approved for the first guest observer period. *Ginga*'s low-Earth orbit and slow maneuverability, however, will restrict coordinated observations using its Large Area Counter. In fact, observers are requested to schedule their multiwavelength observations six months in advance, in order to minimize the frequency of maneuvering. *Ginga*'s All Sky Monitor is being used to alert the astronomical community to some transient high-energy events for followup observations in other wavebands, and to monitor the brighter X-ray sources. However, it has a very low duty cycle (approximately one scan every 1 or 2 days) and low spatial resolution. The Gamma-ray burst detector on *Ginga* is designed to make simultaneous gamma-ray and X-ray measurements of gamma-ray bursts. The only other X-ray telescopes observing in the late 1980s is a package of four European and Soviet instruments on the Soviet *Mir* space station. Collectively termed the *Röntgen* observatory, it detects photons with various degrees of spatial and spectral resolution between 2 and 1300 keV. It is not a guest observer facility, and will probably only participate in worldwide campaigns of exceptional interest, such as that of Supernova 1987A.

Extreme Ultraviolet

In the late 1980s there is no EUV capability. A relatively unexplored band, its potential is great, as suggested by an EUV detector on *Apollo-Soyuz*, and surveys of various classes of objects using the low-energy X-ray detectors on *Einstein* and *EXOSAT* and the far UV spectrometers on the *Voyager* spacecrafts. These observations suggest that the EUV observatories of the future, such as NASA's *Extreme Ultraviolet Explorer* and the United Kingdom's *XUV* wide-field camera on *ROSAT*, will be utilized for fruitful multiwavelength studies of nearby AM Herculis variables and dwarf novae in outburst, the central stars of planetary nebulae, flare stars, hot stars, soft pulsars, white dwarfs, selected AGN, and, of course, the interstellar medium.

Ultraviolet

No international guest observatory has been used by as many observers, for such a wide range of astrophysical objects, as the *IUE*. Launched in 1978 January, it was still producing high-quality ultraviolet spectra more than ten years later, making it the longest-lived astronomical satellite. A series of papers in *Nature* (1978, Vol. **275**) describe the instrumentation, inflight performance, and surveys of major areas. A recent book details its significance in almost every field of astronomy (Kondo *et al.* 1987). As of 1988 January 1, over 60,000 spectra had been taken with the *IUE*. It has observed objects as bright as Venus (at magnitude -4), and as faint as 20th apparent magnitude. Its success in multiwavelength collaborations is owed to a number of factors: its wide simultaneous wavelength coverage with two spectrographs covering the ranges 1150 Å to 1950 Å and 1900 Å to 3200 Å; its reasonable temporal coverage for variable events (can take spectra within the same wavelength range less than 1 hour apart); continuous communication with the satellite, which allows astronomers to use the telescope in real time; and the cooperation by the *IUE* observatory staff in participating in coordinated campaigns and accommodating targets of opportunity such as novae and supernovae. The *IUE* project is also remarkable for the collaborations it engendered between astronomers on different continents, especially between European scientists using the VILSPA observing and reduction analysis facility and U.S. astronomers using the Goddard Space Flight Center facility. Almost every chapter in this book testifies to the importance of the *IUE* in multiwavelength observing in the 1980s. The major drawbacks of the *IUE* are its low sensitivity, which requires very long observations of faint targets, the long read-out time of its detectors, which make rapid time series observations impossible, and the small dynamic range of its vidicon detectors, which prohibit simultaneous UV continuum and emission-line observations of strong emission-line objects. Solar pointing constraints and a slow slewing time are other limiting factors.

The energy range of the *IUE* is extended into the far UV by the ultraviolet spectrometers (UVS) on *Voyager 1* and *2*. These detectors span the wavelength range from 500 Å to 1700 Å, and have been in operation even longer than *IUE* (i.e., since August and September of 1977). However, their usefulness for observations of extra solar system objects really dates from the Jupiter encounters (1979 March 5 for *Voyager 1* and 1979 July 9 for *Voyager 2*). An log of *Voyager* observations prepared by Carone, Holberg, and Polidan (1987) contains over 600 observations of about 300 different objects, all made between March 15, 1979 and December 31, 1985. The instruments on both *Voyagers* are expected to continue operating well into the 1990s. A combined total of 125 pointings per year is anticipated (Polidan, private communication). Most UVS observations have been of subluminous stars, O and B stars, active binaries, and diffuse emission. *Voyager* data is being increasingly used for coordinated observations. The *Voyager* UVS team, operating at the Lunar and Planetary Laboratory at the University of Arizona, allocated over 50% of the 1987/88 observing schedule to guest programs; this amounted to about 50 guest observations in one year (Polidan 1987, private communication). In 1987 alone there were 10 *IUE* programs that depended significantly on coordinated *Voyager* observations. The drawbacks to *Voyager* observations are that they must be scheduled at least three months in advance (thus targets of opportunity are not possible); that a high data rate mode of 4 s resolution is only possible a small fraction of the time (the most commonly available mode is 10 minute resolution, although this will improve after the Neptune encounter of 1989); and that a *Voyager* UVS is committed to observing only one object for any interval of time (there is no switching back and forth between targets). In addition, the spectrometers can only observe targets brighter than $\sim 2 \times 10^{-13}$ ergs cm^{-2} s^{-1} Å^{-1} at 1000 Å. R. Polidan says that this implies limiting V magnitudes of $\sim$16 for a magnetic, hot white dwarf, 12 for a cataclysmic variable or active galactic nuclei, $\sim$11 for an OB star, and $\sim$3.5 for A, F, G, K, or M stars (all assumed unreddened).

Optical

In principle, optical astronomy provides the greatest freedom and flexibility to the multiwavelength observer. When transient events occur, optical astronomers may be alerted at their telescopes through an informal network of colleagues, or more formally through an *IAU* (International Astronomical Union) *Circular*, now available through electronic mail. Unlike the case at space satellite operation centers, the observer at an optical telescope can immediately change his observing program to make coordinated measurements of a transient phenomenon. Telescope equipment alternatives are usually various and changes simple, adding to the flexibility. There is also a growing consciousness among astronomers about the potential value of simultaneous observations of transient phenomena, so that observers may relinquish some part of their programs to accommodate especially urgent requests. With the many optical telescopes available all over the world, it then seems peculiar that there is often so little coordinated coverage of transient events. In part this is due to a lack of rapid communication: It is difficult to find out quickly who is on what telescope and with what equipment. Many opportunities for coordination must be lost for lack of this information. Another problem is that a telescope user, alerted to news of an exciting transitory event, may not know what kind of observations would be valuable.

The more typical kind of coordinated observation made with optical telescopes are the scheduled, carefully planned campaigns, in which all of the participants share a common interest in the target(s) of observation. Some examples of these are considered in Section 3. For the large amounts of observing time required for these efforts, the 1 meter-class telescopes are used most often because they are less in demand, but yet are sensitive enough to observe the more well-known (i.e., brighter) members of classes of star. The most difficult type of coordinated observations are those made with satellites whose schedules are not fixed until weeks or days before the observations. Time allocations on the optical telescopes, on the other hand, are made months in advance. There is little room for flexibility in optical telescope scheduling because of the large numbers of users and their sometimes stringent observing requirements. It is therefore difficult to plan coordinated campaigns, and they are looked upon by the schedulers with anxiety. Ironically, because there are so many optical telescopes apparently available for coordinated observations, each with their own time-allocation committees, this is the band in which it is the most difficult to establish uniform policies for campaigns, (e.g., setting aside a fraction of observing time for coordinated, multiwavelength studies, or coordinating schedules with satellite observatories). Weather is the most significant constraint in making successful coordinated optical observations. This factor requires some redundancy in a campaign; it is important to schedule more than one telescope to ensure simultaneous coverage. This complicates requesting and obtaining telescope time: observatories are not eager to allocate valuable observing time for a contingency plan.

Infrared

The year 1983 saw the launch of the U.S./U.K./Netherlands *Infrared Astronomical Satellite (IRAS)* which provided such a large catalog of infrared sources in its survey of the entire sky at 12, 25, 60 and 100 μm (more than 200,000 sources in the *IRAS* Faint Source Catalog) that it will be useful for comparative multiwavelength studies for years. In fact, since it ceased operating, almost all guest observer usage of the data base is for correlative multiwavelength studies. Besides point sources, the data base can provide important comparisons of extended features such as supernova remnants (see the chapter in this book by R. Chevalier and F. Seward). During its operation, less than 1% of *IRAS* time was used for coordinated multiwavelength observations (Chester 1987, private communication). The ones that were made were chiefly of prominent extragalactic objects. Because of difficulties in scheduling *IRAS* these observations were usually not simultaneous, but done within days or weeks of each other.

The Kuiper Airborne Observatory is the principal, working, far-infrared (submicron to approximately 1 mm) facility. In any given year, more than a dozen instrument teams, and 15–20 guest observers, are awarded observing time on the basis of proposals submitted in response to a NASA Announcement of Opportunity (AO). There is no discretionary time, but a special AO can be released outside of the yearly proposal cycle, as was done for SN 1987A. The C-141 airplane is used for coordinated observations of novae, SN 1987A, AGN, and solar system objects, to name a few examples.

In the near-infrared (1 μm to 30 μm) a number of ground-based telescopes are being used in the late 1980s for coordinated observations. Among the largest of these are the 3 meter NASA/University of Hawaii Infrared Telescope Facility (IRTF) and the 3.8 meter United Kingdom Infrared Telescope (UKIRT), both operating above much of the Earth's infrared absorbing water vapor near the summit of 13,800 foot Mauna Kea in Hawaii. Both of these telescopes are used extensively by guest observers and have been used for coordinated observations with telescopes at all other wavelengths. UKIRT is exclusively a common user facility and provides service observing. Discretionary time may be requested for coordinated observations. In addition to these dedicated IR facilities, there are a large number of optical telescopes which can be used for infrared studies and these are often utilized in coordinated campaigns. The main problem is that IR observers using optical telescopes are usually allocated only bright time, thus constraining the scheduling of coordinated observations.

Submillimeter and Millimeter

Some sources, especially active galactic nuclei, have been observed with millimeter telescopes, particularly the NRAO 12 m on Kitt Peak, and the Owens Valley, California, 10.4 m interferometric array. In general, though, it is difficult to get coordinated millimeter observations, mainly because these telescopes are few in number and demand for them is great for observing star-forming regions. Until recently millimeter telescopes were also used to make modest submillimeter observations. Now, however, there are larger, higher-altitude dishes which have taken over the submillimeter domain. Two telescopes on Mauna Kea, the Caltech 10.4 m telescope and the 15 m U.K./Netherlands/Canada James Clerk Maxwell Telescope (JCMT), promise much more sensitive submillimeter measurements than heretofore possible with smaller-dish or lower-altitude telescopes. The JCMT operates between 0.35 and 2 mm. It has a service program in which sources can be monitored for extensive periods and discretionary time may be requested for targets-of-opportunity (see Section 4). Sweden and the European Southern Observatory also have a 15 m submillimeter dish at La Silla, Chile.

Radio

Nearly every radio telescope has been used at some time for comparative multiwavelength studies, e.g., for studies of the radio counterparts of X-ray sources, and many have been used in simultaneous observing campaigns. The National Radio Astronomy Observatory's Very Large Array (VLA) in New Mexico has been in full-time operation for guest observations since the fall of 1980. B. Clark (1987, private communication) offers the following information on its usage for multi-wavelength studies: For wide spectral coverage on nonvarying or slowly-varying objects, the various wavelengths it supports (i.e., 1.3, 2, 3.8, 6, 20, and 90 cm) can be interleaved during an observation, the time for switching taking only a minute. About one-third of VLA time is used in this manner. For rapidly varying objects (e.g. flare stars and the Sun), the VLA has observed simultaneously at 20 cm and 90 cm, using half the array's bandwidth at each band; or the array has been split into two or three pieces and each piece used at one of the six VLA bands. Probably not more than 2% of the time is used in this way to do simultaneous radio measurements. The VLA is used in coordination with observatories operating at other wavelengths: just under 1% of the time to observe stars together with *IUE*, and about the same percentage of time or a little more in coordination with *EX-OSAT* before it expired. An example of the VLA observing in conjunction with *Ginga* and an infrared telescope is given in the next Section. The VLA can and does award large blocks of time for particularly comprehensive multiwavelength programs. An example is a program to monitor flux changes in quasars and BL Lacertae objects from the near UV to 90 cm. This program in 1987 got a little less than half a percent observing time on the VLA. Using the VLA, quiescent radio emission has now been detected throughout the HR diagram (as reviewed by Bookbinder 1987). VLA surveys of classes of X-ray selected objects have also effectively eliminated some classes of stars as radio sources at the sensitivity limit of the VLA.

Telescopes across the U.S. and Europe have been used in Very Long Baseline Interferometry for campaigns where very high spatial resolution is desired (e.g., the time-varying jets of SS 433 and Cygnus X-3 during outbursts). The VLBI community is very interested in coordinating observations between "the two cultures," i.e., space and ground radio telescopes. Recognizing that the feasibility of launching space-based VLBI elements into Earth orbit is under investigation by space agencies around the world, and that the full scientific benefits of VLBI will result only from observations obtained through the combined and simultaneous use of space-based antennae with existing ground facilities, the 19th IAU General Assembly which met in Delhi in 1985 recommended that "the appropriate national and international authorities concerned with space and ground-based VLBI make every effort to coordinate in a timely way the contributions to this important international program."

The most utilized radio telescopes for coordinated observations are smaller, less sensitive ones in which a large fraction of time is used to continuously monitor a specific class of source. A prime example is the University of Michigan radio telescope, which is used chiefly for coordinated campaigns on active galactic nuclei and some X-ray binaries. These facilities come close to being "dedicated" telescopes, but they are dedicated to specific projects, and are not guest observatories.

3. Campaigns

As the name implies, a multiwavelength observing campaign is an organized effort involving a large number of observers and telescopes. Campaigns are necessitated by the desire to achieve measurements of different wavebands simultaneously, or nearly simultaneously. Such endeavors are made possible because of the availability of guest observer facilities in every waveband. Unsuccessful campaigns are not usually publicized because they bear few if any results; therefore the impression is that campaigns are more successful than they generally are. Prolific campaign organizers claim that anywhere between 1/3 to 2/3 of the campaigns they have managed have been successful enough, i.e., with some overlapping spectral coverage, to publish the results. Here I give examples of a few selected campaign efforts for several different kinds of objects, and mention some of the limitations campaign organizers have found.

The International Halley Watch (IHW)

The 1986 apparition of Comet Halley was sensational enough to engender the enthusiastic cooperation of observers and telescope schedulers alike, without the usual need to justify multi-facility observing. The actual Watch duration extended from 1982 October to early 1989. The IHW included eight observing nets (astrometry, infrared, large scale phenomena, near nucleus studies, photometry and polarimetry, radio studies, spectroscopy, amateur) which included 1000 scientists from 51 countries and 1187 amateurs in 54 countries. It was the largest campaign ever mounted. The IHW tried to standardize the ground-based observing techniques wherever possible and coordinate the observing; it was facilitated with a newsletter to participants. The chief aim of IHW, however, was to publish all the data in a comprehensive archive, thus avoiding the calamity of the 1910 apparition when the major monographs on Halley were not published until decades after the event, and much of the data were not published at all.

R. Newburn and J. Rahe led the Watch and NASA provided funding for the archive. In the eyes of the organizers one problem with the coordination effort was getting the ground-based telescopes to observe simultaneously with space facilities. The ground-based data were necessary because the comet's parameters changed rapidly and the space instruments were only gathering data over a limited time interval and with limited bandwidth. A particular problem was that the schedule of the Kuiper Airborne Observatory, which was important for gathering long-wavelength infrared data, was so uncertain that the time of the observations were not known until the last minute. This precipitated frequent, hurried attempts to try to change the schedules of the coordinating ground-based observatories. Another problem was getting observers to reduce their data and supply the results in a timely way for the published archive.

Hot Stars

In the interval 1984–1987, simultaneous *IUE, Voyager*, and ground-based spectroscopic, photometric, and polarimetric observations of Be star and B supergiants were made to search for links between nonradial pulsations and mass loss, among other objectives. A vehicle used successfully to generate interest and participants in these campaigns is the Be Star Newsletter, edited by G. Peters and distributed to about 300 persons by the European Southern Observatory. These campaigns were largely successful in achieving overlapping coverage. Some of the problems that incurred were that *Voyager* data were lost due to difficulties with guiding the spacecraft after only three hours of observing one of the targets, and that optical data acquired for another campaign were not forthcoming from the optical observer. Ground-based hot star observers find that time allocation committees are often not sympathetic to projects like this which require large amounts of observing time, and represent only a piece of a much larger project. Committees prefer, instead, self-contained proposals requiring small amounts of time to produce a publishable result. A concern of hot star enthusiasts is that telescope schedulers be able to respond to the episodic activity in "ordinary" stars the same way as they did for SN 1987A.

Cool, Active Stars

Some examples of extensive cool star campaigns are given by Córdova *et al.* (1985) and J. Linsky in this volume. Karpen *et al.* (1977) monitored the flare star YZ CMi using the *SAS 3* X-ray satellite, five optical observatories and seven radio observatories over a period of three days. Despite this extensive coverage, triply coincident X-ray, optical, and radio observations were obtained only for 30% of the campaign. Although 31 minor optical flares and 11 radio events were recorded, no X-ray flares were seen. Similarly extensive campaigns were conducted on YZ CMi and Proxima Cen using the *Einstein* X-ray satellite (Kahler *et al.* 1982; Haisch *et al.* 1981). The Kahler campaign involved 31 participants representing 20 institutions. A flare from YZ CMi was detected in all three wavebands, but an X-ray flare from Proxima Cen was not coincident with the numerous optical and radio flares detected. These results indicate the complexity of the flaring phenomenon and illustrate a wide range of flaring behavior in these stars, suggesting that further multiwavelength campaigns are appropriate. Continuous *EXOSAT* and VLA observations of W UMa stars over two of their 0.25 day orbital periods demonstrate that flares can occur simultaneously in both wavebands in contact binaries also (Vilhu *et al.* 1987). The VLA's increased sensitivity compared with forerunner radio telescopes, and *EXOSAT's* unique capability for continuous X-ray coverage over days, made such a detection possible.

X-ray Bursters

Several international X-ray/optical observing campaigns have been undertaken to search for optical counterparts of X-ray bursts, and to determine properties about the environment of the X-ray source. A typical effort was the joint US-European-Japanese observation of 4U 1636-53 during 1979. In two months of observing, only five bursts were observed with overlapping X-ray and optical coverage by the 29 collaborators on the project (Pedersen *et al.* 1982). More examples of burst campaigns are given in the chapters by D. Hartmann and S. Woosely, and by K. Mason.

The biggest difficulty in coordinating burst coverage is that bursts are not predictable. A ground-based observer can wait for days for an optical burst counterpart and not detect anything; alternatively, several bursts can occur in the space of hours. For such long runs, weather, too, becomes important. Perhaps the most significant fact that burst campaign organizers learned through trial-and-error was the necessity of using large telescopes of at least 1.5 m: the optical burst flux is typically $\sim 10^{-4}$ of the X-ray burst flux and the X-ray bursts themselves are not bright because they are viewed at distances of $\sim$10 kpc. For example, the persistent B mag of the X-ray burster 1636–53 is 18.5 and the maximum B mag during a burst is 17.5. Position-sensitive, photon-counting detectors might improve the situation by making it possible to detect burst counterparts with smaller telescopes.

Cataclysmic Variables

In recent years a large number of observations of these binary systems have been done in a multiwavelength mode. This is because the sources are extremely variable over the entire spectrum of their emission, which extends from X-ray to infrared wavelengths (or to radio wavelengths for a few objects). The variability in different wavebands is not simply correlated, making simultaneous observations important. The *IUE* has been used extensively with ground-based optical and infrared telescopes, and *Einstein* and *EXOSAT* to make coordinated observations. CV campaign organizers claim that weather is the main cause of lack of success, with scheduling conflicts between satellites and ground-based telescopes the second most significant reason for difficulties. The latter was usually due to the object being observable only at high air mass or during daytime. Successes were due to active participation, usually on the part of a graduate student who was willing to go to any telescope at any time to do coordinated observations with orbiting telescopes. The frequency with which such observations are made ensure that discoveries are made, in spite of many failures – and sometimes because of them! The rescheduling of an observation because of satellite breakdown has more than once led to more successful multiwavelength coverage by all participating facilities during the rescheduled episode. This emphasizes the serendipity involved in obtaining in coordinated coverage.

One of the most well-organized and thorough multiwavelength campaigns on a dwarf nova, which was observed during both quiescent and outburst states, was the 70 day coverage in 1984 of VW Hydri by *EXOSAT, Voyager, IUE,* and optical telescopes. The *Voyager* data were the sparsest, owing to ground station unavailability for part of the run and lack of sensitivity for detection when the source was in a low state. *EXOSAT* observations were somewhat limited by Earth occultations of the source. The great success of this campaign was due to 1.5 years of planning. For example, the *Voyager* team prepared for the multi-facility run by mounting a smaller, "pre-campaign" consisting of optical and *Voyager* observations of the star in 1983. This experience was invaluable for determining the capabilities of the *Voyager* spectrometers for observing VW Hyi. Frequent communication between participants, as well as good scientific justification for the large amount of time involved, also benefited the campaign. Some of the participants were resident astronomers at the observatories involved, and thus could readily deal with technical problems as they arose. The results of the VW Hyi campaign are described in the chapter on cataclysmic variables by P. Szkody and M. Cropper.

X-ray Binaries with Resolved Radio Components

Cygnus X-3, thought to be a neutron star accreting matter from a small, normal companion star, was discovered in an X-ray rocket flight in 1966. News of a large radio outburst in 1972 was communicated chiefly by telephone to virtually all other major observatories. The outburst, which lasted for several days, was subsequently covered at X-ray wavelengths with the *Copernicus* and *Vela 5B* satellites, in the optical and infrared using the Palomar 5 m telescope, and in the radio by observers worldwide. Ten years later, with the advent of large, sensitive guest observer facilities covering virtually every wavelength, Cyg X-3 was again observed simultaneously in the X-ray, infrared, and radio, this time with much higher signal-to-noise ratio and increased timing resolution. (Optical detection has not been made due to high interstellar extinction). During multiwavelength runs on Cyg X-3 using *EXOSAT*, the IRTF, and the VLA, failure of *EXOSAT's* onboard computer limited X-ray observing time on one occasion; inclement weather reduced IR coverage on another occasion; and viewing constraints (i.e., the source setting early at the VLA site, but still easily viewed with the X-ray and IR telescopes) further limited the multiwavelength overlap. A two day run in 1984 produced only 5.4 hours of simultaneous X-ray and IR coverage (Mason, Córdova, and White 1986). A 1987 August coordinated multiwavelength observation employed the IRTF, the VLA, and *Ginga* with overlap for much of 3 ten-hour intervals on consecutive days. The times of non-overlap were due to (1) the launch of another satellite on one of the observing days, which required use of all ground tracking stations and prohibited recording *Ginga* data for several hours, (2) obtaining VLA time for only two of the three days that IRTF and *Ginga* times were acquired, and (3) problems with the IRTF instrumentation which hampered infrared observing for about half the time. During a large radio outburst in the Fall of 1985, VLBI was also used concurrently with X-ray, radio, and infrared telescopes. With VLBI time-variable, resolved emission was detected which could be interpreted as a double-lobed jet with a speed of 0.3c.

Similarly high velocity jets have been detected with radio observations of SS 433. A recent ambitious multiwavelength campaign on this object was made using VLBI, MERLIN (Multi-Element Radio Linked Interferometer Network), and several optical telescopes worldwide doing spectroscopy and photometry, (i.e., Calar Alto 1.2 m and 2.2 m, La Palma 2.5 m, La Silla 0.9 m and 1.5 m, Flagstaff 1.8 m, and others); the VLA; the Greenbank Interferometer, and radio telescopes in Bologna, Westerbork, Australian (MOST), and the U.S.S.R. (RATAN); and *Ginga* for X-ray observations. This campaign, lasting for 18 days in May/June of 1987, was extremely successful in achieving overlapping coverage, chiefly due to the coordination efforts of R. Vermeulen and R. Schilizzi.

Active Galactic Nuclei

For violently variable quasars and BL Lacertae objects, any one wavelength band does not contain enough diagnostic information to determine the conditions in the emitting plasma and constrain models. To achieve these goals observations are obtained in wavebands from the radio through gamma-rays at nearly the same time. Flux variations in the X-ray band can occur on timescales less than one day. The most common AGN multiwavelength observations are simultaneous single epoch spectra from the radio through the ultraviolet (a few dozen such spectra have been taken), although usually without data in the important submillimeter region (20–1000 μm). About a dozen of these observations also include X-ray data and a few (< 10) have a time series of multiwavelength spectra. J. Bregman, who organized about 15–20 AGN campaigns in the interval 1979 October to 1985 December, communicates the following details. The radio data in the 1–24 GHz region were the easiest to obtain, thanks to the efforts of H. and M. Aller who use the University of Michigan Radio Observatory to collaborate with a wide variety of groups successfully. Radio data were always taken within a couple of days of the target date. Optical data were easily obtained, owing to the wide number of optical telescopes available and the willingness of observers to participate in active galactic nuclei studies. Infrared telescopes and observers are fewer and their schedules (i.e., bright time) often conflicts with the dark time monitoring programs of the optical observers. IR data were obtained within a day or two of the target date. Very few groups can obtain far infrared and submillimeter measurements, so that measurements were made sometimes only within a month of the target date. For lack of enough telescopes and interest, the millimeter band is rarely used for AGN studies but occasionally the NRAO 12 m telescope is used to obtain simultaneous observations within a couple of weeks of the target date. X-ray, UV, and optical measurements were generally made simultaneously on the target date. The limiting instrument in sensitivity was the *IUE*. The primary difficulties in coordination were due to the pointing constraints of the spacecraft which were often very inconvenient for ground-based observers. For campaigns in which the *Einstein* X-ray satellite was involved, the targets frequently had to be observed at large air masses. Despite these problems, the AGN campaigns of Bregman and his collaborators were about 2/3 successful in obtaining overlapping, multispectral coverage.

4. Recommendations

As the examples of the previous Section implied, there is room for improving multiwavelength observations. The following suggestions come from participants in the the two 1987 conferences on multiwavelength observations that were mentioned in the beginning of this chapter, from participants in a special session on multiwavelength astrophysics at the 1988 January meeting of the American Astronomical Society, and from individuals cited in the acknowledgements below. The recommendations are posed in the context of the biggest problems encountered in trying to coordinate campaigns.

Weather

This is the largest limiting factor with ground-based observations, especially in the optical and infrared bands. One way to get around it (as well as to improve image quality) is to put the telescope in Space; for example, *SIRTF* for the infrared, or the *Hubble Space Telescope* for the optical and near-UV. Coordination between these and other large Space Observatories operating at higher frequencies (i.e., *AXAF* for X-rays and *GRO* for gamma-rays) is an exciting idea in principle, but is frought with scheduling headaches because of the different orbits of these,

NASA's "Great Observatories" of the 1990s, and because of the competition for observing time from other programs which do not require coordination between satellites. For the immediate future at least, such facilities will be oversubscribed and it will be difficult to obtain the requisite large blocks of time for coordinated campaigns. A more mundane but less costly method of mitigating against weather-related failures of campaigns is to make sure that there is redundancy in ground-based optical/IR telescopes sites scheduled for any campaign. This, of course, further burdens the already-complicated logistics of organizing the campaign and the difficulty of obtaining yet more interested, willing participants.

Communications

Planning coordinated campaigns is rapidly improving because of the facility of electronic mail. For example, the *IAU Circulars* inform subscribers of new super-novae, comets, and the outbursts of symbiotic stars, recurrent novae, and X-ray transients, to name some examples. Until recently the *IAU Circulars*, which are meant to offer speedy communication, often reached observers too late to make use of the information because of the reliance on the post to transmit the Circulars. Now, however, they are available through electronic mail, greatly speeding up the communication of transient events. Unfortunately, groups at the various observatories, universities, and laboratories, do not all have network capability, and certainly not all are using the same networks. Also, some networks are unreliable so that messages get lost or returned to the sender. This means that even events that are exceptionally dramatic, such as SN 1987A, are not communicated rapidly to all those who could join the observing effort. Every effort should be made to support, improve and extend the best of the existing networks to be universally accessible and reliable. A recommendation of the Taos workshop was the startup of an electronic mail "hotline" which would share news of interesting transient events among those with rapid access to ground-based and orbiting telescopes. Such a hotline could be used to communicate far more news (e.g., observing program details and quick results, as well as electronic mail addresses of observers), than through an *IAU Circular* alone. The excitement over SN 1987A engendered the formation of a SN 1987A electronic Bulletin Board at the Space Telescope Science Institute in Baltimore, Maryland. The bulletin board offers a forum for the exchange of ideas and information, access to relevant *IAU Circulars*, a bibliography of preprints and published articles, a list of planned observing campaigns and their goals, and a directory of investigators including their postal and electronic mail addresses.

A different recommendation for improving communications for multiwavelength campaigns, this one from observatory schedulers, is that the same person is principal investigator on all telescopes for any given multiwavelength program.

Organization/Coordination

Multiwavelength campaign organizers agree that starting far ahead of time, even as much as a year, helps to make a successful program. Having a sound and scientifically interesting program is of great importance in attracting people and support. Keeping the number of participants as small as possible, and limited to those with a keen interest and good knowledge of the subject is important. Getting the participants together informally to discuss goals and methods is necessary. For the large campaigns requiring only small amounts of time from astronomers who do not know much about the objects they are viewing, it helps to choose targets that are among the most well-known members of their class. It is also valuable for the organizers to understand and accommodate the various restrictions relevant to different wavebands and individual telescopes. Some considerations are: (1) satellite targets are often about six hours from the zenith, putting the objects at high air mass, and optical telescopes have different air mass constraints owing to smog and city lights, among other things; (2) infrared observers are almost never scheduled dark time on an optical/IR telescope; therefore, considering the position of the Moon is important for faint targets; (3) casual collaborators would rather observe the multiwavelength targets during the second half of the night (i.e., in the West), after their own observing program is done. All of these examples mean choosing observing dates carefully when organizing a coordinated effort.

Proposal Review and Scheduling

A headache from both the campaign organizer's and the telescope scheduler's points of view, scheduling is the area which generates the most opinions about improvement. The organizer must submit his proposal several months in advance of the observations to a half-dozen or more different review committees, each with different proposal deadlines and schedules that are often out of phase. The organizer must try to determine the specific time period that meets all the constraints of all of the participating observatories, and then justify the request for this time period to all the committees. It is particularly difficult to get time scheduled in advance on satellites, whose final schedules are made usually only a day or two before the observation.

From the point of view of the schedulers, White and Barr (1987) comment that coordinating observations of the *EXOSAT* satellite with other facilities proved to be the most important constraint on scheduling *EXOSAT*, and posed the biggest challenge to mission planning. Sometimes different coordinated programs counted on the same *EXOSAT* time, resulting in conflicting scheduling. At other times a coordinated program would be denied time by the *EXOSAT* review committee but approved by the review committees of the other wavebands, causing much pressure to be put on the *EXOSAT* project scientist. The reverse also happened: *EXOSAT* time was approved for a coordinated program which in the end was denied coverage at other wavebands. Some programs had *IUE* and ground-based time scheduled out of phase, yet both programs relied on simultaneous *EXOSAT* coverage! One attempt at resolving some of these problems was made in 1984 December when schedulers of *EXOSAT*, *IUE*, and *ESO* met at Villafranca, Spain and reached an agreement for more joint consultation before the production of timelines. An understanding was also reached with the VLA scheduler. Although this helped to optimize some observing schedules, many ground-based telescopes in other countries were not involved and the chain of communication was still unsatisfactory. Thus a recommendation would be for all major observing facilities to make their proposal deadlines and scheduling intervals compatible. Another recommendation, already in effect at some observatories, is for more frequent (e.g., quarterly) proposal reviews; this would give more flexibility in designing campaigns. It also, however, puts more of a burden on time allocation committees. A further recommendation is for review committees to award time contingent on the success of the proposers in obtaining time in the other wavebands of interest.

Many campaign organizers are frustrated by having to submit as many as ten observing proposals to do a coordinated observation, calling the approval process "long and tortuous" and the scheduling process "an exhausting ordeal." They would prefer to submit one multiwavelength proposal for review, and not to have to interact with ten different schedulers. A strong recommendation of the 1987 Taos Workshop on "Multiwavelength Astrophysics" was the establishment of a committee that would screen all multiwavelength observing proposals and then pass on its evaluations to the relevant time-assignment committees. The multiwavelength committee would have worldwide representation of astronomers familiar with multiwavelength issues, and representation from all the major observatories. This committee would not have power to allocate telescope time; it would act in an advisory capacity only. Its main functions would be (1) to assist observers (e.g., by providing a directory of observers and observatory names interested in supporting multiwavelength observations, or by analyzing campaign plans and perhaps offering recommendations for their improvement), and (2) to assist time-allocation committees by evaluating proposals that might be beyond their specific area of expertise. An alternate recommendation, this from the 1987 Colloquium on the "Coordination of Observing Projects," was to set up an IAU working group to discuss needs for coordination and exchange of information among schedulers following proposal screening at the level of the individual time-allocation committees.

Obtaining Simultaneous Multiwavelength Coverage

One solution for achieving simultaneous coverage, proposed by Cordova *et al.* (1985), is adding a small, imaging optical/UV monitor to high-energy satellite observatories. Such an instrument could monitor not only the multiwavelength variability of all the active objects discussed in previous chapters, but also could provide astrometry, broadband colors, low-resolution spectroscopy, and imaging of constant sources and fortuitously observed field objects. Córdova *et al.* (1985) note that the concept of providing multiwavelength simultaneous coverage of astrophysical objects in an unbiased way allows new phenomena to be discovered. Wisniewski (1988) points out that the scientific loss from non-overlapping coverage is significant: there are extensive synoptic observations of many sources in, for example, the radio band, but no accompanying optical or other wavelength coverage. The optical Fine Error Sensor on the *IUE* is a successful example of extending the wavelength coverage of a spacecraft. An imaging optical photometer is being seriously considered for the upcoming European Space Agency's *X-ray Multi-Mirror Mission (XMM)*. Several of the proposed U.S. Explorer Class missions for the 1990s would carry a platform with instruments of different wavelength capability, but with comparative sensitivity (see the following chapter for details).

An alternative possibility is to launch a high-energy observatory with sensitive imaging capability and an extremely wide field of view so that most of the nighttime sky would be simultaneously observed. This would eliminate the need for an optical monitor on the spacecraft because simultaneous high-energy data would automatically be available any time a target within the field of view is observed from the ground. An X-ray telescope that would simultaneously image 77% of the night sky has already been designed and is further discussed in the next chapter.

The latter solution would be facilitated by dedicating existing medium-aperture ground-based optical/IR telescopes to coordinated observations. In fact, all satellite observatories would benefit from this. A 1 meter telescope can easily observe an object with an apparent magnitude of 20, given the present state-of-the art CCD and infrared imaging array detectors. The proposal process for these dedicated telescopes would be linked to the orbiting facilities. Wisniewski (1988) advocates a more daring plan: the construction of several 2.5 m optical/IR telescopes to support the large number of new space observatories proposed for the 1990s. Six telescopes, at equally-space longitudes, would be constructed in the northern and southern hemispheres. The cost of such a project, he says, is less than a fraction of one space mission. Wisniewski further recommends that all ground-based telescope dedicated to multiwavelength observing should be implemented with similar photometric and spectroscopic equipment for uniformity of the data.

Another suggestion is to expand remote and service operations on ground-based telescopes because it is difficult for one group to provide the manpower and expertise for all the facilities required for extensive coordinated coverage. The United Kingdom has had a service program in place on its Hawaii telescopes for some time (Smith 1988, private communication). Applications for service observing on UKIRT and the JCMT can be sent through electronic mail to the Royal Observatory in Edinborough, Scotland, and are reviewed every 2–3 months. Up to 18 nights a year can be reserved for service observing on either telescope. This, combined with 10–15% discretionary time on the part of the telescope schedulers, gives flexibility for observing targets of opportunity or doing long-term monitoring of sources. The discretionary time is sometimes used for coordination with space satellite programs which may have sudden agenda.

Adopting an *EXOSAT*-type orbit for all spacecraft would increase overlapping coverage with ground-based observations significantly. Earth-occultation time would then be a less significant part of an orbit, resulting in much longer, continuous coverage of targets. Many important discoveries concerning low-mass X-ray binaries were missed with low-Earth orbit X-ray satellites. Earth occultations lasting for approximately one-half the satellite orbit disguised phenomena occurring on the time scale of the satellite orbital period, which coincidentally is the time scale of the binary periods of the X-ray sources. A geosynchronous orbit is not good enough for X-ray satellites because of the high particle background. A higher-Earth orbit (e.g., 4 days) is preferable, and also allows direct communication with the satellite, allowing fast response to commands for time-critical viewing.

Data Analysis

The ways in which the large amounts of data being produced by orbiting facilities are being processed and made available to guest observers is discussed in the previous chapter by T. Chester. A recommendation from the Taos workshop is that the emphasis should shift from "hard" products like tapes and printed handbooks to electronic products. Electronic networks (such as SPAN, STARLINK, and ASTRONET) are favored for general data analysis because the observer is then able to do data reduction at his home institution. If data bases can be accessed on computer networks, there is little need for data analysis centers that have many data bases, although it is recognized that very detailed analysis of any particular data base will have to be done at the institution which has "curatorship" of the data (e.g., as IPAC has for *IRAS* data, and ESTEC has for *EXOSAT* data). A common software environment is favored for reduction of data from future orbiting observatories (e.g., the Image Reduction and Analysis Facility, IRAF).

The European Space Information System (ESIS) will go a long way towards facilitating data analysis. It will use a distributed archive system which allows users to access archived data bases (like those of *IUE*, *EXOSAT*, and Simbad) using an efficient computer network. A central catalog will contain references and descriptions of the data archived in the remote centers. A Pilot Project is in progress.

Also addressing the problem of the coordination of large data bases was a conference on "Astronomy from Large data bases: Scientific Objectives and Methodological Approaches," sponsored by the Space Telescope-European Coordination Facility in Garching, W. Germany in 1987 October, and two workshops on Astrophysical Data Systems (ADS), sponsored by NASA, and taking place in 1987 August and 1987 November. These conferences were concerned with the access to and dissemination of astrophysical data. The NASA ADS workshops result from this agency's intent to support guest observers to maximally utilize the data from the upcoming Great Observatories. The ADS plans for the future are summarized in the previous chapter.

Funding

For guest-supported facilities, the use of telescopes is funded by the participating observatories and their sponsors. The home institutions of the observers, as well as the observers' research grants must absorb the costs of staff time, travel and computing. Some potentially excellent multiwavelength efforts have been hampered due to lack of funding for staff, guest observers, or computing time to analyze the data. A case in point is the funding of *Voyager* UVS (Ultraviolet Spectrometer) observations and data analysis. The *Voyagers* are planetary spacecraft, but the UVS are doing astrophysics. Thus there is an apparent conflict in NASA about whether the UVS operations should be funded by the Planetary or Astrophysics Divisions, with the result that neither division allocates enough money to expand the present limited operations. The U.S. National Science Foundation does not contribute much funding because it holds that NASA should be funding the active missions. The *Voyager* UVS team contends that were it not for the *IUE* funding of joint *IUE/Voyager* programs, UVS operations probably would not be continuing.

Smaller ground-based facilities often have difficulty with funding day-to-day operations of the telescope. Observers who use these facilities on a routine basis and are often called upon to assist in multispectral campaigns complain about the money its costs them in coolants, transportation, and telescope maintenance to support these efforts. They note that the time involved is often much more than just the time to take, for example, an optical spectrum of the target; the observer must also measure standard stars and make extinction determinations, as well as make allowances in his time allocation budget for bad weather. The costs incurred, then, may be appreciable.

Blockbuster astronomical events can garner some important government financial support. For example, NASA responded to SN 1987A by forming a two-year program to support *IUE, SMM*, airborne infrared, balloon-borne gamma-ray, and X-ray and UV rocket observations of the nearby supernova. The typical multiwavelength campaign is not funded in this way.

A promising development in the support of multiwavelength data analysis is NASA's Astrophysics Data Program (ADP), first announced in 1986. This program funds basic research proposals conducting scientific investigations that use archival data from previous and current astrophysics space missions (specifically, *IRAS*, the *HEAOs*, *EXOSAT*, and *IUE*). Part of the funding is dedicated to proposals for comparative multiwavelength studies using several of the spacecraft. Computational resources are also made available to ADP participants. ESA is funding the ESIS Pilot Project described earlier.

The 1987 reorganization of NASA includes the formation of the Astrophysics Data Systems and Science Operations Branch within the Astrophysics Division. The purpose of the Branch is to support the recommendations of the NASA Astrophysics Data Systems Workshops.

While it is good news that some government support is forthcoming for data analysis, there is still the problem of funding the multiwavelength observations themselves. This will be a much bigger problem with the launch of several orbiting telescopes in the 1990s. These facilities will need ground-based support.

General Considerations for Coordinating Observations: A Summary

Although the goal is for observing strategies to be guided by scientific issues rather than by telescope constraints, it is recognized that this is not always practical, especially when it comes to scheduling orbiting observatories whose lifetimes are short and operations costly. It is recommended, then, that the value and objectives of multiwavelength observing be kept in mind when planning satellite observatory missions and strategies; that some consideration be given to enhancing the value of the space mission with either additional monitors for other wavebands, or supporting dedicated telescopes on the ground (or reserving fractions of time on existing ground-based telescopes for support of the space mission). It is recommended that some discretionary time on all telescopes, in the ground and in space, be set aside for coordinated observations and targets of opportunity. It is also worthwhile keeping in mind that coordination considerations should not be limited to space-plus-ground facilities, but should include coordination of large ground facilities (e.g., the 10 m class optical telescopes of the 1990s) with smaller, supporting telescopes. In addition, the Strasbourg Colloquium participants pointed out that it will be necessary to coordinate space and ground instruments within the same wavelength range (e.g., optical ground support of the Hipparcos astrometric satellite).

Time allocation committees are encouraged to hold a broad, open view towards these projects, which often require long and repeated measurements, and are part of a much larger program not necessarily guaranteeing speedy results.

Schedulers for ground-based telescopes are urged to keep some flexibility in their scheduling, perhaps to set aside dedicated or discretionary time for coordinated observations, and to appreciate the necessity for specific dates for an observation. It is recommended that efforts be made to coordinate timelines and proposal deadlines among all major facilities, and to consider increasing the number of proposal cycles per year.

Observers are aware that, if they desire that science motivate schedules rather than vice-versa, they should amply justify and make clear the scientific aims and technical constraints of their multiwavelength observations. Schedulers emphasize that, to compete with other programs, a multiwavelength proposal must, first, promise good science. For example, no ground-based telescope proposal should receive a favorable bias merely because it includes observations by the latest space mission. Nor should time be granted on an orbiting observatory for a proposal just because it promises simultaneous observations across the spectrum. A multiwavelength proposal should be harder, not easier, to justify, because it must show that observations are important in every wavelength band of interest.

5. Epilogue

The scientific results presented in previous chapters of this book justify a multiwavelength approach, and ensure that multispectral studies will be a dominant focus in astrophysics. These results have also spawned a number of related efforts: there are electronic mail hotlines to inform those with quick access to telescopes of transient events such as supernovae; there are newsletters on specific classes of objects, and updates on space missions and telescope instrumentation, which are sent to all interested parties; there are workshops devoted exclusively toward promulgating the scientific rewards of coordinated observations and exploring ways to improve them. A followup to the 1987 Taos and Strasbourg conferences was a special session on the coordination of multiwavelength observations, held at the 1988 January meeting of the American Astronomical Society. At that session it was voted to form a Working Group of an IAU commission to further explore implementing some of the recommendations summarized here. An increasing number of graduate students are obtaining Ph.D. theses making multiwavelength observations of specific classes of sources. For some astronomers organizing and participating in multiwavelength observing campaigns is their principal, consuming activity. This vocation is an exhausting administrative task for almost all campaign organizers. But as one astronomer remarked, "I can't imagine doing observations that are *not* multiwavelength."

I was inspired by the enthusiasm of the 1987 Taos Workshop participants, and I hope that this article summarizes fairly their concerns and conclusions. I thank the following people for contributing extensive, useful information to this article: Drs. J. Bregman, T. Chester, B. Clark, F. Makino, K. Mason, G. Peters, R. Polidan, F. Seward, M. Smith, P. Szkody, and Y. Kondo. Special thanks go to my co-partners in organizing an ongoing forum for this subject, Drs. T. Courvoisier and M. Malkan. This paper was written with support from the U.S. Department of Energy.

References

Bookbinder, J. A. 1987, *Activity in Cool Star Envelopes*, ed. O. Havnes, B. Pettersen, and J. Schmitt (Dordrecht:Reidel), in press.

Carone, T., Polidan, R., and Holberg, J. 1987 *Lunar and Planetary Laboratory*, internal document.

Córdova, F. A., Mason, K. O., Priedhorsky, W. C., Margon, B., Hutchings, J. B., and Murdin P. 1985 *Astrophys. and Spa. Sci.*, **111**, 265.

Haisch, B. M., *et al.* 1981, *Ap. J*, **245**, 1009.

Kahler, S., *et al.* 1982, *Ap. J.*, **252**, 239.

Karpen, J. T., *et al.* 1977, *Ap. J.*, **216**, 479.

Kondo, Y., *et al.*, editors, 1987, *Exploring the Universe with the IUE Satellite* (Dordrecht:Reidel).

Makino, F. 1987, *Astron. Lett. and Comm.*, **25**, 223.

Mason, K. O., Córdova, F. A., and White, N. E. 1986, *Ap. J.*, **309**, 700.

Pederson, H., *et al.* 1982, *Ap. J.*, **263**, 325.

Swanenburg, B. N., *et al.* 1981, *Ap. J. (Letters)*, **243**, L69.

Vilhu, O., Caillault, J.-P., Neff, J., and Heise, J. 1987, *Activity in Cool Star Envelopes*, ed. O. Havnes, B. Pettersen, and J. Schmitt (Dordrecht: Reidel), in press.

White, N., and Barr, P. 1987, *EXOSAT Express*, No. 18, ESA document.

Wisniewski, W. 1988, *Proc. of Colloq. on Coordination of Observing Projects*, ed. C. Jaschek and C. Sterken (Cambridge Univ. Press), in press.

Future Space Instrumentation for Multiwavelength Astrophysics

Gerard A. Kriss

Center for Astrophysical Sciences

The Johns Hopkins University

1. Introduction

The flowering of space-based astronomy over the past two decades has opened entirely new windows onto the investigation of astrophysical problems. Astronomers have become increasingly aware that most celestial objects emit important components of their radiation over a broad range in wavelength. As the previous chapters in this book show, multiwavelength observations are essential for understanding many astrophysical phenomena. Our knowledge of active galaxies, X-ray binaries, cataclysmic variables, stellar physics, and the interstellar medium has profited enormously from multiwavelength observations. While many multiwavelength studies consist of distinctly separate observations made at different epochs, variability on timescales ranging from milliseconds to months has prompted a greater emphasis on *simultaneous* multiwavelength campaigns. We have learned much from such campaigns, but, as outlined in the previous chapter, successful campaigns are infrequent, and they have come about only through the heroic efforts of dedicated teams of observers and the understanding of many time assignment committees.

Current approaches to multiwavelength astrophysics are inefficient in their use of the available instrumentation and of the time of the personnel involved (see the previous chapter by F. Córdova). Satellites in low earth orbits have typical observing efficiencies of ~35% due to the large angle subtended by the earth and passage through the South Atlantic Anomaly. Coordinating simultaneous observations between two or more satellites in low earth orbit is logistically difficult at best, and physically impossible at worst. Many satellites are constrained to point ~90° from the sun, which means that accessible targets are near the horizon for a ground-based observer. Continuous coverage from the ground requires the use of several observatories at well spaced longitudes. The vagaries of weather also force redundancy in a ground-based support program. A successful multiwavelength campaign may represent first class, pioneering research, but frequently this is true only if all necessary wavebands participate. Taken separately, the observations in any single waveband may yield little new information.

No satellites have been consciously designed to support multiwavelength observations either by their instrument complement or by their method of operation. Similarly, few ground-based observatories have scheduling and operational policies that specifically accommodate the logistical difficulties of coordinated multiwavelength observations. A new generation of satellites and new operational procedures for ground-based support are now being planned that are specifically geared to the special problems of multiwavelength observations.

In this review I will concentrate on topics related to future instrumentation for multiwavelength astrophysics that were covered in two sessions of the 1987 Taos Workshop on Multiwavelength Astrophysics. One session was devoted to the presentation of concepts for future space instruments and of the capabilities of current instruments, and the second was an open discussion of the most important characteristics of potential satellites for multiwavelength astrophysics. The status of current instruments and strategies for coordinating observations between several space and ground-based observatories are discussed in the previous chapter by F. Córdova.

2. Desirable Characteristics of Future Instruments

Future instruments for multiwavelength observations fall into two broad categories. Pointed spacecraft containing several co-aligned instruments sensitive to different wavebands are best suited to studies of known objects. Transient phenomena such as supernovae and γ-ray bursts, however, require instruments with large fields of view since the location of an event is not known *a priori*. For either category, the detectors within each waveband should have similar time resolution and sensitivity, *i.e.*, a typical object for study should yield comparable signal-to-noise ratios per resolution element in each detector system on an appropriate time scale.

Multiwavelength Pointed Platforms

A multiwavelength pointed satellite with low to moderate resolution spectroscopic capability from the X-ray to the optical bandpass would give astronomers the power to attack a broad range of fundamental problems in astrophysics, including

- the nature of the central power source producing the continuum radiation in active galactic nuclei (AGN),
- the structure, kinematics, and physical conditions of material in the broad and narrow line regions surrounding the central power source in AGN,
- the dynamics of accretion disks and the physical processes driving the transport of angular momentum in X-ray binaries and cataclysmic variables,

- the characteristics of matter and radiation in the extreme magnetic and gravitational fields of white dwarfs, neutron stars, and black holes, and
- the physics of flares, spots, and plages in the coronae of stars and their relation to similar phenomena on the sun.

For studies of continuum radiation, resolving powers of ~10 are adequate to avoid strong emission lines. Plasma diagnostics throughout the soft X-ray, UV, and optical bands require resolving powers of several hundred to a thousand, however.

Several concepts for missions presented at the Taos workshop take the multiwavelength platform approach, and a summary of their salient features is presented in Table 1. For each mission I list its bandpass and resolving power (if known) in various wavebands from the γ-ray through the visible, and I also give its launch status. Multiwavelength concepts including infrared instruments were notably absent among the presentations in Taos although several missions devoted exclusively to IR and submillimeter astronomy are being planned. The basic reason for this lack of IR coverage at present appears to be the technological lag for developing IR detectors comparable in performance to optical and UV detectors, plus the high cost and short lifetime of cryogenic systems.

Table 1 Pointed Instruments for Multiwavelength Observations

Mission	γ-ray (MeV)	X-ray (keV)	EUV (Å)	UV (Å)	Visible (Å)	Status[a]
SpEx	...	0.3-12 R=3-60	400-800 R~1000	800-3200 R~1000	2800-5500 R~1000	Concept
HST	...	...	...	1150 to R~2 to 10^4	10000 R~2 to 10^4	1989?
EUVE	...	0.06-0.15 R=1-400	170-760 R=1-1000	...	...	1990?
Lyman	...	...	100-350 R~300	910-1250 R~ 10^4	...	Phase A
SYNOP	...	soft Broad	~175 Broad	1200-3200 R~ 10^4	Broad	Concept
Argus	...	soft	200-800 R=20-800	840-3200 R~1000	3000-7000 R~1000	Concept

[a]The status column shows either a tentative launch date for an approved mission, "Phase A" to denote a concept with funding for more detailed studies, or "Concept" to indicate that it is a proposal still under consideration.

Ideally, a multiwavelength pointed spacecraft should be in an orbit that would permit long, unbroken observations of individual objects. Such continuous multiwavelength coverage is unbiased by preconceived notions of which observations require complete simultaneity, and it minimizes the effects of aliasing when cross-correlating results between the various wavebands. As pointed out by Córdova *et al.* (1985), this allows the discovery of entirely new, unexpected phenomena. A multiwavelength pointed platform in space has the advantages of simplified scheduling, assured simultaneity, and a clear dedication to multiwavelength observations.

High earth orbits have much higher observing efficiencies than low earth orbits, and operations can usually be coordinated from a single ground station. This simplifies the scheduling process, and the small solid angle subtended by the earth permits long observations unbroken by earth occultation. The successful operation of the *International Ultraviolet Explorer* (*IUE*) for 10 years has demonstrated the advantages of a facility in a geosynchronous orbit. A satellite need not be in a geostationary orbit to obtain these advantages. In fact, geostationary orbits have the disadvantage that the particle background is extremely high. The *IUE* orbit is slightly elliptical, and this carries it out of the radiation belts for ~16 hours per day. Even longer unbroken observations and higher observing efficiencies can be obtained in higher orbits. The 90 hour, 190,000 × 350 km orbit of the European X-ray astronomy satellite *EXOSAT* permitted unbroken observations of up to 40 hours. The discovery of quasi-periodic oscillations (QPO's) in low mass X-ray binaries (Middleditch and Priedhorsky 1985) and of the variability of AGN on nearly all time scales (Lawrence *et al.* 1987; McHardy and Czerny 1987) was due in large part to the ability to obtain long, uninterrupted observations.

Placing a satellite in a high orbit is expensive, however, so many of the experiments I discuss later in this review will be in low earth orbits. A multiwavelength experiment which includes an optical telescope does not suffer as much from the scheduling problems of a low earth orbit since the broad wavelength coverage obviates the need for coordination with ground-based facilities. Earth occultations can be avoided by observing targets near the poles of the orbit, but passage through the SAA will limit the time of continuous coverage to less than 90 minutes.

Multiwavelength Survey Instruments

Coordinating multiwavelength observations of transient phenomena is particularly difficult since observers do not know when and where to look. Coordinated observations of X-ray bursters have been successful because of their somewhat predictable behavior (see the chapter by Hartmann and Woosley), but the burst watch campaigns have used much telescope time to record a few bursts. The γ-ray burst sources are particularly enigmatic — no simultaneous optical or γ-ray burst has yet been recorded. Ground-based searches for optical flashes associated with γ-ray bursts are frustrated by the high background rates of meteors and sunlight glinting from satellites. A space-based multiwavelength instrument with wide fields of view in the γ-ray, X-ray, and UV/optical bandpasses can overcome these problems and operate with high efficiency.

The multiwavelength survey instruments are not as strongly affected by orbital characteristics since they view such a large fraction of the sky continuously. The survey instruments have dramatically less stringent pointing requirements than pointed platforms; consequently the spacecraft are simpler. In addition, since their pointing constraints are not rigid, coordination with ground-based observers is eased. Careful attention must still be given to the planning of coordinated observations, but the logistical problems are greatly eased by the wide fields of view. Weather and uncontrollable constraints such as passage through the SAA can still present insurmountable barriers in particular cases.

Table 2 Survey Instruments for Multiwavelength Observations

Mission	γ-ray (MeV)	X-ray (keV)	EUV (Å)	UV (Å)	Visible (Å)	Status[a]
HETE	0.06-1 R∼10	2-25 R∼5	...	800-3100	...	Concept
GRANAT	0.02-30 R∼10	2-150 R∼5-10	...	...	5000-8000 Broad	Nov. 1988
ASTRE	...	0.2-10 R∼5	...	...	...	Concept
MOXE	...	2-20 R∼5	...	...	...	Concept

[a]The status column shows either a tentative launch date for an approved mission or "Concept" to indicate that it is a proposal still under study.

In Table 2, I give a summary of the capabilities of the multiwavelength survey instruments in the same format as Table 1. In the following sections I will give more detailed descriptions of the instruments listed in Tables 1 and 2 in the order of their appearance in the tables. I also give descriptions of single waveband instruments with potential multiwavelength applications, and I conclude with a brief description of future multiwavelength instruments for solar observations.

3. Pointed Instruments for Multiwavelength Observations

SpEx — The Spectroscopy Explorer

SpEx was proposed to NASA in July 1986 as a concept for an Explorer Phase A study by The Johns Hopkins University, Goddard Space Flight Center, the Harvard-Smithsonian Center for Astrophysics, and Los Alamos National Laboratory. *SpEx* is uniquely suited to simultaneous spectroscopic observations of astronomical targets spanning the electromagnetic spectrum from X-ray to visible wavelengths. *SpEx* can be economically assembled from well-matched, existing or soon to be constructed NASA shuttle payloads. The *Broad Band X-ray Telescope* (*BBXRT*) being developed at GSFC for a Space Shuttle flight as a portion of the *SHEAL 2* payload will observe the X-ray bandpass. The *Hopkins Ultraviolet Telescope* (*HUT*), ready for flight on the *ASTRO* missions, will cover the far UV bandpass. The near UV and optical bandpass will be viewed by a single *Near UV/Optical Telescope* (*NUVOT*) that is an optical-mechanical copy of the small (0.40m) telescope currently in use in the sounding rocket program at Johns Hopkins and being adapted for use in a Spartan mission. The near UV and optical detectors and controlling electronics are duplicates of the systems used on *HUT*. The overlapping wavebands of the UV and optical telescopes will permit them to be calibrated with good relative precision. Table 3 gives a summary of the wavelength coverage, spectral resolving power, sensitivity, temporal resolution, and fields of view of each of the telescopes.

The *BBXRT* achieves a large collecting area with light weight by using over 100 conical foil reflectors instead of conventional Wolter 1 mirrors. This comes at the expense of spatial resolution — the half power radius of the conical mirrors is 1.3', approximately constant over the full field of view. The mirrors are constructed of high-polish aluminum foil coated with an acrylic lacquer. Gold is vacuum deposited onto the lacquer surface. Two mirror assemblies each 40 cm in diameter with a 3.8 m focal length are incorporated in *BBXRT*. In the focal plane of each telescope is a segmented Si(Li) detector, similar in principal to the Solid State Spectrometer on the *Einstein Observatory* (Giacconi *et al.* 1978), but with greatly improved performance. A central segment with a field of 3.6 arc minutes is surrounded by four annular segments extending to a radius of 17.3 arc minutes. The detectors are cryogenically cooled to 125 K. Currently this cooling is provided by stored cryogen with a limited lifetime, but a Stirling cycle refrigerator is under development for long-term use in space.

HUT consists of an f/2, 1 m telescope with a Rowland circle spectrograph at the prime focus. The mirror is coated with iridium for good reflectivity in the

Table 3 Performance Characteristics of the *SpEx* Instruments

	BBXRT	*HUT*	*NUVOT*
Bandpass	1 - 40 Å	410 - 1850 Å	1600 - 5500 Å
	(0.3 - 12.0 keV)		
Spectral Resolving Power	3 - 60	300 - 1100	450 - 1950
Peak Sensitivity (cts s^{-1} bin^{-1}/μJy)^a	1.0	1.4×10^{-4}	2.2×10^{-4}
(cts s^{-1}/10^{-12} ergs cm^{-2} s^{-1})	0.41	0.55	12
Wavelength of Peak Sensitivity	15 Å (0.85 keV)	1100 Å	4000 Å
Limiting Flux (μJy)^b	0.003	33	29
(erg cm^{-2}s^{-1}Å^{-1})	4×10^{-16}	8.2×10^{-16}	5.5×10^{-17}
Instrument Background (cts s^{-1} bin^{-1})^a	1×10^{-4}	4×10^{-3}	1×10^{-2}
Temporal Resolution	62.5 μs	1 ms	1 ms
Field of View	3.6' and 17'	9" to 18×120"	9" to 18×120"

[a]10 Å wide bins are used for *HUT* and *NUVOT*, and 1 keV bins for *BBXRT*.
[b]Flux required to obtain a signal to noise ratio of 5 at the peak sensitivity in 10^4 s.

FUV and EUV, and the holographically ruled grating is coated with osmium. The windowless detector has a CsI photocathode directly deposited on a micro-channel plate intensifier. The intensifier is fiber-optically coupled to a Reticon linear photodiode array with 1024 channels. *HUT* can automatically acquire targets with a video camera that views the slit jaws.

The *NUVOT* is an f/4.5, 0.4 m Cassegrain telescope with two Rowland circle spectrographs in the focal plane. The mirrors are coated with Al/MgF$_2$. A beam-splitter provides simultaneous coverage of the 1600 - 3000 Å and the 2750 - 5500 Å bandpasses with a single small telescope. The spectrographs and detectors duplicate *HUT*, except for the photocathode materials. The UV detector has a sapphire window with a CsTe photocathode directly deposited on the microchannel plate. The optical detector has a borosilicate window with a semitransparent bialkali photocathode deposited on its rear surface. A selection of different slits (with matching projected sizes) can be placed at the foci of the *HUT* and *NUVOT* telescopes.

Of the pointed platforms listed in Table 1, *SpEx* has the highest sensitivity over the broadest wavelength coverage at moderate spectral resolution. It is ideally suited to studies of AGN, cataclysmic variables, X-ray binaries, and stellar phenomena. To place its conceived performance in perspective, in 10^4 s *BBXRT* can obtain the *spectrum* of any X-ray source *visible* to *Einstein* in a single orbit. In 1000 s, the *HUT* and *NUVOT* spectrographs can obtain spectra of higher signal to noise and higher spectral resolution than any obtainable by the *International Ultraviolet Explorer* in its longest possible integration time of ~15 hours. As an example of the power of *SpEx*, Figure 1 shows simulated spectra of the Seyfert galaxy NGC 4151 spanning the full range in wavelength covered by *SpEx*. The spectra each represent 2000 s integrations. In the brighter spectrum the non-thermal power law component is six times stronger while all other components retain the same intensity.

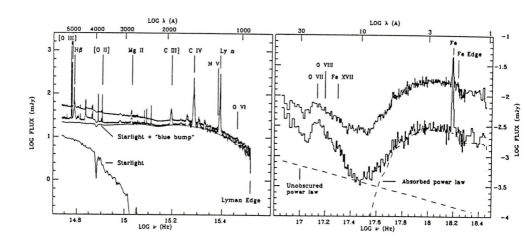

Figure 1 Simulated spectra of the Seyfert galaxy NGC 4151 obtainable in 2000s integrations. In the visible and UV portion the power law, the "blue bump", starlight, and prominent broad and narrow emission lines are all visible. In the X-ray the partially covered power law can be seen, as well as a million degree thermal component.

While *SpEx* is powerful, it is also large. The total size is 2.7 m × 4.2 m × 1.2 m, and it weighs ~3000 kg. It can be launched into a low earth orbit by either a Delta or the Space Shuttle, but ideally it should be in a high orbit. A Titan III-D plus a Centaur upper stage could place *SpEx* in a geosynchronous orbit, albeit at higher cost. Even from a low earth orbit, however, *SpEx* guarantees simultaneous, multiwavelength observations with high sensitivity and moderate spectral resolution that cannot be matched by any other proposed mission.

The Hubble Space Telescope

Of the satellites planned in NASA's "Great Observatories" series, the *Hubble Space Telescope* (*HST*) has the greatest intrinsic capabilities for multiwavelength observations (albeit not truly simultaneous), and the characteristics of its focal plane instruments are already widely known in the astronomical community. Briefly, the focal plane contains the Faint Object Spectrograph (FOS), the High Resolution Spectrograph (HRS), the Wide Field/Planetary Camera (WF/PC), the High Speed Photometer (HSP), and the Faint Object Camera (FOC), which is supplied by ESA. This instrument complement will provide broad and narrow band imaging capabilities at high angular resolution ($<$0.1 arc sec) from 1150 Å to $\sim 1\mu$, and low to high resolution spectroscopy from 1150 Å to $\sim$7000 Å. Only one instrument at a time can view a given celestial object, so truly simultaneous multiwavelength observations are not possible. For objects variable on timescales of hundreds to thousands of seconds, however, it is potentially possible to obtain nearly simultaneous spectra or images over the full bandpass provided by the focal plane instruments. Second generation instruments under development will provide two dimensional spectroscopic capability and near infrared imaging capabilities in the late 1990's.

EUVE — the Extreme Ultraviolet Explorer

The *Extreme Ultraviolet Explorer (EUVE)* is scheduled for launch in 1990. It will survey the sky for six months in three band passes spanning 80 to 900 Å at an angular resolution of 0.1 degree, and then spend another six months conducting followup spectroscopic observations over the 70 - 760 Å range. As described by Bowyer (1983), *EUVE* will carry four grazing incidence telescopes on a spin-stabilized spacecraft. Three telescopes oriented perpendicular to the spin axis will carry out the survey, and the fourth telescope will be aligned with the spin axis in the anti-solar direction.

The three survey telescopes are 40 cm, f/1.24 Wolter-Schwarzschild 1 designs with microchannel plate detectors in the focal plane. Broad band passes will be defined by filters of parylene (80 - 190 Å), aluminum and carbon (170 - 300 Å), titanium and antimony (350 - 540 Å), or tin (500 - 750 Å). Approximate sensitivities for these band passes over the course of the survey are 3 μJy, 0.3 mJy, 0.1 mJy, and 0.1 mJy respectively.

The telescope pointing in the antisolar direction will perform a deep survey of a narrow strip of the sky in the 80 - 350 Å band pass defined by an aluminum/parylene filter. The deep survey telescope is a 40 cm, f/3.4 Wolter - Schwarzschild type 2. When the all-sky survey is complete, pointed observations will be performed using three spectrometers which view the sky through the deep survey telescope (Hettrick *et al.* 1985). Only objects within 45° of the ecliptic plane will be accessible for pointed observations. The spectrometers are slitless, varied-line-spacing, grazing-incidence gratings, each of which picks off one sixth of the light from the telescope focal plane. The "A" spectrometer will cover the 70 - 190 Å band at 0.5Å resolution with a peak effective area of 1.2 cm^2 at 100 Å. The "B" spectrometer will cover the 140 - 380 Å band at 1.0Å resolution with a peak effective area of 0.5 cm^2 at 200 Å. The "C" spectrometer will cover the 280 - 760 Å band at 2.0Å resolution with a peak effective area of 0.6 cm^2 at 400 Å. As on the survey telescopes, the detectors will be microchannel plates with a CsI photocathode and a wedge-and-strip readout (Martin *et al.* 1981).

Lyman — the Far Ultraviolet Spectroscopic Explorer

Lyman is a concept for an Explorer mission which would concentrate on high resolution (R > 10^4) spectroscopy over the 912 - 1216 Å range with an effective area of ~100 cm^2. An option for EUV spectroscopy is also under consideration. ESA has already performed a Phase A study of *Lyman*, and NASA has recently approved a Phase A study of its own to be conducted by a large team of investigators headed by W. Moos of the Johns Hopkins University. Negotiations are underway to make *Lyman* an international effort.

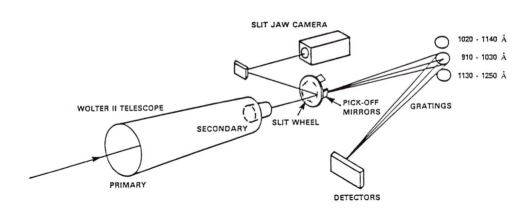

Figure 2 A schematic illustration of the telescope, Rowland spectrograph with three separate quasi-toroidal gratings, and slit jaw camera for *Lyman*.

The concept for *Lyman* proposed to NASA uses an 80 cm Wolter-Schwarzschild II telescope to gather the light. A rotating slit wheel with grazing-incidence pick-off mirrors behind the slits would divert the light to the far UV and EUV spectrographs. A near-normal incidence Rowland spectrograph with three separate gratings of resolving power ~ 3 × 10^4 would be used for the 910 - 1250 Å band. The quasi-toroidal gratings would be optimized for the 910 - 1030 Å, 1020 - 1140 Å, and 1130 - 1250 Å. As shown in Figure 2, a separate slit and pick-off mirror would be associated with each grating. EUV capability is provided by a slit for the 100 - 350 Å band that would divert light to a focusing ellipsoid and a varied-line-spacing grating with a resolving power of ∼ 300. This is similar to the spectrometers used on *EUVE*.

Microchannel plate detectors with opaque CsI photocathodes would be used on the spectrographs. For the high spatial resolution necessary in the 910 - 1250 Å range, a multi-anode readout would be used (Timothy and Bybee 1981). The lower resolution contemplated for the 100 - 350 Å band is well suited to a simpler wedge-and-strip anode.

While *Lyman* covers a wide range of wavelengths, it can view a given object in only one narrow wavelength range at a time. This limits its usefulness for multiwavelength observations to long time-scale phenomena and coordinated observations with other facilities.

SYNOP

SYNOP has been proposed as a geosynchronous satellite facility dedicated to long term, synoptic observations of stars in the X-ray or EUV, the UV, and the visible bandpasses. Observations with *SYNOP* would be coordinated with the ground-based observing program of a 3.5m telescope instrumented with a high resolution spectrograph and a wide-field 1.5m telescope for imaging photometry at the Ca II K line. These telescopes would be new facilities constructed by the National Solar Observatory solely for the support of *SYNOP*. A science working group coordinated by Mark Giampapa of the National Solar Observatory has proposed *SYNOP* as a concept for a Phase A Explorer Study. He has kindly provided the information contained in this summary.

The ultraviolet instrument of *SYNOP* would be a clone of the *International Ultraviolet Explorer*. The telescope and spectrograph would be duplicated, but new, more sensitive detectors would be designed. Apart from the UV spectrographs, *SYNOP* is not yet a fully developed concept. Continuum monitoring would be provided by additional photometers — a UV photometer centered at 1600 Å and a white light photometer for the visible bandpass. The visible photometer would preferably be an array that could be used to image multiple stars. To monitor coronal activity, a soft X-ray telescope or an EUV telescope would be added to the basic *IUE* telescope. The EUV option would use a normal incidence mirror with a multilayer coating to isolate bands around the Mg VI line at 95.5 Å or the Fe IX, X, XI complex around 175 Å. At coronal temperatures of ∼10^6 K, this iron line complex has greater emissivity than the entire X-ray region with λ < 100 Å (Stern *et al.* 1983). The detector could be a CCD sensitized for use in the EUV using ion implantation with laser annealing (Stern *et al.* 1987).

While *SYNOP* is conceived primarily as an instrument dedicated to the study of stellar activity, many other scientific areas such as cataclysmic variables, X-ray binaries, novae, supernovae, and active galactic nuclei could benefit from such a satellite and its intended mode of operation. With improved detectors, all objects now accessible to the *IUE* could be easily studied with *SYNOP*.

Argus — a Multidiscipline, Multiwavelength, Monitoring Spacecraft

Argus is a joint proposal of the Lunar and Planetary Laboratory of the University of Arizona and the Department of Astronomy of the University of Trieste for a multidiscipline experiment consisting of three independent instruments for astronomical observations, solar observations, and Earth plasma studies. Each instrument consists of a cluster of co-aligned spectrometers covering the spectral range from the X-ray to 7000 Å. Long term, uninterrupted observations of targets including the Sun, the Earth, planets, comets, galactic, and extragalactic targets for studies of temporal and/or spectral variations are planned from a Langrangian point orbit or a very eccentric Earth orbit. Details of the *Argus* concept were presented at the Taos conference by R. Stalio of the University of Trieste.

Table 4 Astronomical Platform Spectrometers

Spectr. Channel	λ Range (Å)	Res. (Å)	Aperture (cm)	Sens.[a]	Configuration
EUV2	200-600	10	2x6	4.3	Grazing Incid.
EUV1	400-860	1	28x28	3.1	Objec. Grating
FUV	840-1300	1	40x40	1.6	Objec. Grating
SUV	1150-2200	5	9.2 diam.	1.0	Cassegrain/Slit
LUV	2150-3200	5	7.6 diam.	0.7	Cassegrain/Slit
VIS	3000-7000	7.5	6.0 diam.	0.3	Cassegrain/Slit

[a]Point source continuum flux at full spectral resolution with a S/N = 10 in 1200s in units of 10^{-13} erg cm^{-2} s^{-1} Å^{-1}.

The Earth monitor platform has five monochromatic imagers in the lines of He II λ304, He I λ584, O II λ834, Ly α, and O I λ1304 as well as five spectrographs covering the range 200 to 3200 Å. The solar monitor platform contains a solar spectrograph and a rare gas ionization cell for measurement of the absolute solar flux. The astronomical observatory platform contains an X-ray imager, a wide field Ly α imager, and three telescopic imaging spectrographs (angular resolution 2 - 4") covering the wavelength range 200 - 1300 Å, and three Cassegrain telescopes with slit spectrographs covering 1150 - 7000 Å. The characteristics of he spectrographs on the astronomical platform are summarized in Table 4.

Versions of the EUV1 and FUV spectrometers and the rare gas ionization cell roposed for *Argus* will be tested in two planned shuttle missions, the *International EUV/FUV Hitchhiker* (*IEH*) and the *International EUV/FUV Spartan* (*Janus*). While *Argus* covers a large range in wavelengths, its low sensitivity rould limit its ability to study short time scale variability in all but the brightest elestial objects.

4. Survey Instruments for Multiwavelength Observations

HETE — the High Energy Transient Experiment

ETE is a small, multiwavelength survey satellite especially designed to ascertain he origins of γ-ray bursts and other high energy transient events. It has been roposed to NASA by the Massachusetts Institute of Technology, the Los Alamos Jational Laboratory, Centre National d'Etudes Spatiales, the University of California at Santa Cruz, and the University of Chicago. As described in the chapter n γ-ray and X-ray bursters by D. Hartmann and S. Woosley, the basic nature of '-ray bursters, even whether they are galactic or extragalactic, is still not understood. *HETE* would allow precise positioning and multiwavelength spectral measurements for $\sim$100 γ-ray bursts over its proposed 18 month life. The information iven here was presented by G. Ricker of MIT at the Taos Conference.

An omnidirectional γ-ray spectrometer sensitive from 6 keV to 1 Mev, a wide-eld X-ray transient monitor sensitive from 2 - 25 keV, and an array of ultraiolet transient cameras sensitive from 1800 Å to 3100 Å comprise the instruments f *HETE*. The γ-ray detectors have positional capabilities of $\sim \pi$ ster, time resolution of 4 ms, and energy resolution of $\sim$7% at 662 keV. The X-ray emissions of -ray bursts would allow localization of an event by the X-ray monitor to within -6 arc minutes. The X-ray monitor has a 2 ster field of view, time resolution of $\sim$ 1 ms and an energy resolution of $\sim$15% at 6 keV. The ultraviolet cameras rould provide positions to better than 6 arc seconds. The estimated sensitivities f these instruments (10 σ) are compared to measured and modelled γ-ray and X-ay burst fluxes in Figure 3. While the primary operating mode of *HETE* would rigger on either a γ-ray or an X-ray event, triggers generated by the UV camera rould survey the sky for new classes of transients that emit primarily in the UV.

In Figure 3 the B-band archival flash reported by Schaefer (1981) and the V- and candidate optical flash observed by Pedersen *et al.* (1984) are shown. The redictions for optical and UV flashes from γ-ray burst sources are from the hodels of Melia, Rappaport, and Joss (1986) and Melia (1987). The model of Ielia, Rappaport, and Joss predicts a UV flash lasting $< \sim$10 s. Sensitivities for

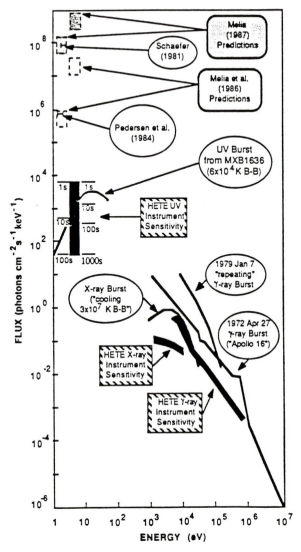

Figure 3 Estimated sensitivities of the *HETE* UV, X-ray, and γ-ray instruments is compare with measured and modelled X-ray and γ-ray burst fluxes.

three different integration times are shown for the UV camera. Those on the le are for bright flashes greatly exceeding the readout noise of the CCD, and tho on the right are for signals comparable to the readout noise. A "classical" γ-ra burst spectrum (total flux above 30 keV of $3.8{\times}10^{-6}$ erg cm^{-2} s^{-1}) is also show The extension down to 1 keV is based on the *Apollo 16* observation of the 197 γ-ray burst (Gilman *et al.* 1980), and the extension above 1 MeV is based on *SM* observations by Nolan *et al.* (1984). The 1979 January 7-type "repeating" γ-ra burst spectrum is from Laros *et al.* (1986). Also shown is a typical X-ray burs

pectrum with a temperature of 3×10^7 K and a blackbody spectrum of 6×10^4 K
tted to the 1979 June 28 optical flash from the X-ray burst source MXB 1636-53
Pedersen *et al.* 1982).

In addition to the multiwavelength capabilities inherent in the *HETE* pack-
ge, participation of ground-based "guest observers" is provided by transmitting
he coordinates and the time histories of detected bursts to ground-based observa-
ories in seconds to minutes after the onset of a burst. A participating observa-
ory would have a self-contained, automated receiver controlled by a personal
omputer. The PC would display burst data and store it on a small hard disk.
he total cost of an automated ground station would be less than $10,000.

The *HETE* spacecraft would have no pointing capability. This greatly
educes its cost. The spacecraft would slowly spin about an axis pointed at the
un, and the attitude would be derived from known X-ray burst sources and from
he UV camera doubling as a star tracker. *HETE* is small enough to be launched
ither on a Scout or from a Get Away Special cannister (GAS-can) on the Space
huttle. The simplicity of *HETE*'s design, its low cost, and its potential to solve
ne of the most enigmatic problems in modern astrophysics drew enthusiastic sup-
ort from the astronomers participating in the workshop discussion on mul-
iwavelength satellites.

The Soviet GRANAT Experiment

he Soviet *GRANAT* mission will be launched in November 1988. *GRANAT* will
e in an elongated, highly inclined orbit (57°) with a period of 96 hours. The
enera-class spacecraft has a 2' pointing capability. The mission centers around
IGMA, a 20 - 2000 keV Anger camera with a coded mask that is a Soviet/French
ollaboration. Other instruments include

ART-P, a Soviet 4-60 keV position sensitive proportional counter with a
coded mask,

ART-S, a Soviet 3-150 keV mechanically collimated Xe proportional counter,

WATCH, a Danish 6-120 keV NaI/CsI detector with a rotating modulation
collimator,

KONUS, a group of 7 γ-ray burst detectors sensitive from 0.02-5 MeV supplied by Leningrad, and

FEBUS, a group of 6 γ-ray burst spectrometers sensitive from 0.1-30 MeV
supplied by CNES/Toulouse.

The *Sunflower* instrument is the heart of *GRANAT*'s multiwavelength capa-
ility. Like *HETE*, *Sunflower* will search for simultaneous optical emission from
-ray and γ-ray burst sources. The following information was provided by G.
icker from presentations by E. Mazets and L. Filipov at the Institute for Space

Research in Moscow on June 11, 1987.

Sunflower consists of X-ray and optical detectors mounted on a gimbaled pla
form that can slew to within a few degrees of a γ-ray burst location within
second of the time of the burst onset. Since the rapid slews of *Sunflower* ca
potentially destabilize the spacecraft at certain sun angles, the duty cycle
Sunflower operation will be low. The X-ray detectors have a 15° field of viev
and they appear to be proportional counters of several hundred square centimetei
of effective area. The optical detector is a Soviet CCD at the focus of a 6 cn
f/1.5 refractive telescope. The bandpass is approximately 5000 - 8000 Å. Tl
CCD is 288×360 pixels with a 20 e⁻ rms readout noise. Optical flashes can b
localized to 4' in a 4°×4° field of view to a limiting magnitude of $m_v = 9$ in
2 - 4 s integration time. A second generation *Sunflower* is planned for th
SPECTRA-1 (X-ray) mission in 1992 and/or the *SPECTRA-2* (γ-ray) mission i
1994 (Mazets and Filipov, private communication to G. Ricker). The new instru
ment would incorporate a UV detector either in place of, or in addition to, th
optical detector.

ASTRE — the All-sky Supernova and Transient Explorer

ASTRE is a concept for a Phase A Explorer study submitted to NASA by a teai
from the Smithsonian Astrophysical Laboratory, the Max Planck Institute, an
the Los Alamos National Laboratory. The summary presented here is extracte
from a preprint by P. Gorenstein, the principal investigator. *ASTRE* is a nov
X-ray camera that images a large fraction of the sky with moderate angular resc
lution (~7') and an effective area of ~100 cm². It would be used as a high energ
transient surveyor as well as a monitor of the long term temporal behavior of X
ray sources on time scales from one day to one year. Although *ASTRE* is larg
(~ 2m x 3m, 2000kg), the spacecraft would require no pointing. This helps t
keep the cost down. Attitude would be determined from the positions of know
X-ray sources. *ASTRE* would be stabilized to view the anti-solar direction, and i
a single day it would see 77% of the *night* sky (as defined by a ground-base
observer). Sources are inaccessible only during earth occultations. Thus an
ground-based observer at a clear site would have X-ray coverage of many poter
tial targets for simultaneous, multiwavelength observations.

ASTRE images in two dimensions with different techniques. One dimensio
focuses X-rays via single and triple reflections from plane mirrors using th
method originally proposed by Schmidt (1975) and described in further detail b
Angel (1979). The perpendicular dimension recovers spatial information from
pseudo-random coded mask combined with a position sensitive proportione
counter. A cylindrically shaped detector with radial mirrors and an azimuthe

ɔded aperture permits *ASTRE* to view 1.5 π ster at an angular resolution of 7'. The diagram in Figure 4 illustrates the camera.

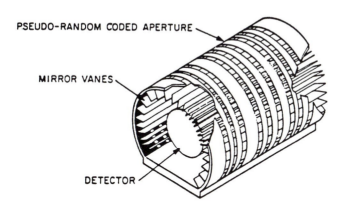

Figure 4 Diagram of the *ASTRE* X-ray camera. The radial mirror vanes focus X-rays from a celestial source at a fixed azimuth on the cylindrical detector. The coded aperture together with the position sensitive detector determines the axial direction of the X-rays.

Because of the focusing provided by the radial mirrors, background rates for his camera are dominated by the cosmic X-ray background, and they are an order of magnitude less than a two dimensional multiple pinhole camera of comparable size. Background rates and flux limits for 5σ detections in an 1800s for a ɔurce with $f_\nu \sim \nu^{-1}$ and no low energy absorption are given in Table 5 for four energy bands covered by the *ASTRE* detector.

Table 5 Background, Effective Areas, and 5σ Sensitivities for *ASTRE*

Energy Band (keV)	Mean Area (cm²)	Background (cts s⁻¹)	Sensitivity (μJy at 2 keV)
0.20 - 0.28	48	6.7	0.2
0.8 - 2.0	80	5.2	0.2
2 - 6	96	4.3	0.4
2 - 10	80	5.3	0.4

ASTRE has good sensitivity at soft X-ray energies below the carbon edge. Its large field of view makes it a good surveyor of *soft* X-ray transient events, a largely unexplored region. This spectral range will probably be dominated by stellar flares, but Klein and Chevalier (1978) and Lasher and Chan (1979) suggest

that a soft X-ray flare should appear in Type II supernovae several hours afte
core collapse. For 83 Type II supernovae per year out to a distance of 100 Mp
($H_o = 55$ km s^{-1} Mpc^{-1}), and using the models of Klein and Chevalier (1978
Gorenstein predicts *ASTRE* would detect 32 events per year at $>5.6\sigma$ confidence
The positional information from *ASTRE* would be sufficient to secure th
identification of the soft X-ray flare with the supernova.

All-Sky X-ray Monitors for Simultaneous Multiwavelength Observations

As stated by Bill Priedhorsky at the Taos conference, X-ray all-sky monitors ar
everyone's *second* favorite experiment for an astronomical mission. They provid
extremely useful information that everyone would like to have, but usually the
are not at the forefront of scientific work, and they are not very sensitive. In th
age of multiwavelength observations, however, all-sky X-ray monitors provide
low-impact, inexpensive method of obtaining automatic support of mul
tiwavelength campaigns in the X-ray bandpass. Priedhorsky discussed two con
cepts for future all-sky X-ray monitors: a set of X-ray pinhole cameras placed o
the NASA Space Station, and an inexpensive free-flyer launched from a Shuttl
GAS-can — the *Monitoring X-ray Experiment (MOXE)*.

A pinhole camera is a simple detector. It is lightweight and compact, con
sumes little power, uses a low bit rate, and requires only modest position resolu
tion. To monitor the whole sky from the Space Station would require only 6 cam
eras each with a 90° field of view. Each camera would be 64 cm square. Th
ensemble would weigh ∼30 lbs., consume a few watts of power, and require onl
200 bits s^{-1} of telemetry. Sensitivity is necessarily low. In one hour, sources o
∼10 μJy would be visible. Extragalactic sources would require integrations o
several days to one week. These sensitivities are suitable for long time-scale moni
toring, but they are not adequate for studying rapid variations in weaker sources.

The *MOXE* concept would use a low-cost spacecraft with no pointing ability
The spacecraft would rotate about an axis directed to the anti-sun, and ∼6 one
dimensional detectors would view the sky through pinholes on the opposite side o
the spacecraft. Alternatively, *MOXE* could carry ∼6 proportional counters witl
fan beam collimators and achieve a comparable sensitivity limit. Spatial resolu
tion would be traded for much higher signal-to-noise ratios on bright objects
Operationally, *MOXE* would be simple. Positions would be determined from
known X-ray sources. *MOXE* would require a low bit rate ($<$ 2 kilobits s^{-1}), i
could be tracked from a single ground station, and data could be made availabl
to the public via a network or phone link. Again, the sensitivity would be low —
∼10 μJy in one day.

5. Single Waveband Instruments with Multiwavelength Applications

Most satellites now planned for launch in the coming decade have limited or no intrinsic capability for multiwavelength observations. Nevertheless, they will be the prime instruments for multiwavelength astrophysics through the usual method of coordinating campaigns that include several ground-based and space-based observatories. Many of these (but not all) experiments will be in low earth orbit, so coordination will be difficult. Table 6 on the following page gives a summary of the capabilities of the single waveband experiments in the same format as Tables 1 and 2. More detailed descriptions of the instruments follow in the order of their appearance in the table.

ROSAT and SPECTROSAT

The German X-ray satellite *ROSAT* is currently scheduled for launch in February 1990 on a Thor-Delta II from Cape Canaveral. It is a collaborative effort between the Max Planck Institute (MPE), the United Kingdom, and NASA. It will be launched into a low earth orbit, but one high enough to last 5 - 10 years before decaying. A summary of *ROSAT*'s capabilities was presented by G. Hasinger at the Taos conference.

The general objectives of *ROSAT* are to 1) perform an all-sky imaging X-ray survey, 2) perform an all-sky XUV survey, and 3) conduct pointed observations. The all-sky survey will take 6 months, and pointed observations will be performed until the end of the mission. The X-ray and XUV instruments on *ROSAT* give it intrinsic multiwavelength capability covering a new region of the spectrum. It should be especially valuable for studies of coronal emission in stars, nearby SNR, cataclysmic variables, bright AGN, and the interstellar medium. Coordinated multiwavelength observations will be comparable in difficulty to those performed with the *Einstein Observatory*. Detailed mission planning will be done ∼6 months before the actual observation yielding a mission timeline accurate to ∼1 hour.

The Wolter 1 X-ray telescope on *ROSAT* has three focal plane instruments — two position sensitive proportional counters (PSPC) and a high resolution imager (HRI). The PSPC has an angular resolution of better than 1' for pointed observations, and it spans 0.1 to 2.0 keV where the mirror cuts off. The peak effective area is ∼200 cm^2 at 1 keV. The XUV imaging telescope is a Wolter-Schwarzschild 1 with microchannel plate detectors. It has a 5° field of view and an angular resolution of 2' for the survey and 1' for pointed observations. Filters of Lexan plus aluminum, beryllium, or carbon can be rotated in front of the detectors to define bandpasses spanning 60 - 200 Å. The effective area peaks at ∼30 cm^2 at ∼125 Å using the beryllium filter. The X-ray telescope and the PSPC's

Table 6   Single Waveband Instruments with Potential Multiwavelength Applications

Mission	γ-ray (MeV)	X-ray (keV)	EUV (Å)	UV (Å)	Visible (Å)	IR (μ)	Status[a]
ROSAT	...	0.1-2 Broad	60-200 Broad	...	...	...	Feb. 1990
SPECTROSAT	...	0.1-2 R~100	60-200 R~100	...	...	...	Mid 1990's
GRO	0.05-3x10⁴ ~10	...	...	...	...	...	Mid 1990
XTE	...	2-200 R~5	...	...	...	...	Mid 1990's
AXAF	...	0.2-10 R~1-2000	...	...	...	...	Late 1990's
SIRTF	...	...	...	...	...	2-700 R~1-1000	Late 1990's
XLA[b]	...	2-20 R~5	...	...	...	...	Concept
XMM	...	0.2-10 R=10-2000	...	1600 to Broad	6000 Broad	...	Late 1990's
ASTRO-D	...	0.3-10 R=3-100	...	...	...	...	1993

[a]The status column shows either a tentative launch date for an approved mission or "Concept" to indicate that it is a proposal still under study.
[b]The *X-ray Large Array* is unique in that it is intended primarily for guiding observations with *gravitational* radiation detectors.

are from MPE, the HRI is from the Smithsonian Astrophysical Observatory, and the XUV instrument is supplied by Leicester.

A follow-on instrument to *ROSAT* called *SPECTROSAT* is planned for the mid 1990's. It is basically a copy of *ROSAT* with transmission gratings added for spectroscopy of X-ray sources at moderate to high resolution. The facets of the transmission grating will have 1000 l/mm. With a planar HRI, the resolving power varies from ~20 at 2 keV to over 200 at 0.5 keV. The expected throughput is thirty times that of the objective grating plus HRI on *Einstein*. The Zerodur blanks for the X-ray mirrors has been purchased, and mass production of the grating facets has begun. Collaborators in the U.S. are being sought for two high resolution imaging detectors and an add-on instrument, preferably xtending the energy range to 10 - 20 keV.

The Gamma-Ray Observatory

The *Gamma Ray Observatory* (*GRO*), another entry in NASA's series of "Great Observatories," will provide a sensitivity some 10 times better than any other previous mission for observations of gamma rays emitted by celestial objects. Four instruments on *GRO* will cover the energy range 0.05 MeV to 3×10^4 MeV: the Energetic Gamma Ray Telescope (EGRET), the Imaging Compton Telescope (COMPTEL), the Oriented Scintillation Spectrometer Experiment (OSSE), and the Burst and Transient Source Experiment (BATSE).

EGRET, built by Goddard Space Flight Center, is a wide field-of-view, high energy gamma ray telescope, based on a spark chamber system, a multi-element time-of-flight coincidence system, and a total absorption scintillation counter, covering 20 to 3×10^4 MeV. It will be able to locate sources, depending on the nature and location of the source and the observing time, in the $0.5°$ to $0.1°$ range, and provide an energy resolution of $\sim 15\%$ in the center of its energy range. EGRET has a maximum effective area of 2000 cm^2 and has an estimated source sensitivity above 100 MeV of $\sim 5 \times 10^{-8}$ cm^{-2} s^{-1} (continuum source) under optimum conditions.

COMPTEL, built by the Max Planck Institute and the University of New Hampshire, is a wide field-of-view (1 steradian) double-Compton telescope covering 1 to 30 MeV with $\sim 8\%$ energy resolution and 7.5 arc min position resolution (1 σ, strong source). COMPTEL's maximum effective area is 50 cm^2 and its point source sensitivity is 3×10^{-5} to 3×10^{-6} cm^{-2} s^{-1} (line) and 5×10^{-5} cm^{-2} s^{-1} (continuum, E > 1 MeV).

OSSE, built by the Naval Research Laboratory, consists of 4 identical shielded and collimated Phoswich scintillation detectors, each mounted on a gimbal allowing it to rotate (in one plane) through $180°$. Each detector has a $3.8° \times 10°$ field of view. The energy range is 0.1 to 10 MeV, with an energy resolution of 8% at 0.66 MeV. OSSE has a maximum total effective area of 2310 cm^2 and an estimated source sensitivity of 2×10^{-5} cm^{-2} s^{-1} (line) and 3×10^{-5} cm^{-2} s^{-1} (continuum, 0.1 to 10 MeV).

BATSE, built by Marshall Space Flight Center, is designed to monitor continuously a large segment of the sky for the detection and measurement of gamma-ray bursts. For burst monitoring, it has an energy range of 0.05 to 1.0 MeV, a time resolution of <1 millisecond, and a maximum effective geometric factor of 15000 cm^2 sr. For burst spectroscopy, BATSE has an energy range of 0.05 to 20 MeV, a resolution of 7.3% at 0.66 MeV, and a maximum effective area of 127 cm^2 on each of 4 detectors.

The X-ray Timing Explorer

The *X-ray Timing Explorer*, being prepared for launch in 1991, will study rapid time variability in galactic and extragalactic X-ray sources. As described by Gruber (1984), *XTE* consists of an array of large area proportional counters (LAPC), the High Energy Timing Experiment (HEXTE), and a scanning shadow camera (SSC) which serves as an all-sky X-ray monitor. Eight xenon-methane proportional counters which view the sky through mechanical circular collimators of 1° FWHM form the LAPC. The counters are sensitive from 2 - 60 keV with an energy resolution of 16% at 6 keV. The total geometric area of the array is 10400 cm^2. While such a large area counter is extremely sensitive, the 1° spatial resolution makes the LAPC confusion limited at 0.1 μJy. Two arrays of 6 NaI/CsI phoswiches comprise the HEXTE. Like the LAPC, they are mechanically collimated to a circular field of 1° FWHM. Each array has an area of 1000 cm^2. The phoswiches are sensitive from ~20 - 200 keV with an energy resolution of 18% at 60 keV. The SSC will monitor the X-ray sky for bright transients and flares in known X-ray sources to serve as a guide for observations with the LAPC and the HEXTE. Two one dimensional coded mask X-ray cameras with a net area of 30 cm^2 will scan the entire sky once every 90 minutes to a flux limit of 50 μJy and find positions to 0.1°. The SSC data will also be useful for studies of long term variability in bright galactic X-ray sources.

AXAF — the Advanced X-ray Astrophysics Facility

The *Advanced X-ray Astrophysics Facility (AXAF)* is another of the satellites planned by NASA as part of the "Great Observatories." It is currently in an extended Phase B until a new start is approved by NASA. Glass for the high resolution mirror assembly has been purchased, and a launch in the mid to late 1990's is expected for the observatory. The AXAF facility is described by Weisskopf (1985). Six concentric Wolter 1 mirror pairs image X-rays from 0.1 to ~ 8.5 keV over a 1° field at an angular resolution of 0.5 arc sec. The mirror assembly has a geometric area of 1700 cm^2. Five instruments have been studied for the AXAF focal plane — a high resolution camera (HRC), an array of CCD detectors (ACIS), a segmented solid state spectrometer, an X-ray calorimeter, and a Bragg crystal spectrometer. In addition, a high resolution objective grating assembly has been developed.

The HRC will be an advanced microchannel plate detector similar to that flown on the *Einstein Observatory*. A CsI photocathode will be deposited on the microchannel plate, and a crossed-grid charge detector will read the array (Murray and Chappell 1985). Several CCD detectors are being studied for potential use on

AXAF (Garmire *et al.* 1985). The most promising is a virtual phase Texas Instruments chip which is an array of 584×390 30μ pixels. This chip currently gives 150 eV resolution from 0.2 to 10 keV with a peak effective area of ~1000 cm^2 at 1.5 keV. The solid state spectrometer is a segmented Si(Li) detector similar to the *Einstein* SSS and to the detector used in BBXRT (as described earlier). The X-ray calorimeter is an entirely new concept (Holt *et al.* 1985) that may achieve a resolving power of nearly 700 at the 7 keV iron lines with a non-dispersive instrument. The calorimeter is a cryogenically cooled monolithic Si structure. X-rays are absorbed in a layer of Bi evaporated on the surface. The temperature rise induced in the silicon is detected by an Al/As doped silicon thermometer.

The Bragg crystal spectrometer for AXAF is described by Canizares, Markert, and Clark (1985). It consists of 10 curved diffractors mounted on a turret. Each diffractor has a quasi-toroidal geometry. The spectrometer will give resolving powers of 200 - 2000 over the 0.5 - 8.0 keV band, and 50 - 70 over the 0.14 - 0.50 keV band. Low background imaging proportional counters mounted on the Rowland circle will detect the diffracted X-rays. In a typical observation which scans across a single emission line, the effective collecting area ranges from 4 to 60 cm^2. The corresponding minimum detectable line flux is 4 - 30 $\times$ 10^{-6} phot cm^{-2} s^{-1}.

The objective grating spectrometer on AXAF consists of three grating assemblies. The high energy gratings are placed behind the the inner three mirrors, and they are optimized for 4 - 8 keV X-rays. The medium energy gratings are behind the outer three mirrors, and they are optimized for 0.4 - 4.0 keV X-rays. These two grating assemblies are described by Canizares, Schattenburg, and Smith (1985). The gratings are oriented so that the high and medium energy dispersed spectra form a shallow "X" in the focal plane. The resolving power ranges from ~150 - 1000 with a peak effective area of ~250 cm^2 at 2 keV. The high energy grating has a resolving power of ~150 with an effective area of ~30 cm^2 at the 7 keV iron lines. The gratings can be used with either the HRC or the ACIS imaging detectors. With the HRC the low energy limit is 0.2 keV, and with ACIS it is 0.4 keV. The low energy grating assembly described by Brinkman *et al.* (1985) is separate from the other grating assemblies and cannot be used simultaneously. When used with the HRC, it covers the range 0.1 - 6.0 keV with a peak effective area of ~40 cm^2 at 1 keV. The resolving power ranges from 50 to 500.

The Space Infrared Telescope Facility

The *Space Infrared Telescope Facility (SIRTF)* is being studied by NASA as part of the "Great Observatories" program for a launch in the mid to late 1990's. A 1 m class telescope will produce diffraction limited images over the 2 - 700 μ band from a 900 km orbit (Brooks *et al.* 1986). The focal plane instruments consist of the Infrared Array Camera (IRAC) (Fazio *et al.* 1986), the Multiband Imaging Photometer System (MIPS) (Rieke *et al.* 1986), and the Infrared Spectrometer (IRS) (Houck *et al.* 1986). These instruments are still under development, and this summary reflects design goals rather than actual performance.

The IRAC will use three different arrays to image the 2 - 5 μ, 5 - 18 μ, and 18 - 30 μ bands with a variety of broad and narrow band filters. The MIPS will use separate arrays optimized for the 3 - 7 μ, 13 - 30 μ, 30 - 55 μ, 50 - 80 μ, 80 - 120 μ, 120 - 200 μ, and 200 - 700 μ bands. Most of these bands will have a selection of filters available. Sensitivities (1σ in 1000 s) for the IRAC and the MIPS will range from ~0.5 μJy at the short wavelength end to ~300 μJy at the longest wavelengths. The IRS will perform spectroscopy at resolving powers of 50 and 1000. The low resolution mode will cover the 2 - 200 μ band with a sensitivity (1σ in 1000 s) of ~ 0.1 - 1 mJy. The high resolution mode will cover 4 - 200 μ at sensitivities of 1 - 5 mJy.

The X-ray Large Array

The *X-ray Large Array* (*XLA*) is a 100 square meter array of X-ray detectors proposed by the Naval Research Laboratory, Stanford University, and the University of Washington, and it has been the subject of a pre-Phase A study by the NASA Marshall Space Flight Center in 1987. It would be constructed in orbit at the NASA Space Station and then operated attached to the Station. The large aperture provides microsecond photometry and millisecond time-resolved spectroscopy of bright X-ray sources at energies from 0.5 - 25 keV. The implied data rates (up to at least 50 Mbps) are accommodated with direct onboard storage on optical mass storage media. The 2000 brightest sources are observable without significant source confusion, and this provides access to most of the known classes of X-ray objects.

The *XLA* also opens up the novel category of coordinated observations in X-rays and gravitational waves. Low mass X-ray binaries, the brightest X-ray sources in the Galaxy, are thought to be weakly magnetized neutron stars accreting matter from a close companion star. Angular momentum transferred by the accretion stream from the companion star will spin up the neutron star to millisecond periods. The system will then exhibit the Chandrasekhar-Friedman-Schutz general relativistic instability and radiate angular momentum in gravitational radiation (Wagoner 1984). If such binaries pulsate, the large aperture of the *XLA* would be required to detect the low pulsed fraction of the X-ray continuum. If a pulse period is discovered in X-rays, a gravity wave antenna tuned for the associated gravity wave signal can be constructed. Estimates of the expected signal strength suggest either bar detectors or laser interferometers should be capable of detecting the signal. The pulsar can be simultaneously monitored in both X-rays and gravity waves. This provides a new view of these accreting systems since most of the energy provided by accretion is radiated in X-rays while most of the angular momentum is carried away by the gravitational waves.

ESA Planned Missions Including the X-ray Multi-Mirror Mission (XMM)

The ESA space science program as presented by N. White at the Taos conference calls for three types of missions. Big-budget, cornerstone missions (~$400 M) are planned for every four years starting with a solar and geophysics satellite in 1994, *XMM* in 1998, a mission for submillimeter spectroscopy in 2002, and a comet sample return mission in 2006. Medium budget missions (~$200 M) and small budget missions (~$100 M) will also occur every four years. The first of these will be *Hipparcos*, an astrometric satellite to make 0.002 arc second position measurements of 10^5 stars, to be launched in 1988. The *ISO* mission, planned for 1992/3, will perform 3 - 200 μ imaging, photometry, spectroscopy, and polarimetry. The planned mission lifetime is 18 months. A new medium mission selected from either *Lyman*, *GRASP*, or *Quasat* will be launched in 1996. *Lyman* (described earlier) will explore the far UV (900 - 1200 Å) with high resolution spectroscopy, possibly with U.S. collaboration. *GRASP* stands for Gamma Ray Astronomy with Spectroscopy and Positioning. *Quasat* will be an orbiting VLBI experiment. Of these missions, *XMM* has the greatest potential applicability to multiwavelength observations.

The *XMM* mission is based on the premises that spectroscopy will produce the next big leap in X-ray astronomy, that the K edges of carbon through iron must be covered (0.2 - 7.0 keV), and that many detected photons are necessary. X-ray telescopes with high angular resolution are heavy and expensive, so *XMM* trades off angular resolution for collecting area. Its emphasis on spectroscopy with high signal-to-noise complements the high angular resolution of *AXAF*. The primary scientific objectives of the mission are the nature and evolution of all classes of active galactic nuclei and their contribution to the X-ray background, the physical conditions near accreting compact objects, the study of hot gas in stellar coronae, supernova remnants, and clusters of galaxies, and the physics of the interstellar medium.

The design goals for *XMM* call for broad band spectroscopy (R ~10) over the 0.2 - 10.0 keV bandpass with effective areas of 10,000 cm^2 and 5000 cm^2 at 2 and 8 keV respectively and an angular resolution of 30". Under these constraints, the mission will support medium resolution spectroscopy (R ~ 250) with maximized effective area below 3 keV. High resolution spectroscopy (R > 1000) will be concentrated around the oxygen and iron K lines.

The model payload has four telescope modules. Each mirror module will be a nested set of 58 Wolter I telescopes of focal length 7.5m. All modules will contain CCD butted arrays and gas scintillation proportional counters (GSPC). A single *XMM* module with a CCD detector has an expected sensitivity of $\sim 10^{-14}$ erg cm^{-2} s^{-1} in a 10,000 s observation. This is comparable to the *AXAF* sensitivity for the ACIS detector. Two modules will have fixed reflection gratings, and the other two modules will have Bragg spectrometers. X-ray calorimeters are still under study with the main emphasis on instruments with a long lifetime.

The bore-sighted star tracker on *XMM* will double as an optical monitor. This gives *XMM* an intrinsic multiwavelength capability. The monitor is similar to *AXIOM*, the design proposed by Córdova *et al.* (1985) for *AXAF*. It will be a 0.3 m mirror with a 500×500 pixel photon counting microchannel plate detector. The detector will cover an 8' field of view at 1" resolution. A selection of broad band filters will span the wavelength range 1300 - 6000 Å. Similar to *EXOSAT*, *XMM* will be in a high elliptical orbit with a 48 hour period. This will give greater than 40 hours of visibility for any given object on each orbit and ease the logistical difficulties of coordinating ground-based observations.

Planned Japanese Missions Including ASTRO-D

Japanese space experiments are coordinated by the Institute of Space and Aeronautical Science (ISAS), a collaboration of Japanese universities. The next high energy ISAS payload following *Ginga* will be *ASTRO-D*, which will be launched in 1993. The instrument complement is still under study, and the Japanese are seeking a collaboration with NASA for the mission. The prime purpose of *ASTRO-D* will be X-ray spectroscopy, and a tentative payload includes four grazing-incidence conical mirror assemblies similar to those used for BBXRT. Two of the telescopes will have imaging gas scintillation proportional counters in the focal plane, and the other two will have either pin diode arrays or CCD's for the detector system.

For the mid to late 1990's, ISAS is developing concepts for space VLBI, an additional X-ray payload, and IR and UV satellites (Tanaka 1987). The Japanese are also considering an attached payload for the space station that would perform optical, X-ray, and γ-ray observations of γ-ray burst sources. This experiment would be part of the Japanese Equipment Model (JEM) on the Space Station.

USSR Space Experiments for Multiwavelength Astrophysics

Building on its extensive experience in γ-ray astronomy, the satellites being planned by the USSR are heavily weighted toward γ-rays and high energy X-rays. While several of the Soviet missions carry intrinsically multiwavelength experiments with the specific purpose of high energy burst detection and location (*e.g.*, the previously mentioned *Sunflower* instrument on *GRANAT* and the *SPECTRA* series), the ability to coordinate observations with ground-based facilities or other satellites for general-purpose, multiwavelength astrophysics may be difficult. In addition to the usual logistical problems, observers on Soviet satellites typically have even less control over observing schedules than on U.S. or European satellites. For example, pointing the *Kvant* experiment on the *Mir* space station requires orienting the whole station, and the X-ray observations must be integrated into the framework of *Mir* operations. Political problems in East-West relations also have the potential to hamper collaborative multiwavelength campaigns.

In Table 7 I list the planned Soviet missions, their planned launch dates, and a brief description of their objectives. The currently operating *Kvant* X-ray experiment on the *Mir* space station is discussed in the preceding chapter by F. Córdova. All planned Soviet missions include instruments supplied by other European collaborators. The instruments for missions beyond *GRANAT* are not yet firm, and the Soviets are actively seeking Western collaborators for many of them.

Table 7 Future Soviet Satellites for Astrophysics

Mission	Launch Date	Objectives
GAMMA-1	1988?	X-ray and high energy γ-rays
GRANAT	Nov. 1988	X-ray, γ-ray, and burst location
RADIOASTRON	1991	Orbiting VLBI at 1.35, 6.2, 18, and 92 cm
Name Unknown	Unknown	100μ to 2mm spectroscopy
SPECTRA-1	1992	Imaging X-ray and burst location
SPECTRA-2	1994	γ-ray spectroscopy and positioning

Launch of *GAMMA-1* was planned for 1987, but it is suffering repeated delays. It will be in a low earth orbit, and the spacecraft pointing accuracy is $0.5°$. The main instrument is a 50-1000 MeV time-of-flight spark chamber with a tungsten mask. The 50×50 cm, 12 gap chamber has a $30°$ field of view with a 1.5m long time-of-flight path to a Cerenkov detector and a four layer calorimeter. It is a French and Soviet collaborative effort. Additional experiments on board are Pulsar X-2, a Soviet 2-20 keV X-ray telescope, and DISK, a 0.2-20 MeV NaI shielded telescope with a modulated "anti-collimator."

The orbiting VLBI mission, *RADIOASTRON*, is in the planning stages with the Soviets still actively seeking Western collaborators. A 10m dish with 4 circular polarization feeds will give a sensitivity of ~1 mJy in 18 hours. The planned orbit of 9000 × 75,000 km at an inclination of 61° has a period of 24 hours. The spacecraft will have a pointing accuracy of ~1°. The European VLBI consortium is supplying the 6.2 cm receiver, and the Polish are supplying the 92 cm receiver. Australia and the CSIRO are being approached for the 18 cm receiver, and the 1.35 cm receiver has been sought from NRAO. No approval for supplying either of these receivers has yet been made. In addition, collaborations for use of the Deep Space Network (DSN) and for use of European and Australian VLBI antennas are still being negotiated.

A concept for an infrared/submillimeter telescope is still in the planning stages. The Soviets will supply the spacecraft and a 1 m telescope with the primary mirror cooled to 100 K. The French are supplying the focal plane instruments with cooling to 0.5 K. Spectrometers with resolving powers of 10 and 1000 are planned.

The two *SPECTRA* missions will be in high, elongated orbits similar to *GRANAT* with similar pointing capabilities. The first mission scheduled for 1992 will concentrate on X-ray astronomy. Tentatively, the Danes will supply a conical metal foil telescope with 140 nested foils. The focal length is to be 8m with a 70 cm maximum diameter. The goal is to obtain an effective area of 800 cm^2 at 15 keV with 1' resolution. The Max Planck Institute has proposed a low energy imaging X-ray telescope. Focal plane detectors for either telescope have not been defined. The second *SPECTRA* mission will concentrate on γ-ray astronomy. Tentatively, the European Space Agency will supply GRASP, Gamma Ray Astronomy with Spectroscopy and Positioning, and MPE will build a 0.05 - 10 MeV Ge spectrometer and a 15 - 500 keV NaI detector with a coded mask.

6. Multiwavelength Instruments for the 1991 Solar Maximum

The *Solar Maximum Mission (SMM)*, originally launched for the 1978 solar maximum, may still be functioning through the 1991 solar maximum. In addition, the Japanese Institute of Space and Aeronautical Science plan to fly the *SOLAR-A* satellite to observe the sun with a launch scheduled for the fall of 1991 (Tanaka 1987). The experiments on board include a hard X-ray imaging experiment, a soft X-ray telescope, a bent crystal spectrometer for high resolution Doppler spectroscopy, and a wide-band spectrometer.

The hard X-ray telescope will image the solar disk with a resolution of ~7" using fan beam collimators and a Fourier synthesis technique. The detector system is an array of 64 NaI(Tl) scintillators with 3 cm^2 of effective area each over the 10 - 100 keV band.

The soft X-ray telescope is a collaboration between Lockheed (through NASA) and the Tokyo Astronomical Observatory. It will use a novel grazing incidence telescope consisting of two hyperboloids. Compared to a conventional Wolter 1 design, this design sacrifices on-axis resolution for high resolution over a wide field. The entire solar disk will be imaged at energies of $\sim$ 0.25 - 3.5 keV at an angular resolution of $\sim$3". An integrating CCD will be used as a detector. A filter system is being designed for the telescope that will provide unique temperature diagnostics for single component thermal spectra.

The crystal spectrometer will image the lines of S XV, Ca XIX, Fe XXV, and Fe XXVI at resolving powers of 3000 - 8000. Position sensitive proportional counters will be used as detectors. The wide band spectrometer will cover the 2 kev - 50 MeV energy range with a proportional counter in the soft X-ray, an NaI scintillator in the hard X-ray, and a BGO scintillator in the γ-ray.

NASA plans to support a series of long duration ($>$15 day) balloon flights carrying a variety of hard X-ray, γ-ray, and neutron detecting instruments. Called *MAX 91*, this balloon campaign was outlined by C. J. Crannell at the Taos workshop. Design objectives for a hard X-ray/γ-ray imaging experiment are to image the whole solar disk at 1" resolution with a spatial Fourier filter made of centimeter thick layers of tungsten. The detector would have a time resolution of better than 10 ms, an effective area of $\sim$500 cm^2, and cover the energy range 20 keV to 200 MeV. Spectroscopic experiments for hard X-rays and γ-rays are also planned. The active shielding for the detectors would detect neutrons and particles with energies $>$ 10 MeV. Further details can be found in the chapter

7. Conclusions

The quest for multiwavelength observations attends the opening of each new astronomical window. At first this is merely the optical identification of sources detected in the new wavelength window, but as entirely new classes of astronomical objects are discovered, unraveling the complex physical processes at work often *requires* multiwavelength observations. We are at the forefront of a new era in astronomy in which multiwavelength observations will play a major role in solving many outstanding astrophysical problems. In addition, the simultaneous, multiwavelength observation of celestial objects in an unbiased fashion could open a new "observational window" that would allow the discovery of new phenomena. Many of the instruments already approved for completion in the 1990's have minimal intrinsic multiwavelength capabilities. Careful attention to mission planning and operations, however, can lead to significant advances in multiwavelength astrophysics. The key to truly opening the "multiwavelength window" lies with new concepts such as *HETE* and *SpEx* which are specifically dedicated to multiwavelength observations.

I am indebted to many astronomers working on multiwavelength instrumentation for their contributions of technical details for this review, especially L. Acton, F. Córdova, C. Crannell, A. Davidsen, M. Giampapa, P. Gorenstein, G. Hasinger, F. Marshall, W. Moos, W. Priedhorsky, G. Ricker, R. Stalio, N. White, and K. Wood. The discussion of the desirable characteristics for a multiwavelength satellite was organized and led by T. Snow. I thank F. Córdova, K. Long, and W. Priedhorsky for their helpful comments on the manuscript.

References

Angel, J. R. P. 1979, *Ap. J. (Letters)*, **233**, 364.

Bowyer, S. 1983, in *Advanced Space Instrumentation in Astronomy*, ed. R. M. Bonnet (New York Pergamon), p. 157.

Brinkman, A. C., *et al.* 1985, *Proc. SPIE*, **597**, 232.

Brooks, W. F., Melugin, R. K., Lee, J. H., and Lemke, L. 1986, *Proc. SPIE*, **619**, 2.

Canizares, C. R., Markert, T. H., and Clark, G. W. 1985, *Proc. SPIE*, **597**, 241.

Canizares, C. R., Schattenburg, M. L., and Smith, H. I. 1985, *Proc. SPIE*, **597**, 253.

Cash, W. 1982, *Applied Optics*, **21**, 710.

Córdova, F., Mason, K., Priedhorsky, W., Margon, B., and Hutchings, J. 1985, *Ap. Sp. Sci.*, **111**, 265.

Fazio, G. G., *et al.* 1986, *Proc. SPIE*, **619**, 60.

Garmire, G. P., *et al.* 1985, *Proc. SPIE*, **597**, 261.

Giacconi, R., *et al.* 1978, *Ap. J.*, **230**, 540.

Gilman, D., Metger, A. E., Parker, R. H., Evans, L. G., and Trombka, J. I. 1980, *Ap. J.*, **236**, 951

Gruber, D. E. 1984, in *X-ray Astronomy '84*, ed. M. Oda and R. Giacconi, p. 503

Hettrick, M. C., Bowyer, S., Malina, R. F., Martin, C., and Mrowka, S. 1985, *Applied Optics*, **24**, 1737.

Holt, S. S., *et al.* 1985, *Proc. SPIE*, **597**, 267.

Houck, J. R., *et al.* 1986, *Proc. SPIE*, **619**, 69.

Klein, R. J., and Chevalier, R. A. 1978, *Ap. J. (Letters)*, **223**, L109.

Laros, J. G., *et al.* 1986, *Nature*, **322**, 152.

Lasher, G. J., and Chan, K. L. 1979, *Ap.J.*, **230**, 742.

Lawrence, A., Watson, M. G., Pounds, K. A., and Elvis, M. 1987, *Nature*, **325**, 694.

Martin, C., Jelinsky, P. Lampton, M., Malina, R. F., and Anger, H. O. 1981, *Rev. Sci. Instrum.* **52**, 1067.

McHardy, I., and Czerny, B. 1987, *Nature*, **325**, 696.

Melia, F. 1987, *Ap. J.*, in press.

Melia, F., Rappaport, S., and Joss, P. C. 1986, *Ap. J. (Letters)*, **305**, L51.

Middleditch, J., and Priedhorsky, W. 1986, *Ap. J.*, **306**, 230.

Murray, S. S., and Chappell, J. H. 1985, *Proc. SPIE*, **597**, 274.

Nolan, P. L., Share, G. H., Matz, S. M., Chupp, E. L., Forrest, D. J., and Rieger, E. 1984, in *High Energy Transients in Astrophysics*, ed. S. E. Woosley (New York: AIP), p. 399.

Pedersen, H., *et al.* 1984, *Nature*, **312**, 46.

Pedersen, H., Paradijs, J. V., Motch, C., Cominsky, L., Lawrence, A., Lewin, W. H. G., Oda, M. Ohashi, T., and Matsuoka, M. 1982, *Ap. J.*, **263**, 340.

Rieke, G. R., *et al.* 1986, *Proc. SPIE*, **619**, 73.

Schaefer, B. 1981, *Nature*, **294**, 722.

Schmidt, W. K. H. 1975, *Nuc. Instr. and Meth.*, **127**, 285.

Stern, R., Kimball, R., and Davidsen, A. 1987, private communication.

Stern, R., *et al.* 1983, *S. P. I. E.*, **445**, 347.

Tanaka, Y. 1987, presented at the NATO Advanced Study Institute on "Hot Thin Plasmas in Astrophysics," Corsica, France.
Timothy, J. G., and Bybee, R. L. 1981, *Proc. SPIE*, **93**, 265.
Wagoner, R.V. 1984, *Ap. J.*, 278, 345.
Weisskopf, M. 1985, *Proc. SPIE*, **597**, 228.

INDEX